Asian Change in the Context of Global Climate Change

Nearly two-thirds of the world's population live in Asia, and many countries in that region are currently undergoing very rapid industrial, agricultural and economic development. The Framework Convention on Climate Change constrains developed countries with regard to their future emissions of greenhouse gases, but recognizes the special needs of developing countries. There is growing appreciation of the ways in which developing countries in the Asian region contribute both to global changes (by altering biogeochemical pathways and cycles) and are themselves affected by those changes. This volume uses the intellectual efforts and findings of the IGBP community to provide the first integrated analysis of the interactions between global change and Asian change, with particular attention given to the role of China.

JAMES GALLOWAY is a full Professor in, and Chair of, the Department of Environmental Sciences at the University of Virginia. His research focuses on biogeochemistry and includes the study of natural and anthropogenic controls on chemical cycles at the local (watershed), regional and global scales.

JERRY MELILLO is Co-Director of The Ecosystems Centre of the Marine Biological Laboratory in Woods Hole, Massachusetts. His research in biogeochemistry includes work on global climate change, the ecological consequences of tropical deforestation and the sustainable management of forest ecosystems.

The International Geosphere–Biosphere Programme was established in 1986 by the International Council of Scientific Unions, with the stated aim:

> *To describe and understand the interactive physical, chemical and biological processes that regulate the total Earth system, the unique environment that it provides for life, the changes that are occurring in this system, and the manner in which they are influenced by human activities.*

A wide-ranging and multi-disciplinary project of this kind requires well-defined priorities and goals to be effective. There are six key questions that the IGBP seeks to answer. These are:

- How is the chemistry of the global atmosphere regulated, and what is the role of biological processes in producing and consuming trace gases?
- How will global changes affect terrestrial ecosystems?
- How does vegetation interact with physical processes of the hydrological cycle?
- How will changes in land-use, sea level and climate alter coastal ecosystems, and what are the wider consequences?
- How do ocean biogeochemical processes influence and respond to climate change?
- What significant climatic and environmental changes occurred in the past, and what were their causes?

The International Geosphere–Biosphere Programme Book Series brings new work on topics within these themes to the attention of the wider scientific audience.

Titles in the series

1 **Plant Functional Types**
T. M. Smith, H. H. Shugart & F. I. Woodward (Editors)
0521 48231 3 hardback
0521 56643 6 paperback

2 **Global Change and Terrestrial Ecosystems**
B. H. Walker & W. L. Steffen (Editors)
0521 57094 8 hardback
0521 57810 8 paperback

3 **Asian Change in the Context of Global Climate Change**
J. N. Galloway & J. M. Melillo (Editors)
0521 62343 X hardback
0521 63888 7 paperback

INTERNATIONAL GEOSPHERE–BIOSPHERE PROGRAMME BOOK SERIES

Asian Change in the Context of Global Climate Change

Impact of natural and anthropogenic changes in Asia on global biogeochemistry

Edited by

James Galloway

Jerry Melillo

CAMBRIDGE UNIVERSITY PRESS

PUBLISHED BY THE PRESS SYNDICATE OF THE UNIVERSITY OF CAMBRIDGE
The Pitt Building, Trumpington Street, Cambridge CB2 1RP, United Kingdom

CAMBRIDGE UNIVERSITY PRESS
The Edinburgh Building, Cambridge CB2 2RU, UK http://www.cup.cam.ac.uk
40 West 20th Street, New York, NY 10011–4211, USA http://www.cup.org
10 Stamford Road, Oakleigh, Melbourne 3166, Australia

First published 1998

Printed in the United Kingdom at the University Press, Cambridge

Typeset in Ehrhardt 10.5/14pt

A catalogue record for this book is available from the British Library

Library of Congress Cataloguing in Publication data

Asian change in the context of global climate change / edited by
James N. Galloway and Jerry M. Melillo.
p. cm. – (International Geosphere–Biosphere Programme book series ; 3)
Includes index
ISBN 0 521 62343 X – ISBN 0 521 63888 7 (pbk)
1. Climatic changes – Environmental aspects – Asia.
2. Biogeochemistry – Asia. I. Galloway, James N. II Melillo, Jerry M.
III. Series
QC981.8. C5A863 1998
577.27′6 – dc21 97-32157 CIP

ISBN 0 521 62343 X hardback
ISBN 0 521 63888 7 paperback

Contents

Contributors

Z. S. An
Xi'an Laboratory of Loess and Quaternary Geology, Academia Sinica, PO Box 17, 710054 Xi'an, China

J. Boonjawat
Biochemist, Department of Biochemistry, Chulalongkorn University, Bangkok 10330, Thailand

C-T. A. Chen
Institute of Marine Geology and Chemistry, National Sun Yat-Sen University, Kaohsiung, Taiwan, ROC

T. Chuluun
Natural Resource Ecology Laboratory, Colorado State University, Fort Collins, CO 80523, USA

Z. Dai
Research Center for Eco-Environmental Sciences, Chinese Academy of Sciences, PO Box 2871, Beijing 100085, China

C. B. Fu
Center for Global Change Research, Chinese Academy of Sciences, Beijing, China

J. N. Galloway
Department of Environmental Sciences, University of Virginia, Charlottesville, VA 22901, USA

M. Heimann
Max-Planck-Institut für Meteorologie, Bundesstr. 55, 20146 Hamburg, Germany

D. Hu
Institute of Oceanology, Chinese Academy of Sciences, Qingdao 266071, China

S. Kempe
Geological and Paleontological Institute, Technical University of Darmstadt, Germany

J-W. Kim
Department of Astronomy and Atmospheric Sciences, Yonsei University, Korea

A. Lerman
Department of Geological Sciences, Northwestern University, Evanston, IL 60201, USA

C. Li
Institute for the Study of Earth, Oceans, and Space, University of New Hampshire, Durham, NH 03824, USA

F. T. Mackenzie
Department of Oceanography, School of Ocean and Earth Science and Technology, University of Hawaii, Honolulu, HI 96822, USA

J-P. Malingreau
EC Joint Research Centre, Ispra, Italy

J. M. Melillo
The Ecosystem Center, Marine Biological Laboratory, Woods Hole, MA, USA

H. U. Neue
Soil Chemist and Head, Department of Soil Sciences, Centre for Environmental Research Leipzig-Halle, Hallesche Strasse 44, D-06246 Bad Lauchstädt, Germany

D. S. Ojima
Natural Resource Ecology Laboratory, Colorado State University, Fort Collins, CO 80523, USA

Y. Saito
Geological Survey of Japan, Tsukuba, Ibaraki 305, Japan

W. A. Salas
Institute for the Study of Earth, Oceans, and Space, University of New Hampshire, Durham, NH 03824, USA

E-D. Schulze
Lehrstuhl Pflanzenökologie, Universität Bayreuth, D-95440 Bayreuth, Germany

C. Silapathong
National Research Council of Thailand, Bangkok, Thailand

D. L. Skole
Department of Geography, Michigan State University, East Lansing, MI 03824, USA

J. Sun
Research Center for Eco-Environmental Sciences, Chinese Academy of Sciences, PO Box 2871, Beijing 100085, China

L. G. Thompson
Byrd Polar Research Center, The Ohio State University, 1090 Carmack Road, Columbus, OH 43210, USA

S. Tsunogai
Marine and Atmospheric Geochemistry Laboratory, Graduate School of Earth Science, Hokkaido University, Sapporo 060, Japan

L. M. Ver
Department of Oceanography, School of Ocean and Earth Science and Technology, University of Hawaii, Honolulu, HI 96822, USA

C. J. Vörösmarty
Institute for the Study of Earth, Oceans, and Space, University of New Hampshire, Durham, NH 03824, USA

X. Xiao
The Ecosystem Center, Marine Biological Laboratory, Woods Hole, MA, USA

D. Ye
Institute of Atmospheric Physics, PO Box 2718, Beijing, 100080 China

M. Yoshino
5-1-8-202 Komazawa Selagaya-ku, Tokyo 154, Japan

X. S. Zhang
Institute of Botany, Chinese Academy of Sciences, Beijing, China

Z. C. Zhao
National Climate Center, China Meteorological Administration, Beijing, China

Y. H. Zhuang
Research Center for Eco-Environmental Sciences, Chinese Academy of Sciences, PO Box 2871, Beijing 100085, China

Preface

Droughts, forest fires and air pollution have recently focused world attention on environmental problems of South East Asia. These acute, relatively local changes occur within a wider spatial and temporal framework: they have potential global impacts, and represent just some of the interactions between ecosystems, human impacts and developmental processes occurring in Asia, the largest and most populous continent.

The diversity, scale, and complexity of those systems precludes a single–volume account of their dynamic responses to current and future perturbations – and we have only just begun to collect and bring together the necessary scientific information to develop that understanding. In this regard, a start has been made. One of the most significant advances occurred at Beijing in October 1995, when over 400 researchers from 18 Asian and 42 non–Asian countries met to share their ideas and results at a Symposium on the role of Asia in global change, held in conjunction with the Fourth Scientific Advisory Council for the International Geosphere–Biosphere Programme. This book, developed from the plenary presentations at the Symposium, documents the state of some of that knowledge and addresses some key areas for additional work. In addition to the oral sessions, the discussions at Beijing benefited from more than 60 poster presentations; their abstracts have been separately published (*Natural and Anthrogenic Changes: Impacts on Global Biogeochemical Cycles, Book of Abstracts*; IGBP Stockholm, 1995).

Many people helped to make this publication possible. A special note of appreciation goes to the Chinese National Committee for the IGBP, the Symposium hosts; the local Organizing Committee, chaired by Chen Panqin; and the Symposium Organizing Committee, comprising

Hajime Akimoto (Japan), Chen Panqin (China - CAST), Paul Cheung (Singapore), Kasem Chunkao (Thailand), B. L. Deekshatulu (India), Hu Dunxin (China - CAST), Dennis Ojima (USA; Chair, poster session) and Riga A. A. Suprapto (Indonesia). Financial assistance for the production of this book was provided by the IGBP, through sponsorship from the ENRICH program (European Network for Research in Global Change) of the European Commission. Such support is gratefully acknowledged, as is the editorial assistance by Bruce Godfrey, Tamara Gardner, Judith Peatross and Shuang Huang (University of Virginia), publication assistance by Phillip Williamson (University of East Anglia) and the assistance with Symposium arrangements by Suzanne Nash (IGBP Stockholm) and Mary Ann Seifert (Marine Biological Laboratory, Woods Hole).

J. N. Galloway
J. M. Melillo

Charlottesville, April 1998

Asian change in the context of global change: an overview

J. N. Galloway, D. S. Ojima and J. M. Melillo

Introduction

Background

The Earth's biogeochemical cycles of carbon, nitrogen, phosphorus, and sulphur (CNPS) exhibit significant natural variability on time scales of relevance to humans. Furthermore, these cycles are being increasingly affected by human activities, including combustion, agriculture, and industry. These activities have accelerated the mobilization of CNPS from inert (e.g. N_2) and sequestered (e.g. fossil carbon) forms into chemical forms that can affect critical processes of our biogeochemical environment, such as ecosystem productivity, and atmospheric energy adsorption and photochemistry. Historically, changes in the natural CNPS cycles have occurred in the more developed countries of the western portion of the Northern Hemisphere. Over the past few decades, however, combustion, agriculture and industry in Asia have grown to the level that mobilization rates in some Asian countries are now among the highest in the world and are having significant effects on natural cycles. In addition, projections are for significant growth in Asian energy use, agriculture and industry over the next few decades, leading to further changes in natural cycles. These changing source patterns of CNPS are the driving force for our curiosity about the relationship between Asia and global change, and raise three questions: What is the impact of Asia on Asia?; What is the impact of global change on Asia?; What is the impact of Asia on global change?

Hence, it was fitting that the Fourth Meeting of the Scientific Advisory Council for the IGBP met in Beijing, and that the Science Symposium title was "Natural and Anthropogenic Changes: Impacts on Global Biogeochemical Cycles". The symposium and this resulting book have provided a forum to begin to address the three questions above. We stress 'begin'; the region is so large, the changes and potential consequences are so substantial, that this book is only a small selection of the work that has been done and that needs to be done to answer the questions posed.

The remainder of the introductory section of this paper gives an overview of past meetings of the Scientific Advisory Council and details some of the IGBP programs in Asia. It also provides an outline of the organization of the Symposium and this volume. Subsequent sections examine current and future CNPS mobilization rates in Asia relative to other regions and provide an overview of Asian change in the context of global change.

SAC I–III

The meeting of the International Geosphere – Biosphere's Scientific Advisory Council (SAC) in Beijing in 1995 was the fourth SAC meeting held since IGBP was founded in 1988. As an organization of the International Council of Scientific Unions, IGBP was established to develop and coordinate an international scientific programme dedicated to advancing our understanding of the causes and consequences of global change. As part of its initial plan of action (IGBP Report 4), IGBP holds periodic SAC meetings, each at a different location around the world.

The SAC meetings have been valuable events for the refinement of IGBP research direction, presentation of recent global change research results, and interaction with national research groups. The SAC meetings have provided a forum for national committees to meet as groups and develop regional or cross-regional studies. They have also provided a forum for ICSU bodies and other organizations to discuss similar research aims and collaborative research themes, and to review the progress of the IGBP research projects. The first SAC meeting in Stockholm, Sweden of October 1988 was critical to the IGBP in defining the scope of research for the IGBP and setting up the initial coordinating panels and working groups to define the research agenda for what will later become the IGBP Core Projects.

The SAC II meeting in Paris in 1990 presented the deliberations of the coordinating panels and working groups in developing the scientific plan as presented in the IGBP report 12, "The Initial Core Projects". The Paris meeting was critical to the global change community in order to present the initial set of research areas defining the established Core Projects and to outline the implementation and operational strategy of BAHC, GCTE, IGAC, JGOFS, LOICZ, and PAGES as well as the Framework Activities of GAIM, IGBP-DIS, and START. These initial two SAC meetings set the stage for what is now a major global change research program.

Following these developments was a period of implementing research in all of these areas and developing new Core Projects. The SAC III in January 1993 in Ensenada, Mexico demonstrated that the IGBP had implemented many of the research agendas described in the previous SAC, and that it was ready to push new frontiers. The initial presentation of the established LOICZ Core Project was made in Ensenada. SAC III also established research relationships more concretely among the IGBP, WCRP and the IHDP of the International Social Science Council (ISSC). These cross-programs are central to the further development of the System for Analysis, Research, and Training (START) and the Land Use/ Cover Change (LUCC) Core Project. The importance in fostering these two research activities is evident today as we review the role that START and LUCC play in defining a number of research studies including them and represented at the SAC IV.

The regional focus of SAC IV on Asia has highlighted the extent to which research has proceeded in the region to address a complex set of global change issues, and to highlight the human dimension of this research effort. In previous meetings of the SAC, the role of human dimensions of the global change was not as explicitly represented. However, during the past several years the Land Use/ Cover Change (LUCC) Core Project has been established under the joint sponsorship of the IGBP and the International Human Dimensions Programme and has progressed rapidly. The involvement of scientists from social and natural sciences focused on the understanding of drivers of land use and land cover changes as an important link to understanding the drivers of global change and the impact of global change is a welcomed development. In many areas of the Asian region, LUCC projects are now ongoing, including regional efforts in South East Asia, Temperate East Asia, Hindu-Kush and South Asia. These studies in LUCC are taking advantage of two new research mechanisms of the global change community, the IGBP transects and the START regional facilities. Together, these two mechanisms have facilitated the interaction among different Core Projects and research institutes to develop integrated global change research efforts. The Asian research community has been successful in developing global research in all areas of the IGBP Core Projects as indicated in the research presented in the special poster sessions of the SAC IV. (Abstracts available IGBP, 1995.)

IGBP in Asia

The participation of Asian countries in the IGBP has been strong from its inception. The development of the science plan and the implementation strategy of the IGBP has involved and benefited from the involvement of Asian scientists from a variety of scientific disciplines. The IGBP has also benefited from the donation of the resources from Asian research institutes and other agencies. The Asian region is characterized by a rich mosaic of land- and water-scapes that are utilized for a variety of purposes by many people. This diversity of earth system properties and anthropogenic forcing factors found in the Asian region provide an ideal research environment for global change research. The research studies presented at the SAC IV meeting, especially in the poster sessions, indicate the strength of the ongoing research activities on the atmosphere,the oceans and estuaries, the fresh waters, and the land of the Asian region. Examples of the research include the studies on cloud formation over the Pacific Ocean, land use change in Southeast Asia and in temperate East Asia, biomass burning in India, deforestation in Southeast Asia, human impacts on land use, hydrological interactions between terrestrial ecosystem processes, and remote sensing of land cover change. The research studies have been important in understanding how changes in the Earth system are driven by intensive human activities. The IGBP national committees of the region have been extremely active in organizing research projects to address the breadth of global change studies. Recently, international efforts have been developed to foster the interaction among countries. The Asian Pacific Network (APN), for instance, has funded activities that have encouraged young scientists to participate in workshops related to global change and provided support for joint Core Project meetings within the region. The growing interest and support being provided by Asian organizations are most welcome, and the regional context of the research efforts is providing a meaningful perspective of the impact of global change on key system components and the factors affecting climate change. This research effort will also provide a long-term benefit to the people of the region in developing long-term sustainable natural resource use in the area.

Symposium objectives and strategy

Topics at the SAC IV Science Symposium covered all aspects of IGBP and included processes that mobilize, transform, transport and sequester CNPS as well as environmental processes that are affected by the increased concentrations of active CNPS species. In addition to a focus on the current impacts of Asia on global cycles, plenary and poster presentations examined future scenarios of CNPS cycling. The Symposium was organized into four sessions. In the first, change in the historical perspective was presented through the use of historical records of climate and land-use changes. The second session addressed current land use and CNPS mobilization rates, and the third, a Plenary Session, covered a broad view of the effects of changes for biogeochemical cycles. The final overview session focused on the implications of global change on the Asian system. This volume is the result of that symposium. It is primarily composed of the plenary papers presented at the Symposium. A previous publication presented the abstracts of all the poster presentations (IGBP, 1995).

This book is organized around the biogeochemical cycles and their controlling processes. Following a historical perspective, the book presents a series of chapters on the terrestrial system, three chapters on the relationships among global change, biogeochemical cycle alterations and the dynamics of rivers, estuaries and the ocean, and concludes with a review of the impact of global change on Asia and a summary of the symposium. The rest of this chapter broadly discusses Asian impacts on CNPS mobilization now and in the future, and gives an overview of significant points in the book.

CNPS mobilization

Human-induced changes of environmental systems on the global scale are driven by increases in the human population and per capita resource use. The resources can be lumped into three categories: water, food, and energy. Arguably these are listed in decreasing rank order of their importance to human survival. The increased demand for water has altered its composition, location, storage, and flow rate for many hydrological systems. The demand for food has required extensive conversion of forests to pasture and cropland, and the addition of substantial quantities of commercial N and P fertilizer. The increases can be large. For example, N fixation by food and energy production is approximately equal to natural fixation in terrestrial ecosystems (Ayers *et al.*, 1994; Mackenzie, 1994; Galloway *et al.*, 1995). The increased demand for energy has resulted in extensive deforestation and fossil fuel and biomass combustion. These activities are the primary sources for C and S, and a secondary source for N.

Since the turn of the century, and until recently, most anthropogenic CNPS mobilization has occurred in North America and Europe. Consequences of the mobilization have been observed in those regions and on a global scale for materials with substantial distribution (e.g. CO_2, N_2O). However, in the last few decades, Asia with over 50% of the world's population has become a significant CNPS source due to population growth, economic growth and accompanying resource use (Fu *et al.*, this volume). Asia is now at about par with North America and Northern Europe in the mobilization of CNPS (Fig. 1.1).[1] In 1990, mobilization of CNPS in Asia reached about 25%, 40%, 35%, and 35%, respectively, of the global total. Asian sources of anthropogenic P are primarily as a consequence of food production. This feature is the same for N, even when fossil fuel production of NO_x is taken into account. In 1990, about 8 Tg N were injected into the Asian environment due to combustion, compared to about 36 Tg N due to fertilizer application, with China contributing a significant fraction (Galloway *et al.*, 1996*a*).

1. The 2020 estimates are based upon the 1992a scenario (Leggett *et al.*, 1992; W. Pepper, 1996, personal communication). S mobilization rates include fossil fuel combustion, biomass burning and industrial emissions (Leggett *et al.*, 1992; W. Pepper, 1996, personal communication). The 2020 estimates are based upon the IS92a scenario (Leggett *et al.*, 1992; W. Pepper, 1996, personal communication). N mobilization rates include NO_x emissions from energy combustion (Leggett *et al.*, 1992; W. Pepper, 1996, personal communication), biomass burning (Leggett *et al.*, 1992; W. Pepper, 1996, personal communication) and N fertilizer application (FAO, 1996). The 2020 estimates for the first two N fluxes are based upon the IS92a scenario (Leggett *et al.*, 1992; W. Pepper, 1996, personal communication). The 2020 estimates for N fertilizer application are based upon the Constant and Sheldrick (1992) estimates for N fertilizer growth rates between 1989/90 and 2000/01 which we extended to 2020. The only significant change over those used by Constant and Sheldrick was that they estimated that for Europe (including FSU) there would be a slight decrease in fertilizer use rates up to 2001. The annual rates of increase were 3.1%, 0.2%, 2.9%, 2.9%, and 0.2% for Africa, North America, Latin America, Asia and Europe, respectively. Our estimate to 2020 was that there would be a slight increase in fertilizer use. P mobilization rates are for fertilizer application only (FAO, 1996). The 2020 estimates for P fertilizer application are estimated using the same percentage increases as for N fertilizer.

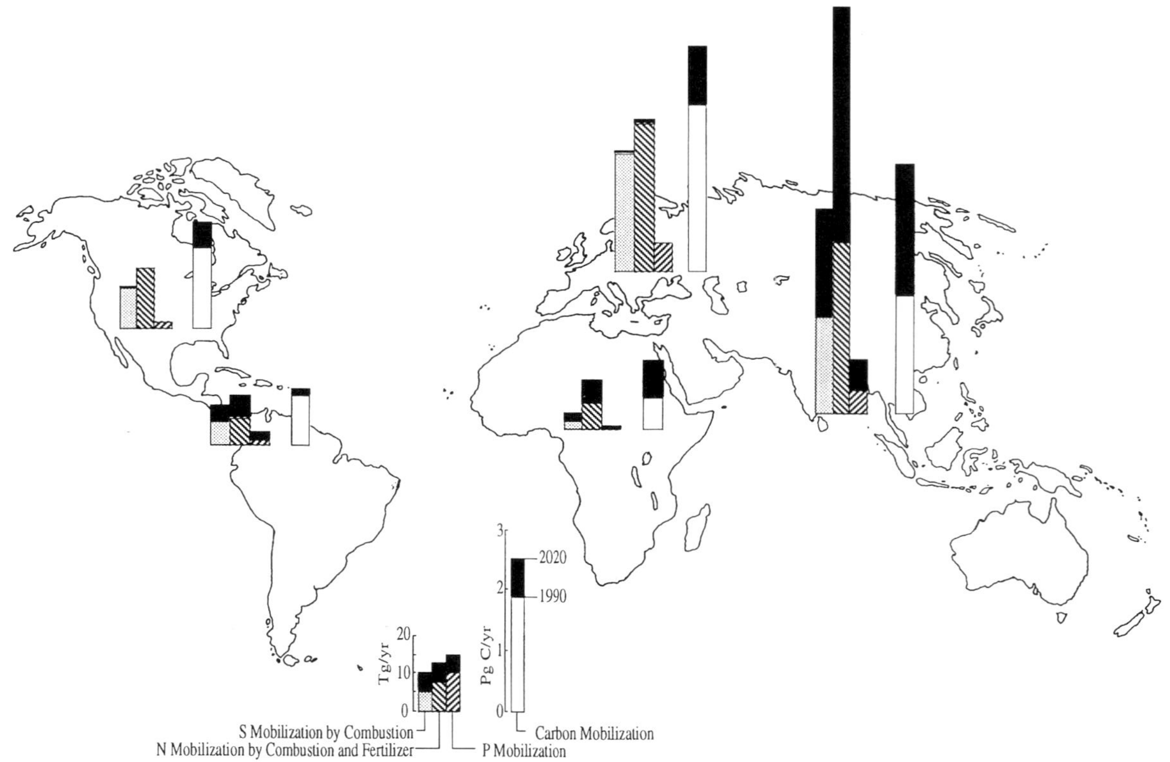

Figure 1.1 *The mobilization of C, S, N and P in 1990 and 2020. The regional aggregations depicted in the figure are based on the IPCC compilations as shown in parentheses: North America (USA); Europe (OECD-W, CPEUR), Africa (AFRICA), Central and South America (LAMERICA), Asia (OECD-A, CPASIA, M-EAST, SEASIA). Carbon mobilization rates for 1990 and 2020 include combustion, deforestation and cement production (Leggett* et al.*, 1992; W. Pepper, 1996, personal communication).*

A critical aspect of the current Asian mobilization rates of CNPS is the rate of recent increase. From 1980 to 1990, N mobilization in Asia from fertilizer use alone approximately doubled (*c.* 20 to *c.* 40 Tg N $year^{-1}$; Galloway *et al.*, 1995). Between 1985 and 1995, C mobilization from combustion increased by about 35% (*c.* 0.8 to *c.* 1.1 Pg C $year^{-1}$; Leggett *et al.*, 1992; W. Pepper, 1996, personal communication). These rates and magnitudes are among the highest, if not the highest, in the last century.

The reactivity, bioavailability, solubility and/or volatility of the injected CNPS species generally permit extensive distribution to secondary reservoirs and accumulation of environmentally reactive species. Following injection into the atmosphere, carbon accumulates in the atmosphere, oceans and continents. Sulfur accumulates in soils and the ocean. Nitrogen accumulates in the atmosphere and in terrestrial, freshwater and marine ecosystems. Lacking a gas phase, P accumulates in terrestrial, freshwater and marine ecosystems.

There are environmental responses to the increased availability and accumulation of each element. The primary effect of N fertilization of agricultural ecosystems is an increase in crop productivity. Secondary effects include, among others, ecosystem acidification, unplanned increases in productivity of downwind and downstream ecosystems, reduction of stratospheric ozone and respiratory stress in humans. Most anthropogenic sulfur is initially emitted into the atmosphere where it increases atmospheric turbidity and acidity with secondary effects on the radiation balance and acidification of poorly buffered terrestrial and aquatic ecosystems. Carbon mobilization contributes to the greenhouse effect and can increase productivity in terrestrial ecosystems. Following injection into the agricultural systems, P is quickly taken up by either the plants or soil surfaces. Exchanges of P among reservoirs are limited, with the primary exchange occurring between terrestrial and aquatic ecosystems via suspended particulate matter in streams and rivers. Primary and secondary effects are increased productivity, intended or not.

In addition to these responses caused by individual elements, there are also extensive environmental changes caused by the interactions among the elements, such that mobilization of one element can have profound impacts on the other elements. The interaction is driven by the fact that CNPS are all components of living biomass (e.g. marine phytoplankton, $C_{106}N_{16}S_2P$; similar equations can be written for land plants, animals, etc.). N and P fertilize forests, resulting in increased retention of C. N and P also fertilize coastal ecosystems resulting in increased production of CO_2. N increases the oxidizing capacity of the atmosphere, resulting in faster chemical transformations in the atmosphere of N, S and C. Ecosystem acidification by S and N results in increased mobilization of P and perhaps C. Hence it is crucial to our understanding of the human controls on Earth systems to determine not only the anthropogenic mobilization rates of specific elements (Fig. 1.1) and the initial environmental responses, but also the degree of distribution of the mobilized element throughout the Earth's reservoirs, and the interactions and feedbacks with the biogeochemical cycles of other elements.

The rapid increase in CNPS mobilization observed for Asia over the last few decades will continue. The mobilization of N by fertilizer use in Asia is projected to double from about 40 to about 80 Tg N year^{-1} over the period 1990 to 2020 (Galloway *et al.*, 1995). Rates of N mobilization by fossil fuel combustion and biomass burning are projected to increase from about 8 to 18 Tg N year^{-1} over the same period (Leggett *et al.*, 1992; W. Pepper, 1996, personal communication). The total increase in N mobilization in Asia is thus estimated to be from about 45 to 100 Tg N year^{-1} from 1990 to 2020. These increases are large relative to what will occur in other regions. In fact, the amount of N mobilized in Asia in 2020 will be significantly greater then in all the rest of the world (Fig. 1.1). Significant increases in C mobilization are also projected to occur, primarily from fossil fuel combustion. In 2020, the total amount of C mobilized (from fossil fuel combustion, cement production and forest clearing) is projected to be *c.* 4 Pg C year^{-1} compared to *c.* 2 in 1990 (Fig. 1.1). P mobilization in Asia will also increase significantly due to the demand for increasing food productivity. In 1990, Asia consumed about 35% of the world's P fertilizer supply. By 2020, we estimate that it will consume about 50%. In 1990, S mobilization from combustion, biomass burning, and industrial activities in Asia was second only to Europe. By 2020, we estimate that it will account for half of the S emitted to the world's atmosphere.

The magnitude of these projections is highly uncertain; the direction of change is not. Population and economic growth in Asia will increase CNPS mobilization. With over 50% of the world's population and a rapidly growing economy, the amount of increased CNPS mobilization will be significant. Given the critical roles that CNPS have in almost all biogeochemical processes, the regional and global impact of the increased CNPS availability will not only be significant, but may be controlling.

Symposium overview

Historical perspective

The continuing passage of Asian regions through the culture milestones, together with increasing population, has resulted in increasing resource use. There are several historical studies of environmental changes in Asia over the past millennia that have accompanied these cultural changes (e.g. India; Gadgil & Guha, 1992). Indeed, in some instances, paleo data for some regions are rich in information for several millennia (NRC, 1992*a*). This volume examines one specific aspect of past change in Asia – the paleo monsoon.

The weather patterns in the Asian region are tightly coupled to the monsoonal patterns coming off the Pacific and Indian Oceans and influenced by the Tibetan Plateau. Understanding the factors affecting the monsoons of the region has been an active research field in Asia. The rich paleoclimatic records for the Asian region provide a basis for studying the spatial and temporal changes of the monsoonal rainfall and temperature regimes in the region during the past 18 000 years and for studying the factors modifying local to global changes in climate, biogeochemistry, and human development (Winkler & Wang, 1993). The data sources are diverse, being derived, for example, from ice cores, peat and lake sediments, loess soil analyzes, tree ring data, and the historical written records of the region. The paleo-records are valuable archives used to better understand changes in regional weather patterns and tele-connections with the global climate system (An *et al.*, 1991*a,b*; An & Thompson, this volume).

The wealth of paleoclimate data provides a basis for discerning possible connections between the changing climate patterns, both at the regional and global scale. Analysis of tree ring data provides evidence of the importance of the monsoonal patterns in the moisture stress of vegetation in the Tibetan, Qinling, and other semiarid regions of Asia. These data are used to reconstruct temperature and precipitation patterns for these regions (Wu *et al.*, 1988; Wu, 1992; Liu *et al.*, 1996). The use of loess soil analysis in various regions of China has provided evidence of regional and temporal oscillation between cold/dry periods and cool/humid periods associated with the Younger Dryas phenomenon. The use of these geological data, including lake levels, pollen records, and eolian dust deposits, indicate that maximum summer monsoonal rainfall occurred at different times for different regions of East Asia (An *et al.*, 1996).

At the global level, paleo-records indicate similar climate patterns during the Little Ice Age and Medieval Warming Period. An and Thompson (this volume) cite tree ring analysis from samples collected from the Tibetan Plateau to determine the regional maximum cooling period of the Little Ice Age to be about the middle of the 17th Century with temperatures approximately 1 °C cooler than present. The analysis suggests that certain North Hemisphere processes were working synchronously during the past 500 years. From historical records of planting and crop production from China, indications of climate patterns can be discerned. An and Thompson (this volume) point out that certain key crops, such as Citrus and a fiber crop, were recorded to be growing in agricultural areas of China north of their current distributions and that the suggested temperature regime may have been as much as 1 °C warmer than at present. These findings are consistent with climate patterns derived from other historical documents, plant phenological records, and ice core records from China, and indicate that certain of the climate phenomenon are consistent throughout the Northern Hemisphere and may also be global in nature.

The ice core records of Tibetan Plateau provide an excellent research opportunity to understand the global climate dynamics during the past 20 000 years. Analysis conducted by Thompson and others indicate that a global warming has taken place during the past 200 years. Comparison of Tibetan ice core analyzes with ice cores from Peru, Antarctica, and Greenland suggest that globally a cooler and drier conditions prevailed during the last glacial stage, and that during the past 15 000 years several major drought and wet periods generally coincide and that temperatures reached a warming maximum approximately 8000 to 5000 years BP (Medieval Warming Period), followed by a cooling indicated of the Little Ice Age, and that in recent deposits a warming is taking place (An & Thompson, this volume).

The use of the various paleo-records to reconstruct regional climate patterns for Asia is progressing rapidly with the collaborative efforts of the scientific community. These efforts will greatly facilitate the regional and global connections controlling the critical monsoonal dynamics of the region.

Terrestrial ecosystems

The Asian region encompasses examples of all the major terrestrial ecosystems recognized on this planet, ranging from cold deserts to humid rainforests and includes barren ice fields and rich coastal forest areas. Much of the region has undergone massive transformations under the pressure of human efforts to harness energy and mineral resources, harvest food and fuelwood, and create settlements for over 3.4 billion people living in Asia. Although human habitation in this region has been present for many thousands of years, recent increases in human populations during the past century have intensified the land use in the region markedly. It is this expansive human–environmental landscape that many of the IGBP core projects utilize to frame their research studies in this region in order to gain a better understanding of the impact of climate changes on ecosystems, feedbacks of these systems to global environmental change, and the role humans play in modifying the rates of changes in ecosystem processes.

The diversity of ecosystem types found in the region reflects the large environmental gradient which exists over the vast area, extending from the tropical regions at the equator to the taiga regions in the high latitudes. The development of these terrestrial ecosystems over the past 20 000 years has served as a biological record of climate-related changes as recorded in pollen records, sediment cores, tree cores (An & Thompson, this volume). The analysis of these ecosystem changes to climate is fundamental to understanding the sensitivity of ecosystems to future climate change. Movement of forest and grassland transition zones has been dynamic over the past 10 000 years, and even today relict forest stands can be found in predominantly grassland steppe areas.

How these terrestrial ecosystems respond to climate change is dependent on a number of factors, including fire and grazing disturbance patterns (Malingreau & Zhuang, this volume; Ojima *et al.*, this volume), factors influencing ecosystem structure such as self-thinning (Schulze & Heimann, this volume), and human activities (Neue & Boonjawat, this volume; Skole *et al.*, this volume). The various ecosystems in the region have different sensitivities to climate change. The semiarid ecosystems tend to be very sensitive to changes in water availability (Ojima *et al.*, this volume). In analysis of semiarid steppe ecosystems in the Asian region, Xiao and others (Xiao *et al.*, 1995*a*) found that above-ground grass production was correlated more to growing season rainfall input than to annual rainfall amounts. This dependence on growing season rainfall was further illustrated in analysis of potential climate change impacts using different general circulation model (GCM) scenarios. The results of this analysis indicated that small shifts in the seasonal rainfall pattern could greatly affect the plant production and the storage of soil C and N in these steppe ecosystems of China, Mongolia and Russia (Xiao *et al.*, 1995*b*; Ojima *et al.*, 1996).

Intensification of human use in the Mongolian steppe region of China, Mongolia, and Russia, due to agricultural activities, increased grazing pressure, and recent fire events have resulted in increased soil erosion, loss of soil fertility, and in extreme cases in desertification of certain landscapes with moving dunes being observed. In many parts of Asia, livestock numbers have doubled or tripled in numbers during the past century (Richards & Flint, 1994; NRC, 1992*b*) resulting in severe overgrazing problems. This intensification of use and the inherent variable rainfall patterns of this region mark this region as highly vulnerable to environmental degradation and loss of natural resources and biodiversity. The ability of the people within the region to maintain a sustainable level of agricultural production of animals and crops will be further tested in the coming decades.

The role fire plays in modifying terrestrial ecosystem processes has been studied for many years; however, the ways in which fire may alter terrestrial ecosystem feedback to the global environment is now becoming more clearly understood. Initial focus of the role of biomass burning has been in the tropical savanna regions of the world, but recent observations of extensive fires in the boreal regions of Russia, Mongolia, and China indicate that these regions, too, may play an important role in global atmospheric chemistry (Schulze & Heimann, this volume; Malingreau & Zhuang, this volume). An important research question for the high latitude region is related to the fire return cycle for this region and the nature of the ecosystem recovery pattern. These two factors affect the net carbon storage and atmospheric exchange of carbon from an ecosystem (Schulze & Heimann, this volume).

In areas of temperate and tropical regions of Asia, fire is more commonly used for removal of crop residue, in shifting cultivation practices, and fuel wood consumption. The extent of the fire areas is not well known; however, research efforts are under way to better document the occurrence and spread of fires and the recovery of vegetation after an area is burned (Malingreau & Zhuang, this volume). The research on fire in Asia is an important element in the development of a better understanding of the factors influencing regional atmospheric chemistry and factors influencing changes in land cover.

In tropical areas of the Asian region, other factors influencing land cover change are being studied intensively (Skole *et al.*, this volume). Deforestation of Southeast Asian tropical forests appears to be in the same order of magnitude as had occurred in the Amazon Basin or approximately 1.4 million ha per year (compared to 1.6 to 2.2 million ha per year in the Amazon) during the mid-1970s thru mid-1980s. The most extensive activity was found to occur in the north and eastern parts of Thailand, where expansion of agriculture and logging have occurred and shifting agriculture has intensified (Skole *et al.*, this volume). This research activity is also attempting to better identify the extent to which secondary forests are being impacted by the intensification of land use. These changes in forest turnover and subsequent changes in forest species and structural characteristics will have an impact on the carbon and water cycling within this region (Schulze & Heimann, this volume; Skole *et al.*, this volume).

Another major transformation which has affected the Asian landscape for many centuries has been the development of rice cultivation practices. Over the centuries the dependency of the Asian people on rice as their major food grain has resulted in the development of an intricate set of rice production systems. Asia is the major rice producing region of the world and accounts for approximately 90% of the area planted as irrigated rice, globally. In addition, the Asian rice production systems are highly productive resulting from the development of double-crop and triple-cropping systems in the region (Neue & Boonjawat, this volume). These factors have maintained the balance between rice consumption and demand in the region. Recent concern over the increasing global atmospheric methane concentrations have focused attention on the cultivation practices influencing methane production and emissions from rice production systems. Research efforts by the IGBP-IGAC RICE Activity have been able to identify water management practices, soil fertility properties, and rice cultivar factors that may be modified to reduce methane production and emissions while maintaining high production levels (Neue & Boonjawat, this volume).

Other major land use changes taking place and rapidly transforming natural resource allocation with the Asian region. Urban expansion is rapidly growing in

Southeast China and other Asian countries, major water works are under development that will alter water resources, and political and social changes are altering market constraints that will alter land use patterns throughout Asia. What the Asian landscape will be in the next century is at the core of much of the current research effort on land use/cover changes. The regional research groups are engaged in collaborative efforts within the global change research communities to better understand the role of Asian ecosystems in changes in the global environment.

Aquatic ecosystems

In Asia, as elsewhere, humans have had a significant impact on aquatic ecosystems. Physically, channels have been diverted; lakes drained or expanded; impoundments created or enlarged. Chemically, sediment and nutrient loads have changed, usually in the upward direction. These types of physical and chemical changes have had consequences for the biological communities specifically and the ecosystem generally. The scale of the changes ranges from the local pond, to the large rivers and oceans. Given our focus on global scales, this book examines the interaction of Asia with aquatic ecosystems for the river basin (Vörösmarty *et al.*, this volume), coastal ocean (Hu *et al.*, this volume), open ocean (Chen & Tsunogai, this volume) and the biogeochemical interactions among these three systems (Mackenzie *et al.*, this volume). The role of Asian rivers in suppling water and sediment to coastal oceans is impressive: 35% of river flow and about 50–70% of the sediment load (Degens *et al.*, 1991; Hu *et al.*, this volume) are discharged from Asian rivers. The Yellow River (Huanghe) has changed its course 26 times as a consequence of human and climatic forcing over the last 3000 years (Vörösmarty *et al.*, this volume; Hu *et al.*, this volume); it even changed the coastal sea into which it discharged (Hu *et al.*, this volume). Just as humans affect rivers, so does the reverse occur. In 1931, the Yellow River flood was responsible for over 4 million deaths (Vörösmarty *et al.*, this volume).

Human settlement in a watershed converts soil to sediment by erosion. During the past 20 years, it is estimated that 480 billion tonnes of topsoil have been lost from farmland globally by erosion (Mackenzie *et al.*, this volume). By way of example, large-scale human migrations of people into the Yellow River watershed occurred in the Han Dynasty (206 BC–AD 220) and the Tang Dynasty (AD 618–AD 907). The migrations were followed by extensive agricultural activities and loss of vegetation cover in the Loess Plateau. As human populations increased, erosion and sediment river load increased due to forest clearing and agricultural practices. Currently, the sediment load of the Yellow River is ten times greater then it was prior to the Han Dynasty (Hu *et al.*, this volume). There is another human impact that lessens sediment loss via rivers – river management (e.g. dike and dam construction). The management of rivers has at least a 5000–6000-year history in

China (Vörösmarty *et al.*, this volume) and in the last few decades has greatly increased. In 1950, there were 8 dams in China, in 1982 there were 18 600, representing 55% of the global total. The increase in dams has resulted in the decrease in sediment load. In the last 40–50 years, the Yellow River sediment discharge has decreased rapidly (Hu *et al.*, this volume). The impact of climate change on this already significant human influence is poorly known. Water scarcity and excess at the regional level have been predicted under climate change scenarios, but as Vörösmarty *et al.* (this volume) discuss, the ability of current GCMs to predict future river flow under climate change scenarios is problematic. Due however, to the potential changes, it is critically important to improve our understanding of the relationship among people, climate and rivers.

Just as humans have had direct and indirect effects on water and its suspended solid phase, there are also effects on the dissolved constituents. The increase in anthropogenic mobilization of N on the Asian continent from zero in the pre-human past to about 45 Tg N year^{-1} in 1990 (Fig. 1.1) is reflected in the riverine N flux. Over the period 1963–1980, the NO_3^- concentration increased by about a factor of four in the Yangtze River (Changjiang). There were also increases in NH_4^+; over the period 1980–1987, concentrations increased by about a factor of two (Zhang *et al.*, 1995). In the associated estuary of the Yangtze, the inorganic N concentration increased by about a factor of three from the 1960s to the 1990s. The main sources of the inorganic N for the Yangtze River are fertilizer, drainage, soil erosion and precipitation (Hu *et al.*, this volume). Increases of NO_3^- and NH_4^+, similar to those observed in the Yangtze, have been also been noted in the Yellow River (Zhang *et al.*, 1995). Although data are very limited, given the extent of phosphorus mobilization in Asia by fertilizer use, it is reasonable to assume that phosphate concentrations are also increasing. The concentrations of these important nutrients will continue to increase. While the rates are unknown, the increases are likely to be substantial. By 2030, assuming an Asian population of 5 billion and similar per capita resource use as currently exists in industrial nations, the N and P discharged via rivers will be four times what the industrial nations discharge today (Mackenzie *et al.*, this volume).

There are very limited data on the fate of the N and P injected into Asian coastal and shelf waters, and some of what exists is not readily available. Evidence from other regions suggests that the consequences of CNPS injection into Asian saline waters will be complex. An analysis of the North Atlantic Ocean and its basin showed that, of the 2.4 Tg N injected into saline waters via rivers, most is denitrified in estuaries and the shelf. In fact, unless there was a major river (e.g. the Amazon), denitrification on the shelf is so strong that N must be supplied from the open ocean to support it. For phosphorus, about 25% is transported across the shelf to the open ocean (Nixon *et al.*, 1996; Seitzinger and Giblin, 1996; Galloway *et al.* 1996*b*). Unfortunately, a similar assessment has not been done for the Asian river waters entering the Asian marginal seas and the Pacific and Indian Oceans. It is a complex issue, for along with the nutrient supply from rivers must be considered the upwelling of nutrient-rich waters form the Asian Seas (Chen & Tsunogai, this volume). Until we are able to more fully understand the natural and anthropogenic controls on the biogeochemistry of Asian marine waters, the ultimate fate and thus impact of riverine anthropogenic N, P and C are uncertain.

The transfer of carbon from Asia to the oceans as a consequence of human activities has increased by direct and indirect mechanisms (Chen & Tsunogai, this volume). Direct input occurs via riverine injection; globally, the flux of particulate organic carbon has doubled (Wollast & Mackenzie, 1989; Mackenzie *et al.*, this volume). Indirect input of anthropogenic carbon occurs following CO_2 release to the atmosphere and subsequent uptake by the ocean. Over the course of time, some regions of the oceans have changed from a net CO_2 source to a net sink, owing to the rising atmospheric CO_2 concentrations (Mackenzie *et al.*, this volume; Chen & Tsunogai, this volume). The marginal seas of the North Pacific are perhaps important sinks for some of this carbon; Chen and Tsunogai (this volume) estimate that they have taken up over 1×10^{15} g of excess carbon.

Humans have had extraordinary impacts on the discharge of water, N, P and C from Asia to the coastal oceans. Water is removed for irrigation, soil, N, P and C are lost from arable land, sewage is transported from populated regions to downstream systems, etc. Keeping in mind that all of these direct impacts will increase significantly in the future, the addition of climate changes will compound the effects on an already stressed system. Unfortunately, our understanding of these direct and indirect human impacts on the biogeochemistry of Asian rivers and coastal areas is at the early stages. Our knowledge must increase, but it will be in the context of a constantly changing system as the direct and indirect impacts of human on Asian aquatic ecosystems increase.

Conclusions

As all chapter authors have illustrated, and Fu *et al.* (this volume) discuss explicitly, Asia currently has, and will increasingly have, significant impacts on Asia. In addition, due to the magnitude of CNPS mobilization, and extensive global distribution, Asia has and will have impacts on global scale processes. Fifty years ago, Asian contributions to many anthropogenic fluxes were small relative to other regions of the world. Now, and increasingly in the decades to come, they will be the highest; in some instances, greater than the sum of all other sources.

In the light of this situation, there are several actions that will improve our ability to determine the impact of Asia on global change.

- A greater involvement of Asian scientists in global change research, especially from those countries with large land areas, extensive coastlines, and/or large populations.
- A series of investigations at the large watershed and airshed scales on the fluxes and fates of CNPS mobilized in Asia.
- Increased efforts on exchange of environmentally relevant data.
- A stronger presence of human dimensions scientists to aid in the understanding of the concerns and constraints of different cultures on global change issues.

References

An, Z. S., Kukla, G., Porter, S. C. & Xiao, J. L. (1991*a*). Magnetic susceptibility evidence of monsoon variation on the Loess Plateau of central China during the last 130 000 years. *Quaternary Research*, **36**, 29–36.

An, Z. S., Kukla, G., Porter, S. C. & Xiao, J. L. (1991*b*). Late Quaternary dust flux on the Chinese Loess Plateau. *Catena*, **18**, 125-32.

An, Z. S., Porter, S. C., Kutzbach, J. E., Wu, X. H., Wang, S. M., Liu, X. D., Li, X. Q. & Zhou, W. J. (1996). Asynchronous Holocene Optimum of the east Asian summer monsoon in central and eastern China. Submitted to *Quaternary Science Reviews*.

Ayers, R. U., Schlesinger, W. H. & Socolow, R. H. (1994). Human impacts on the carbon and nitrogen cycles. In *Industrial Ecology and Global Change*, ed. R. H. Socolow, C. Andrews, R. Berkhout & V. Thomas, pp. 121–55. New York: Cambridge University Press.

Degens, E. T., Kempe, S., Richey, J. E. (1991). Summary: biogeochemistry of major world rivers. In *Biogeochemistry of Major World Rivers*. ed. E. T. Degens, S. Kempe & J. E. Richey, SCOPE. Chichester: John Wiley.

FAO (1996). United Nations FAO, *Annual Fertilizer Review*. Rome, Italy: Food and Agriculture Organization of the United Nations.

Gadgil, M. & Guha, R. (1992). *This Fissure Land: An Ecological History of India*, 274 pp. Oxford: Oxford University Press.

Galloway, J. N., Schlesinger, W. H., Levy II, H., Michaels, A. & Schnoor, J. L. (1995). Nitrogen fixation: anthropogenic enhancement-environmental response. *Global Biogeochemical Cycles*, **9**, 235–52.

Galloway, J. N., Howarth, R. W., Michaels, A. F., Nixon, S.W., Prospero, J. M. & Dentener, F. J. (1996*a*). Nitrogen and phosphorus budgets of the North Atlantic Ocean and its watershed. *Biogeochemistry*, **35**, 3–25.

Galloway, J. N., Zhao, D., Thomson, V. E. & Chang, L. H. (1996*b*). Nitrogen mobilization in the United States of America and the People's Republic of China. *Atmospheric Environment*, **30**, 1551–61.

IGBP (1995). Natural and anthropogenic changes: impacts on global biogeochemical cycles, *Book of Abstracts*. Beijing, China: IGBP.

Leggett, J., Pepper, W. J. & Swart, R. J. (1992). Emissions Scenarios for the IPCC: an update. Assumptions, methodologies, and results. Prepared for the Intergovernmental Panel on Climate Change, Working Group I, May 1992. Published as Appendix A3. In *Climate Change 1992*: The Supplementary Report to the IPCC Scientific Assessment, Report of the IPCC Scientific Assessment

Working Group, ed. J. T. Houghton, B. A. Callander & S. K. Varney. Cambridge: Cambridge University Press.

Liu, Y., Wu, X. D., Leavitt, S. W. & Hughes, M. K. (1996). Stable carbon isotope in tree rings from Huangling, China and climatic variations. *Science in China* (Series D), **39**, 152–61.

Mackenzie, F. T. (1994). Global climatic change: climatically important biogenic gases and feedbacks. In *Biotic Feedbacks in the Global Climatic System: Will the Warming Feed the Warming*, ed. G. M. Woodwell & F. T. Mackenzie, pp. 22–46. Oxford: Oxford University Press.

NRC (1992*a*). China and Global Change: Opportunities for Collaboration. Panel on Global Climate Change Sciences in China. National Research Council. Washington, DC: National Academy Press.

NRC (1992*b*). Grasslands and Grassland Sciences in Northern China. A report of the Committee on Scholarly Communication with the People's Republic of China. National Research Council. Washington, DC: National Academy Press.

Nixon S. W., Ammerman, J. W., Atkinson, L. P., Berounsky, V. M., Billen, G., Boicourt, W. C., Boynton, W. R., Church, T. M., DiToro, D. M., Elmgren, R., Garber, J. H., Giblin, A. E., Jahnke, R. A., Owens, N. J. P., Pilson, M. E. Q. & Seitzinger, S. P. (1996). The fate of nitrogen and phosphorus at the land–sea margin of the North Atlantic Ocean. *Biogeochemistry*, **35**, 141–80.

Ojima, D. S., Parton, W. J., Coughenour, M. B., Scurlock, J. M. O., Kirchner, T. B., Kittel, T. G. F., Hall, D. O., Schimel, D. S., Garcia Moya, E., Gilmanov, T. G., Seastedt, T. R., Kamnalrut, A., Kinyamario, J. I., Long, S. P., Menaut, J -C., Sala, O. E., Scholes, R. J. & van Veen, J. A. (1996). Impact of climate and atmospheric carbon dioxide changes on grasslands of the world. In *Global Change: Effects on Coniferous Forests and Grasslands*, ed. A. I. Breymeyer, D. O. Hall, J. M. Melillo & G. I. Ågren, pp. 271–311. Chichester: John Wiley.

Richards, J. F. & Flint, E. P. (1994). Historic land use and carbon estimates for south and southeast Asia 1880–1980. Carbon Dioxide Information Analysis Center, Oak Ridge National Laboratory, Environmental Sciences Division Publication No. 4174, Oak Ridge, TN.

Seitzinger, S. P. & Giblin, A. E. (1996). Estimating denitrification in the North Atlantic continental shelf sediments. *Biogeochemistry*, **35**, 235–60.

Xiao, X., Wang, Y., Jiang, S., Ojima, D. S. & Bonham, C. D. (1995*a*). Interannual variation in the climate and above-ground biomass of *Leymus chinense* steppe and *Stipa grandis* steppe in the Xilin river basin, Inner Mongolia, China. *Journal of Arid Environments*, **31**, 283–99.

Xiao, X., Ojima, D. S., Parton, W. J., Chen, Z. & Chen, D. (1995*b*). Sensitivity of Inner Mongolia grasslands to climate change. *Journal of Biogeography*, **22**, 643–8.

Winkler, M. G. & Wang, P. K. (1993). The late-quaternary vegetation and climate of China. In *Global Climates Since the Last Glacial Maximum*, ed. H. E. Wright, Jr., J. E. Kutzbach, T. WebbIII, W. F. Ruddiman, F. A. Street-Perrot & P. J. Bartlein, pp. 221–64. Minneapolis: University of Minnesota Press.

Wollast, R. & Mackenzie, F. T. (1989). Global biogeochemical cycles and climate. In *Climate and Geo-Sciences*, ed. A. Berger, S. Schneider & J. C. Duplessy, pp. 453–73. Dordrecht: Kluwer Academic Publishers.

Wu, X. D., Lin, Z. Y. & Sun, L. (1988). A preliminary study on the climatic change of Hengduan Mountains area since AD1600. *Advances in Atmospheric Sciences*, **5**, 437–43.

Wu. X. D. (1992). Dendroclimatic study in China. In *Climate Since AD1500*, ed. R. S. Bradley & P. D. Jones, pp. 432–5. London: Routledge.

Zhang, J., Yan, J. & Zhang, Z. F. (1995). Nationwide river chemistry trends in China: Huanghe and Changjiang. *Ambio*, **24**, 275–9.

Paleoclimatic change of monsoonal China linked to global change

Z. S. An and L. G. Thompson

Introduction

The Asian monsoon plays a significant role in the global hydrological and energy cycles. To a large extent, it controls the climate and environment over populous China. Our knowledge of the Asian monsoon has increased greatly in the last decade. It consists of two relatively independent subsystems: the East Asian monsoon and the India monsoon (Gao *et al.*, 1962), and the Tibetan Plateau monsoon phenomena (Tang, 1979). The thermal and dynamic influences of the Tibetan Plateau not only play a significant role in the evolution and formation of the Asian monsoon circulation but affect global change as well. In addition to the current role of Asia in relation to the earth system, we have to look into past records of climate change for the perspective of the human environment.

Historical, archaeological, tree ring, ice core records, as well as other proxy data for the last 2000 years, indicate that the climate of monsoon China was characterized by the alternation of dry and wet periods through historical times. Palaeomonsoon archives recorded in loess deposits, deserts, lakes, glaciers, vegetation patterns, mammalian fossils, coast lines, continental shelf deposits, and treeline variations clearly delineate the alternation of winter and summer monsoon predominance during the last 150 000 years (An *et al.*, 1991*a*).

The paleomonsoon record of China reflects the linkages of the Asian monsoon system to the global nature of climatic change. The correlation of climatic variations between monsoonal China and the Polar-Atlantic during the last glaciation and the Little Ice Age implies that climates in these regions are linked through the westerlies. In addition, the similarities in climatic history over the last 1500 years recorded in ice cores from the Tibetan Plateau and Peru suggest low frequency teleconnections between the Northern and Southern Hemispheres.

Climatic change in the historical period

The climate of central-eastern China is characterized by seasonal alternations of the East Asian winter and summer monsoons (Gao *et al.*, 1962) (Fig. 2.1). The

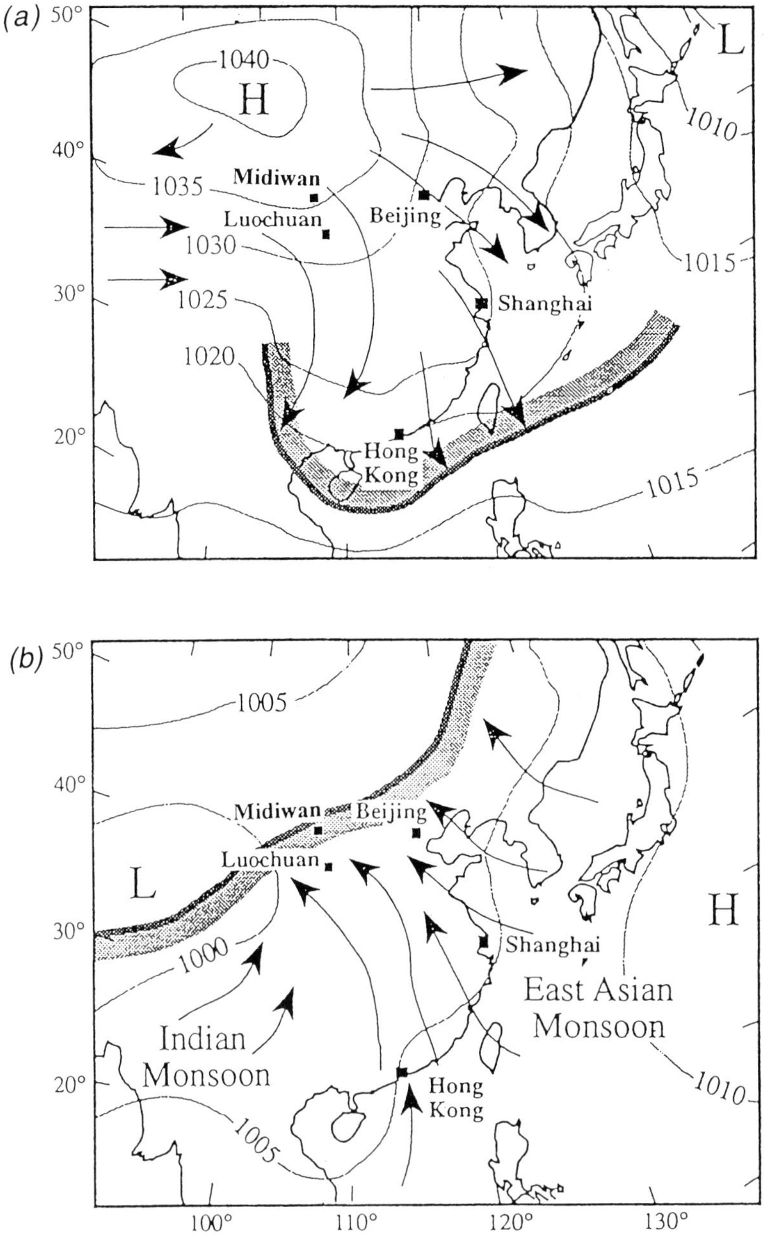

Figure 2.1 *Winter and summer monsoon regimes affecting China (Gao* et al., *1962). (a) January surface pressure (mbar), dominant wind vectors (arrows), and average southern limit of the polar front (bold line). Northwesterly winds associated with the Mongolian High and westerly winds from central Asia entrain the bulk of the eolian dust delivered to the Loess Plateau. (b) July surface pressure, dominant wind vectors, and average landward limit of the monsoon front. Summer monsoon precipitation is delivered to the Loess Plateau by the onshore flow of western Pacific maritime air which dominates the East Asian summer monsoon regime (Porter & An, 1995).*

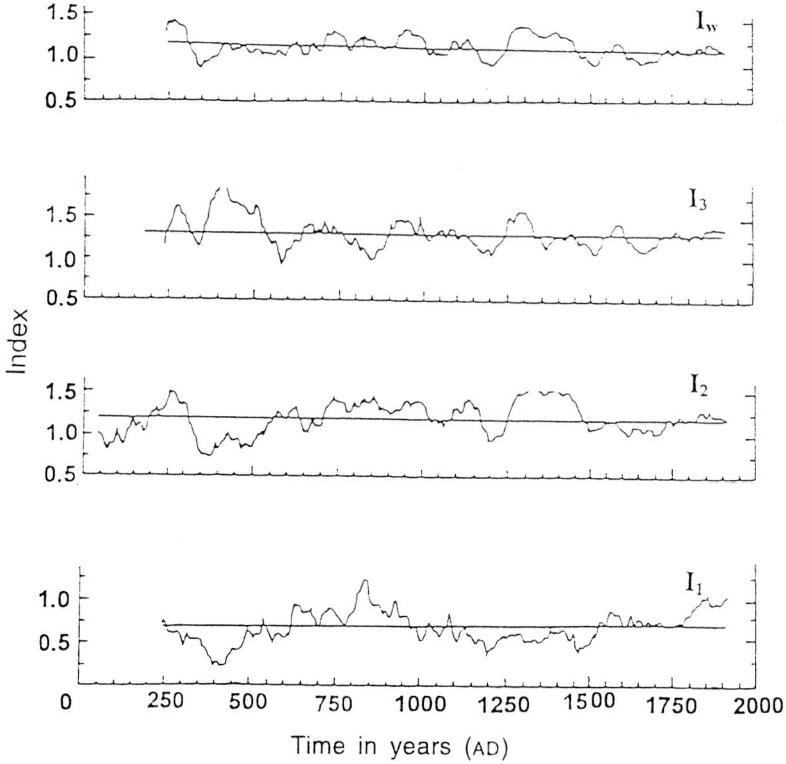

Figure 2.2 *Variations of the moisture indices for the semi-arid area (I_1), semi-wet area (I_2), wet area (I_3) and the whole area indices (I_w) have been smoothed by a 15-point moving average. The index I is defined as: $I=2F/(F+D)$, where F and D are the numbers of reported flood and drought events during a five-year interval. The moisture index curves during the past 2000 years in different regions reflect the climatic fluctuation of dryness and wetness with important temporal and spatial differences of various climatic zones over monsoonal China (Gong & Hameed, 1991). The equation for calculating the moisture index I used by Gong and Hameed (1991) is not very exact, but it is available to estimate the climatic fluctuation of dryness and wetness. This index seems to be biased towards floods or wet periods and should be ≥ 1.*

records of moisture conditions in China over the last 2000 years show alternating wet and dry periods of long duration, with important regional differences between semi-arid, semi-wet and wet zones (Gong & Hameed, 1991) (Fig. 2.2). Zhang P.Y. *et al.* (1994) propose a stage subdivision for climatic evolution history in eastern China for this same period, and suggest that the present structure of the monsoon climate might be similar to that in AD 880–1230. Recently, regional high resolution records of wet and dry periods for the last 1000 years have been published (Zhang, 1995), and much research has been conducted on droughts and floods over the last 2000 to 500 years (Zheng *et al.*, 1975; Zhang, 1983; Xu & Zhang, 1983). The Meiyu (plum rains) activity over the areas of the middle/low reaches of the

Yangtze River (Changjiang) is an important weather and climate phenomenon. The 18th century summer precipitation in eastern China, which was related to the Meiyu activity, was reconstructed based on the *Clear and Rain Records* (Zhang, 1991; Wang & Zhang, 1992) (Fig. 2.3). These results reflect the climatic periodicity as well as regional differences. The spatial and temporal distributions of monsoon precipitation in China seem to be related to the migration of the precipitation band associated with monsoon front changes.

Zhu (1973) was the first to use historical documents and phenological records to reconstruct the history of temperature fluctuations through the last 5000 years in eastern China. This history is consistent with snowline variations in Norway. The variations in air temperature over China are similar to those in the Northern Hemisphere over the last 50 years (Wang & Zhao, 1995), and show cyclicity of 36,

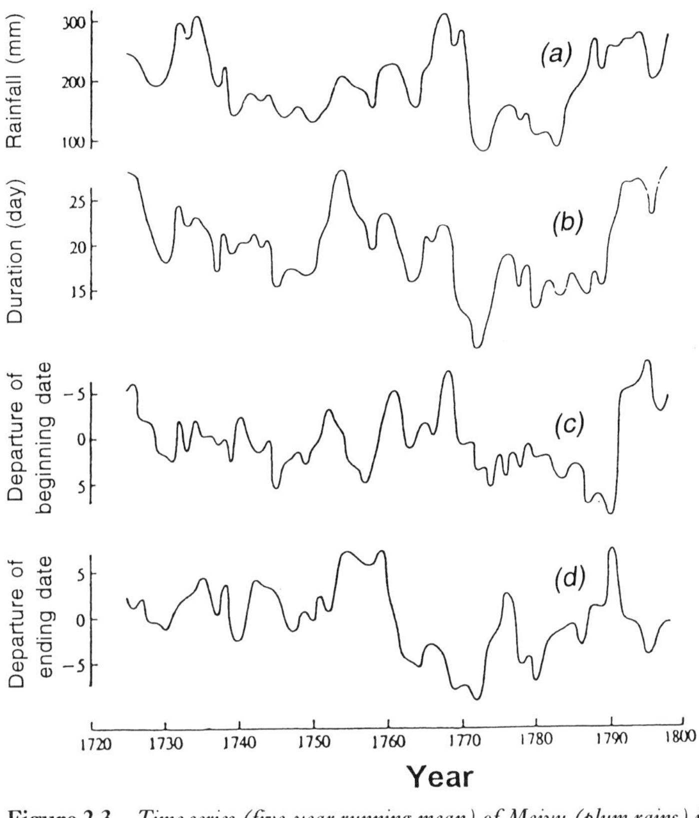

Figure 2.3 *Time series (five-year running mean) of Meiyu (plum rains) indices during the 18th century in the middle and lower reaches of the Yangtze River. (a) Meiyu rainfall (mm), (b) Meiyu duration (days), (c) departure(days) from Meiyu beginning date mean, (d) departure (days) from Meiyu ending date mean. Variations of the beginning dates, ending dates and duration of Meiyu show the period with 2–3 years. The quasi-periods with 4–5 years and 9-year, and a 32-year quasi-period was extracted from Meiyu rainfall time series based on the power spectrum analysis (Zhang & Wang, 1991).*

22 and 2–3 years over the last 100 years (Wang, 1978; 1981). Global teleconnections over the last few hundred years are also seen in the comparison of ice core records from the Dunde ice cap of China, the Quelccaya ice cap of Peru, and Landsberg's Northern Hemisphere temperature reconstructions. All these records also indicate a warming trend in the 20th century (Thompson, 1992).

Since the 1980s, high resolution (annual and seasonal) tree ring records have been used to reconstruct palaeoclimate in monsoonal China. Temperature and precipitation in Tibet and Qinling have been reconstructed for the last several hundred years based on tree-ring widths (Wu *et al.*, 1988; Wu, 1992). The seasonal variations of $\delta^{13}C$ in tree rings are mainly controlled by relative humidity during the growing season and are related to the arrival time of monsoon fronts (Li *et al.*, 1996). June temperature departures in the semi-arid regions are recorded in stable carbon isotopes of tree rings for the last 100 years (Liu *et al.*, 1996). The May–July moisture indices reconstructed from tree ring data from the central Qilian Mountains display a 5–6 year cycle (Wu *et al.*, 1996, in press) which is probably related to the El Niño Southern Oscillation (ENSO). The precipitation history in north-central China reconstructed from tree ring density chronology reveals that a period of severe drought occurred in the mid-20th century (Hughes *et al.*, 1994).

Paleoclimatic events

There are a few indications of abrupt climate change in monsoon China during the historical period (Zhang P. Y. *et al.*, 1994). The Little Ice Age (LIA) and the Medieval Warm Period can be identified from proxy climate records. Data from tree rings and historical documents show that the LIA in the Tibetan Plateau began toward the end of 12th century, and that the coldest period occurred in the middle of the 17th century, when the temperature was about 1 °C lower than the present mean (Wu & Lin, 1981). Documentary evidence has been used to reconstruct decadal mean temperature anomalies in north and east China during the LIA (Wang, 1992), and three cold periods (1450s to 1510s; 1560s to 1690s; 1790s to 1890s) were identified. The decadal mean temperature anomaly in the coldest phase during the LIA was about −1.0 °C and the average temperature anomalies during the three cold phases was −0.3 °C, −0.5 °C and −0.4 °C, respectively. The climatic fluctuations of the LIA in China can be compared with those in Europe and North America (Zhang, 1991), as well as Japan (Suess, 1968), and suggest a possible palaeoclimatic teleconnection in the Northern Hemisphere.

Evidence of the Medieval Warm Period has been observed from the distribution maps, which have been made from ancient records of agriculture involving *Citrus* trees and *Boehmeria niver* (Ramie), an important fiber plant, during the 8th, 11th and 13th centuries. When the historical planting distributions are compared with present, it is noted that the northern boundaries of agriculture of these plants in the 13th century lay to the north of modern boundaries. The 13th century limits suggest that the climate in eastern China at that time may have been the warmest in the last 1000 years, with mean temperature estimated at 0.9–1.0 °C higher than today (Zhang, 1993).

The spatial and temporal distribution of precipitation over the last 12 000 years, which is controlled mainly by the summer monsoon, has been established based on such geological data as lake levels, pollen records and eolian dust deposits (An *et al.*, 1996). Moreover, summer precipitation increments and monsoon strength index have been established using numerical modeling (using GCMs). The results of both geological data and numerical modeling indicate that Holocene full-Optimum, as identified by precipitation maxima, is obviously diachronous in central and eastern China. That is, the Holocene Optimum marked by East Asian summer monsoon precipitation reached its peak at different time intervals in different regions (12 000 years ago in northeast and southwest China, 9000 years ago in northwest and north China, 6000 years ago in the middle-lower reaches of the Yangtze River and 3000 years ago in south China) (An *et al.*, 1996) (Fig. 2.4). The spatial and temporal distributions of the monsoon precipitation over central and eastern China indicate that the intensity of the East Asian summer monsoon is controlled by the variation of solar radiation, which is determined by orbital forcing. The progressive southward and eastward movement of the monsoon precipitation peak is consistent with the general weakening of the summer solar radiation on the Northern Hemisphere during the Holocene (Kutzbach & Gutter, 1986).

The natural geography of East Asia contributes to considerable regional diversity which is characteristic of the East Asian monsoon climate. For example, during the Younger Dryas (YD, 11 000–10 000 years ago) cold and dry conditions were prevalent in the western section of the Tibetan Plateau (Gasse *et al.*, 1991). The Baxie profile (35°34' N, 103°35' E) on the Chinese Loess Plateau and a sediment core from Qinghai Lake (37°10' N, 100° E) provide evidence of a strengthening of the East Asian summer monsoon (An *et al.*, 1993) and an increase in lake water volume (Kelts *et al.*, 1989), respectively, during this interval. High-resolution paleoclimate proxy records at Midiwan (37°39' N, 108°37' E), along with radiocarbon dating of swampy sediments and an eolian sand–paleosol sequence from the desert–loess transition zone spanning the YD reveal the monsoon climatic oscillations from cold/dry to cool/humid, then a return to cold/dry conditions between 11 200 and 10 000 ^{14}C years. These records not only show the global nature of cold climate in the YD but also indicate the significant variability of monsoon climate on a centennial time scale in East Asia (Zhou *et al.*, 1996) which can be viewed as a response to the interaction between inland cold air masses and warm moist oceanic air masses. These paleoclimatic records provide analogs for understanding the temporal and spatial transitional processes from

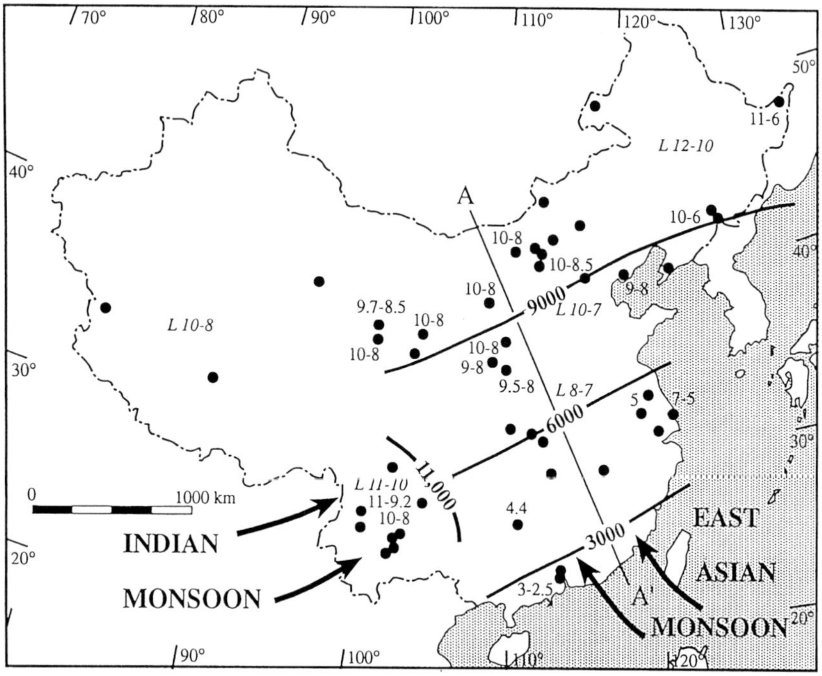

Figure 2.4 *Map of China showing position of East Asian Monsoon precipitation maximum through time based on paleoclimatic proxy data. Maximum precipitation/effective precipitation occurred c. 12 000–10 000 years ago in northeastern China, c. 10 000 – 7000 years ago in north-central and northern east-central China, c. 8000 – 5000 years ago in the middle and lower reaches of the Yangtze River, and c. 3000 years ago in southern China. The monsoon maximum dating to c. 11 000 years ago in southwestern China is related to the northeastward penetration of the Indian summer monsoon. The filled circles represent the data points with the ages of the precipitation peak reflected by biogeological data since the last 12 000 years (An* et al., *1996, in preparation).*

cold to warm conditions in East Asia. Therefore, attention should be paid to the regional differences in global climatic change events for modeling and prediction.

Ice core records related to Paleoclimates of the Northern and Southern Hemispheres

Many ice cores have been obtained from polar and mountain regions in the Northern and Southern Hemispheres (Fig. 2.5). The time series from many deep ice cores recovered from polar regions extends back beyond the Last Glacial Stage (LGS), which ended approximately 15 000 years ago. Ice cores from the Dunde ice cap on the northeast edge of the Tibetan Plateau have yielded the first record of LGS climate from outside the polar regions (Thompson *et al.*, 1989). This,

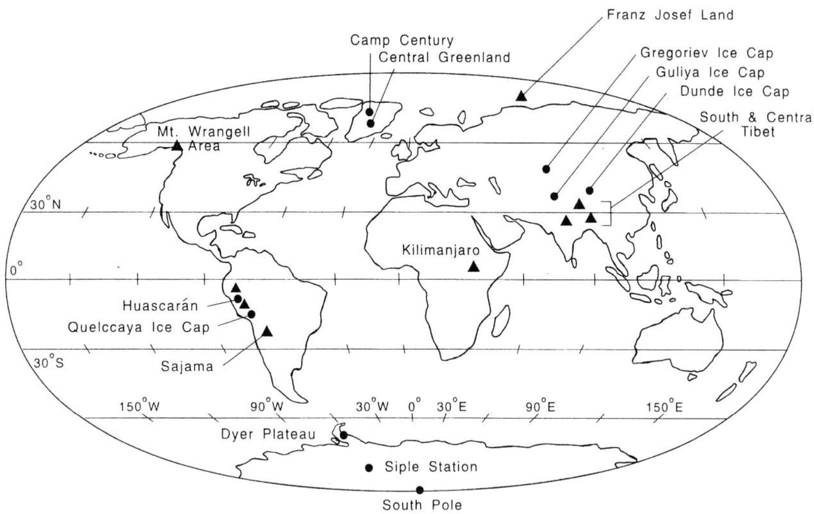

Figure 2.5 *Global ice core sites. The filled circles and filled triangles represent present sites and future sites, respectively.*

along with the subsequent recovery of glacial stage ice from Huascarán, Peru (Thompson *et al.*, 1995*a*) allows for direct comparison with polar cores from Dome C, Antarctica and Dye 3, Greenland (Fig. 2.6). All four records illustrate the high dust concentrations and more depletion of ^{18}O which characterize ice deposited during the LGS. Thus, for the first time, long-term polar ice core records can be compared with those from low latitudes. This comparison demonstrates a global cooling and decrease in water vapor (as inferred from increased dust concentrations and the depletion of ^{18}O in ice cores) during the LGS. Thus, the comparison of the low latitude/high altitude ice core records with their counterparts from the polar regions allows for the examination of global change on many spatial and temporal scales.

Two hundred year averages of $\delta^{18}O$ from five ice core sites (GRIP, Dunde, Guliya, Quelccaya and Huascarán) clearly show that a warming has occurred in the last 200 years in the low latitude sites (Fig. 2.7). This warming is particularly strong in the Dunde and Huascarán records, which suggests that mountain glaciers can be much more sensitive to annual, decadal and centennial time scales because they are much smaller than the polar ice caps. Because of their great size, the polar ice caps often influence their own climate and may not respond as readily to short-term climate variability.

Distinct annual dust layers are preserved in ice cores from the Dunde (eastern Tibetan Plateau) and Guliya ice cap (far western Tibetan Plateau), as well as in cores from the tropical Huascarán and Quelccaya ice cap (Thompson *et al.*, 1989;

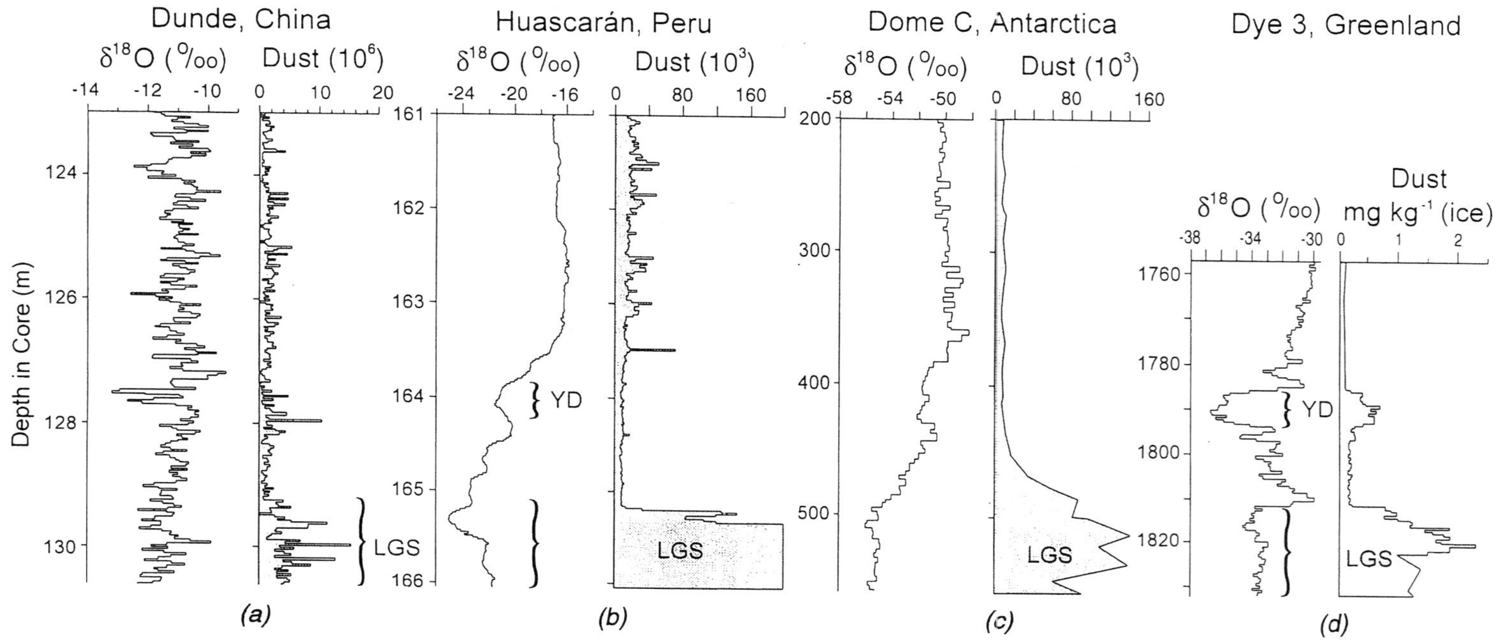

Figure 2.6 *The Late Glacial Stage-Holocene (LGS) Transition recorded by $\delta^{18}O$ and dust records from (a) Dunde, China, (b) Huascarán, Peru, (c) Dome C, Antarctica, and (d) Dye 3, Greenland, for portions of those cores containing LGS ice, the glacial–interglacial transition, and the YD. Dust concentrations for Huascarán, Dunde and Dome C represent the number of particles with diameters in the size range between > 0.63 μm and >16.0 μm per milliliter sample. For comparison of Huascarán dust with Dye 3 dust (in milligrams per kilogram of ice), the Huascarán Holocene average is 0.16 mg kg^{-1} and the LGS average is 32.2 mg kg^{-1}. The data of Huascarán, Dome C and Dye 3 are from Thompson* et al. *(1995a).*

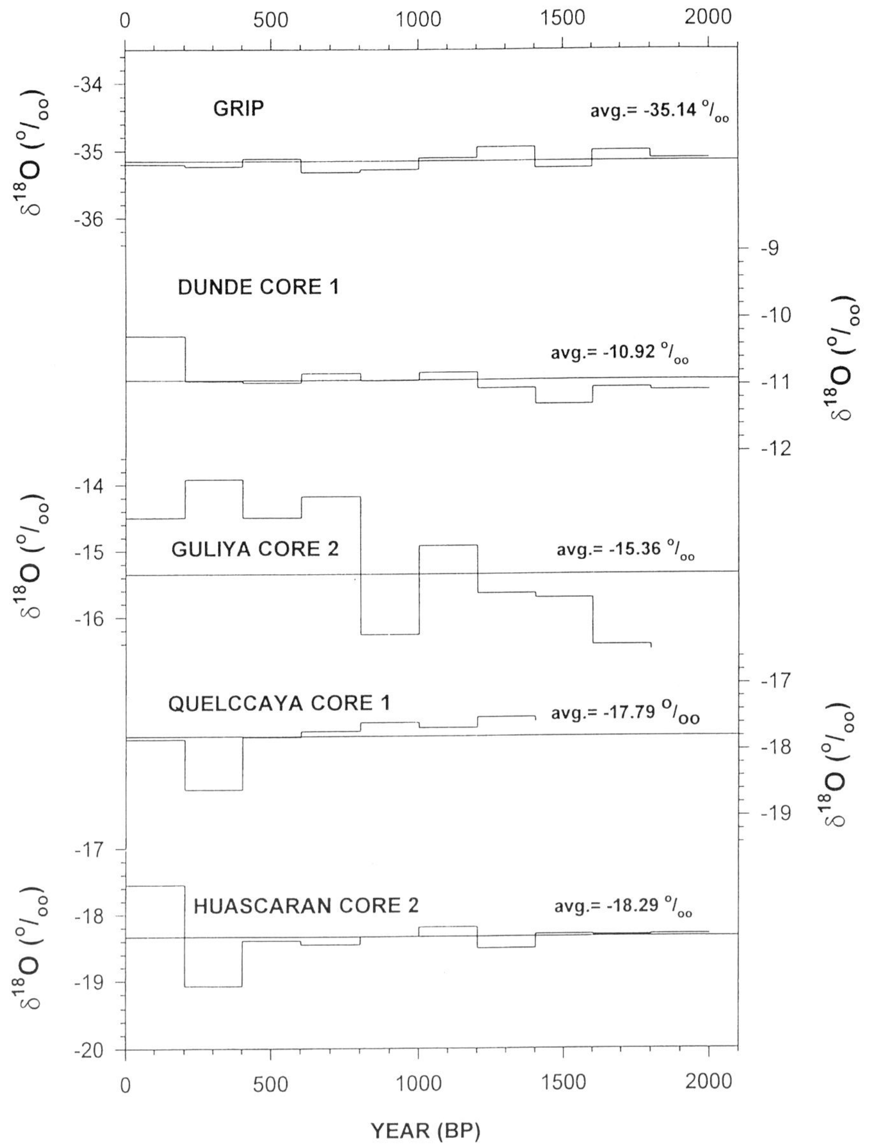

Figure 2.7 *200 year averages of $\delta^{18}O$ for the past 2000 years for Central Greenland (GRIP), Dunde, Guliya, and the tropical Quelccaya and Huascarán ice cores. The warming trend during the last 200 years is apparent in the Dunde, Guliya and Huascarán ice cores. These data indicate that $\delta^{18}O$ in mountain glaciers is a much more sensitive indicator of climatic change than in the polar ice caps.*

1995*a*,*b*; Yao & Thompson, 1992). The Tibetan Plateau and the Bolivian–Peruvian Altiplano have similar mean elevations (*c*. 4000 meters). On average, 70 to 80% of the annual snowfall in these areas arrives during their respective wet seasons.

Net accumulation during the last 2000 years on the Guliya and Quelccaya ice caps are compared with the historical moisture index from the semiarid area of China (Gong & Hameed, 1991) (Fig. 2.8). The similarities of precipitation variability between Chinese and Peruvian records are striking, given that the sites are 20 000 km apart on opposite sides of the Pacific Basin. During the 1500 years of overlap, the major periods of drought and wetness in South America and inland Asia appear contemporaneous, suggesting a past interaction of the Asian monsoon circulation with climatic events such as ENSO. Now, for the first time, a longer temporal perspective is available from ice core records, which suggests that lower-frequency teleconnections in climate fluctuations on decadal and centennial time scales may exist as well.

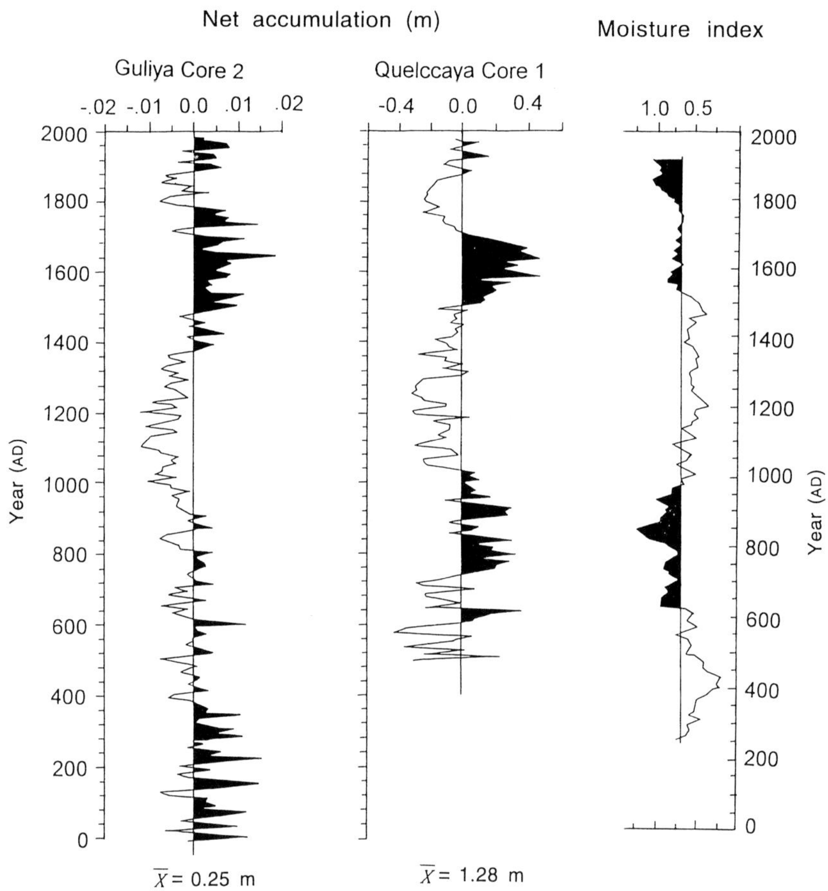

Figure 2.8 *The 2000-year record of net accumulation (calculated as decadal averages) on Guliya reveals contemporaneous trends with net accumulation on the Quelccaya ice cap in Peru (Thompson* et al., *1995b) and moisture index in the semiarid area of China (where higher numbers indicate wetter conditions) (Gong & Hameed, 1991). During the last 2000 years, the major periods of drought and wetness in Peru and China appear consistent, which implies that lower-frequency teleconnections in climatic fluctuations existed.*

Over the last 20 000 years, the climate of the Andes of Peru and the Tibetan Plateau of China show remarkable similarities. For example, the LGS was much colder and dustier, and the vegetation cover was much less extensive in both regions. The climate was the warmest from 8400 to 5200 years BP, then cooled gradually in the Late Holocene. A strong warming trend occurs in many of the recent records from China and the Peruvian Andes.

Chinese loess and global change

Insoluble atmospheric aerosols such as dust or loess can be used to trace atmosphere circulation, variations in wind strength and wind direction, and show evidence for drought conditions. Loess is believed to be a wind-derived, calcareous and porous yellow silty deposit (Liu, 1985) (Fig. 2.9). Loess deposits in China are distributed around the middle reaches of the Yellow River (Huanghe) and constitute the Loess Plateau, which extends over an area of about 300 000 km^2 in central China (Liu, 1985). The loess–paleosol sequence in the Loess Plateau has long been regarded as a well-preserved record of Late Pliocene–Quaternary paleoclimate cycles which can be compared with records from deep sea sediments (Liu, 1985; Kukla & An, 1989). These sequences can be viewed as a proxy record of Asian summer monsoon variations over the past *c.* 2.5 million years (An *et al.*, 1990, 1991*a*, *b*).

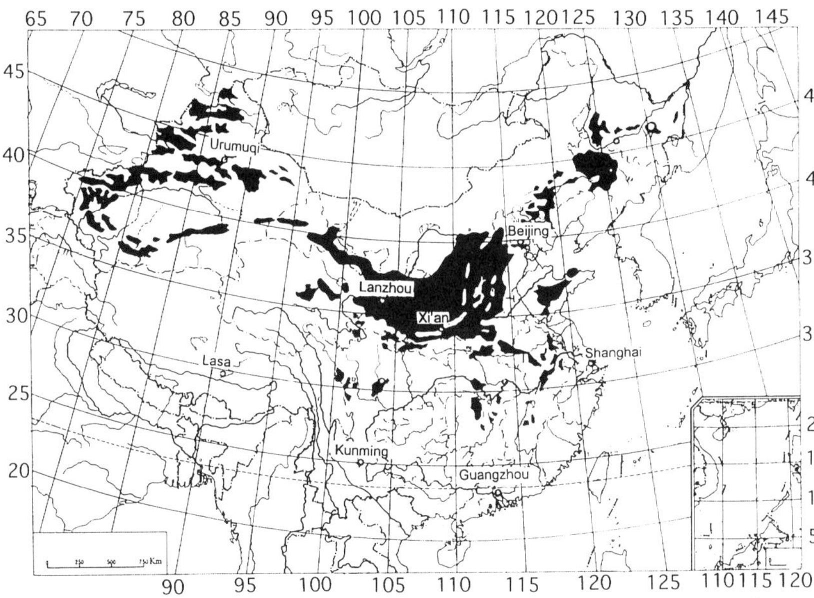

Figure 2.9 *Sketch map showing the loess distribution in China. The developed loess lies between 34° and 45° N, along the middle reaches of the Yellow River, where it attains its greatest thickness. Middle reaches of the Yellow River form the center of loess sedimentation in central China (Liu, 1985).*

Recent work has focused on the study of loess–paleosol sequences during the last glacial cycle. The pedogenic magnetic susceptibility flux (PSF) of loess and paleosols which reflects summer monsoon intensity (An *et al.*, 1991*b*; An & Sun, 1995) and eolian flux (EF) which reflects winter monsoon intensity (An *et al.*, 1991*c*; Xiao *et al.*, 1995) can be compared with the summer insolation at 65° N (Berger & Loutre, 1991) and the oxygen isotope record (Martinson *et al.*, 1987) over the last 140 000 years (Fig. 2.10). This demonstrates the possibility of placing variations of the Asian monsoon into the context of global-scale climate variations. Moreover, comparison of PSF with EF curves displays the existence of three patterns of summer and winter monsoon climates, which we refer to as complementary, harmonious development, and synchronous variation patterns (An & Sun, 1995).

The higher eolian flux and lower pedogenic magnetic susceptibility flux in the last glacial (74 000–12 000 BP) reveal the stronger winter and weaker summer monsoons. During the last interglacial (129 000–74 000 BP), the summer monsoon was strong, while winter monsoon was weak. This complementary relation is quite remarkable within the intervals of the last full glacial (21 000 and 65 000 BP). The greater ice volume reflected by the oxygen isotope curve enhanced the winter monsoon (higher EF) and weakened the summer monsoon (lower PSF) during the last glacial, while the lesser ice volume weakened the winter monsoon and caused the variation of configuration between sea and land in East Asia which led to strengthen the summer monsoon (An *et al.*, 1991*a*). The complementary pattern suggests that monsoon climate is directly related to changes in global ice volume induced by variations in solar radiation during the last glacial cycle.

During the last interstadial (59 000–24 000 BP), the stronger winter and summer monsoons show a pattern with the harmonious development of the winter and summer monsoon, which is consistent with the middle values of the summer insolation and $\delta^{18}O$ (see Fig. 2.10). The harmonious pattern may imply the response of the East Asian monsoon circulation to the interaction between the global ice volume and solar radiation during the last interstadial.

A notable in-phase feature for the fluctuation of EF and PSF curves occurs in the interstadial of the last glacial and the last interglacial, showing an approximately sychronous variations of the winter and summer monsoons. The sychronous variations of winter and summer monsoon climates are closely related to the seasonal cycle of solar radiation (Kutzbach & Gutter, 1986), and accordingly indicate the obvious influence of solar radiation on the East Asian monsoon during the last interglacial and interstadial.

Grain-size and eolian flux data from Chinese loess and intercalated accretionary palaeosols are evidence for climate signals, which are similar to those in the remote North Atlantic region and Greenland area, and record the high frequency variability of the East Asian winter monsoon climate on a millennial time scale in the last

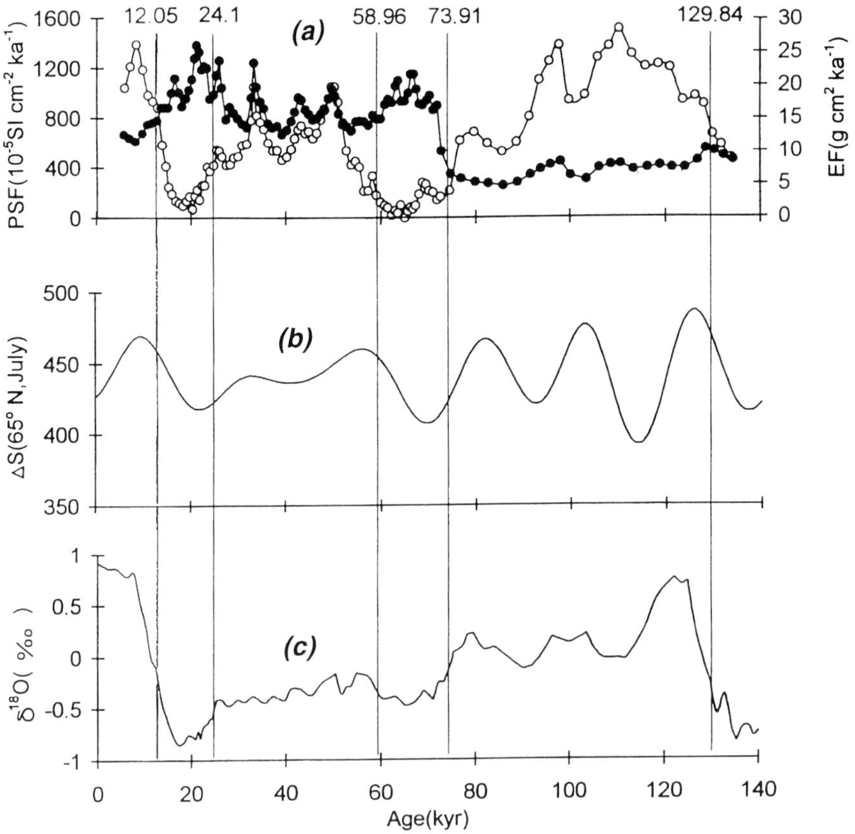

Figure 2.10 *Variation sequences of winter and summer monsoons reflected by (a) the eolian flux (EF, closed circles) and the pedogenic susceptibility flux (PSF, open circles), respectively in the Luochuan section (An & Sun, 1995), compared with (b) summer insolation variations at 65° N (ΔS) (Berger & Loutre, 1991) and (c) oxygen isotope record (Martinson* et al., *1987) over the last glacial cycle. Variations of the summer monsoon on the Chinese Loess Plateau seem to be consistent with the summer isolation variations at 65° N and ice volume variations recorded by oxygen isotope curves during the last glacial cycle. The comparison of the E F and P S F curves in Luochuan section during the last 130 000 years suggests that the complementary, harmonious development, and synchronous variation patters of summer and winter monsoon climates probably existed in the East Asian monsoon climatic changes which were closely related to the solar radiation and ice volume variations.*

glaciation (Porter & An, 1995; An *et al.*, 1995). Grain-size maxima indicate that the changing strength of the East Asian winter monsoon occurs at ages that match those of the last six Heinrich events, (Porter & An, 1995) (see Fig. 2.11). This may suggest that the climates of the North Atlantic and East Asia were linked by the westerly winds. Compared with the present climate regime, the Arctic Polar front and the position of Siberian High during the last glaciation shifted southward significantly, just as it did during the Little Ice Age (Yoshino, 1978).

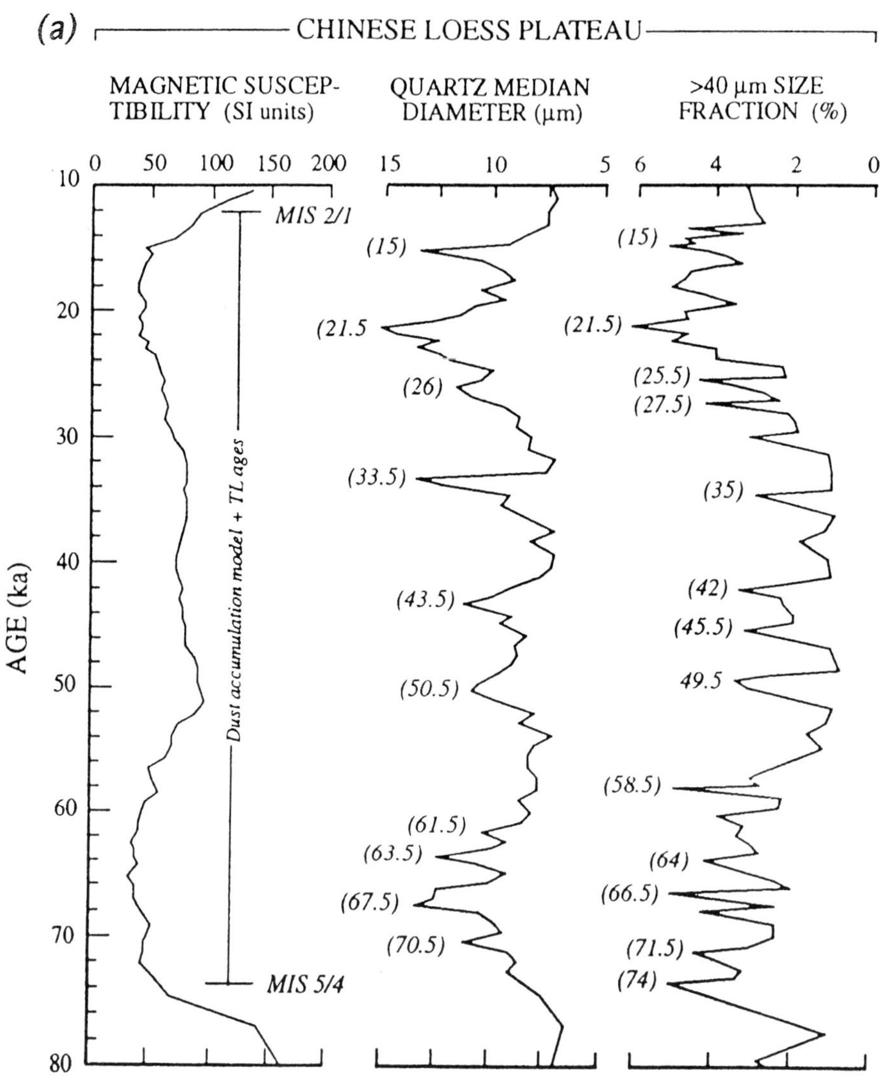

Figure 2.11(a) *Time series comparing magnetic susceptibility and grain-size variations in the Luochuan loess/palaeosol succession with Figure 2.11(b). Loess time series are dated by a sedimentation model keyed to stratigraphic "windows" representing the marine isotope stage 5/4 and 2/1 transitions. High-frequency events recorded by grain-size data from Chinese loess-paleosol sequences in the last glaciation show similar climate signals remote from the North Atlantic region. These climatic variations associated with the winter monsoon regime may be linked to the North Atlantic via the westerlies. However, comparable high-frequency events have not been detected in the magnetic susceptibility curve which reflects the summer monsoon variation (Porter & An, 1995).*

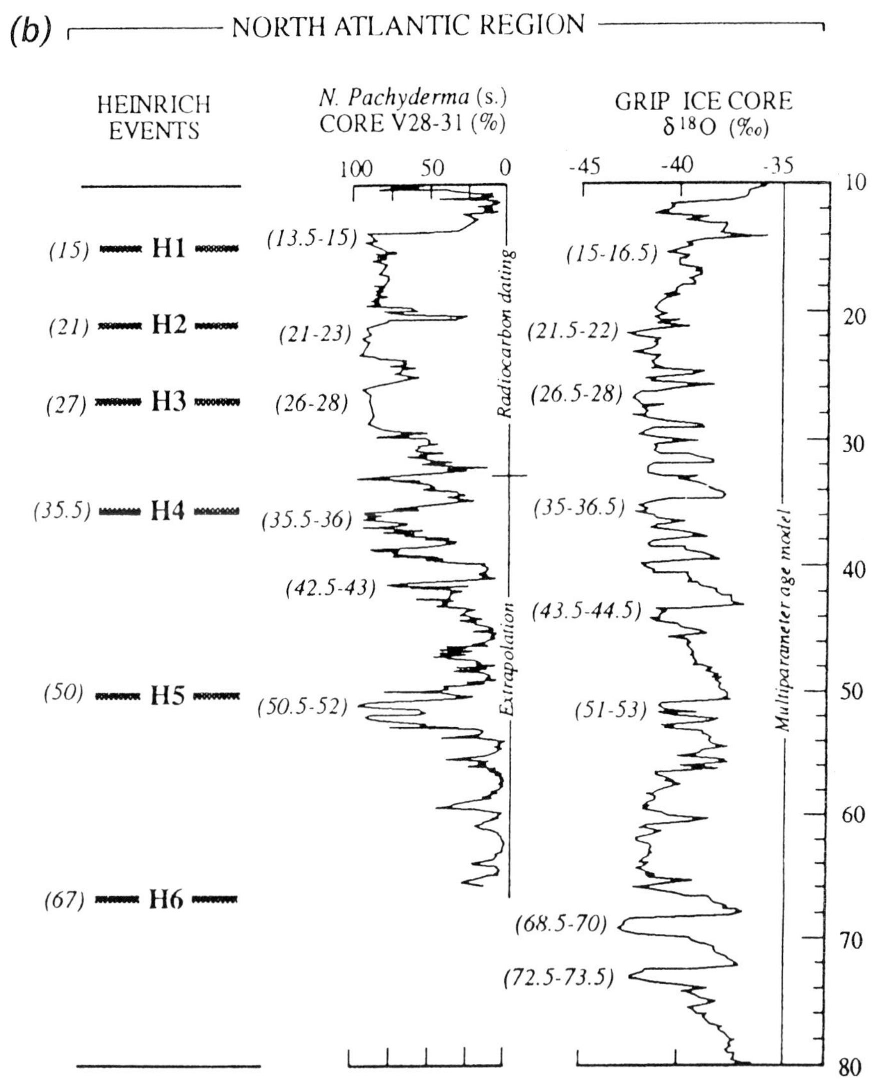

Figure 2.11(*b*) *Time series comparing North Atlantic Heinrich events,* Neogloboquadrine pachyderma *(s.) variation, and Greenland ice sheet* $\delta^{18}O$ *variations during the last glaciation (80 000 – 100 000 years).*

Chinese deserts are major sources for mineral dust and various trace elements in the air, ocean and loess of the mid-latitudes from 25° N to 40° N (Uematsu *et al.*, 1985; Prospero *et al.*, 1989; Zhang *et al.*, 1996). Estimates of atmospheric dust deposition to Asia/Pacific regions indicate that the Chinese deserts inject approximately 800 Tg of mineral dust into the atmosphere each year, about 30% of which (240 Tg $year^{-1}$) is redeposited on to the deserts and 20% of which (140 Tg $year^{-1}$) is transported over regional scales, primarily within continental China (Zhang *et al.*, 1997, in press). The remaining 50% of the dust (420 Tg $year^{-1}$, GESAMP, 1989) is subject to long-range transport to the North Pacific and beyond (X. Y. Zhang, personal communication).

On the basis of concentrations of the desert-derived elements (Al, Fe, Mg and Sc) from the 12 Chinese desert sites, an elemental tracer system was established to proportion the dust input to the loess and other reception areas by using a chemical element balance procedure. The results reveal that the Chinese desert sources can be divided into three major areas called Northwest High Dust Deserts, Northern High Dust Deserts and Northern Low Dust Deserts (Fig. 2.12), which exhibited significant variability in the contributions to the eolian deposit in the center of the Loess Plateau. The pattern uncovered were examined for evidence of climate signals since the last glacial age, and comparisons with several studies in

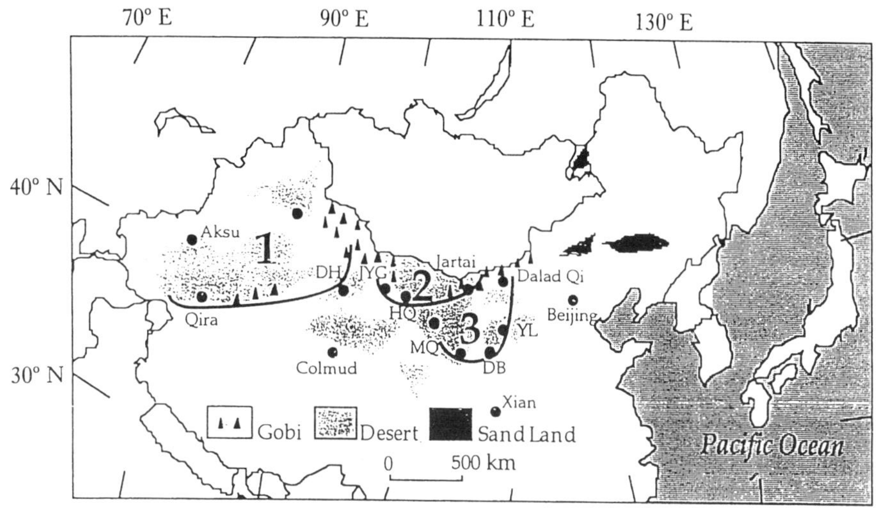

Figure 2.12 *The three major sources (1) northwestern high dust deserts, (2) northern high dust deserts, (3) northern low dust deserts, for Chinese dust are inferred from the elemental signature based on the ratio of four major elements of mineral dust (Al, Fe, Mg, and Sc) to aluminum (Zhang & An, 1995). The signature is proved to be recognizable during dust transport until the larger than 2 μm particles were removed. By using a cluster analysis, the twelve individual desert signatures were grouped together into three major regional signatures (significantly different according to the Fisher Protected Least Significant Difference test). Thus, by using the chemical element balance (CEB) procedure, mixed dust particles in the receptor area are resolved into their source components.*

the North Atlantic indicate that long-term variations in large-scale atmospheric circulation patterns of the past can be reflected by the temporal changes in the source contributions.

From a work of Late Quaternary records of the atmospheric input of eolian dust to the Luochuan loess/paleosol sequence of the central Loess Plateau, one can see that the dust fluxes (17 g cm^{-2} 10^3 $year^{-1}$) during the last glaciation was more than twice as much than during the last interglaciation, which suggests that the dust loadings show significant difference between the glaciation and interglaciation (An *et al.*, 1991*c*; Zhang X. Y. *et al.*, 1994). High deposition rates and long-range transport of mineral dust to East Asia and the North Pacific Ocean from Chinese deserts, especially in the glaciation, must play an important role in biogeochemical cycles of trace substances in the Northern Hemisphere, and may influence the global albedo.

Sensitivity of the Tibetan Plateau to global climate change

On an annual basis, the land–atmosphere system of the Tibetan Plateau as a whole serves as a heat source, especially in the spring and summer, although in the winter it is a cold source (Ye *et al.*, 1979). It thus has an important bearing on the strength and direction of the Asian monsoons. Uplift of the Tibetan Plateau and the resultant thermal effects, as well as its role in blocking and splitting the middle lower layers of the westerlies, has to a great extent controlled the formation, development and variation of the paleomonsoonal circulation of China (An *et al.*, 1991*a*,*b*). The regional weather patterns over the Tibetan Plateau are driven primarily by the sensible heat flux and the latent heat release during precipitation.

Accordingly, the Tibetan Plateau is very sensitive to climate change compared to other places at the same latitudes (Wu & Lin, 1981). Global cooling in the 1950s and 1960s and warming in the 1970s and 1980s were also observed in China. These recent variations appear 5–8 years earlier over the Plateau than in eastern China (Tang, 1989). On a longer time scale, the beginning of the Altithermal, the warmest period in the Holocene, began earlier on the Tibetan Plateau than in other regions of China (Liu *et al.*, 1995). During the last glacial maximum, the expansion of glaciers on the Plateau occurred earlier than along the margins (Li & Shi, 1995) . These data suggest that major climatic change appeared early on the Plateau. If so, the variations in Tibetan Plateau climate may be useful in helping to predict future trends of temperature variations in monsoonal China, and perhaps even in the Northern Hemisphere. This is particularly significant, given that the recent $\delta^{18}O$ records from the Dunde ice cap show that 1937 to 1987 was the warmest 50-year period in the last 12 000 years in that region of the Plateau (Thompson *et al.*, 1989; 1993).

Anthropogenic influence on the Loess Plateau and Yellow River reaches

Recently, in eastern China the impact of human activity is assuming greater influence. The southeastern retreat of the southern boundary of the steppe on the Loess Plateau since about 3000 years BP seems to imply that the destruction of vegetation cover, along with the resulting soil loss, has resulted mainly from human activity (Wu *et al.*, 1994). Using data on the formation of the Loess Plateau from loess–paleosol sequences, pollen analyses, historical geography, and long-term experimentation on soil erosion rates, it has been demonstrated that the current serious erosion is mainly due to human destruction of the capacity and resistance of the soil to erosion (Ren & Zhu, 1994). According to aerial photographic studies of a 7.7 km^2 experimental plot (Jin *et al.*, 1990), 0.18% of the total area of flat ground is being lost every year.

The frequency of flood, drought and river disasters in the lower reaches of the Yellow River and elsewhere in the Shaanxi Province has increased since AD 600 (Wu *et al.*, 1994) (Fig. 2.13). These events can be considered to be responses to anthropogenic impacts. Comparing the volume of Holocene loess-like deposits on the continental shelves of the Bohai, Yellow and East China Sea with modern deposition rates, it is estimated that the present sediment load of the Yellow River is at least three times greater than it was earlier in the Holocene (Ren & Zhu, 1994).

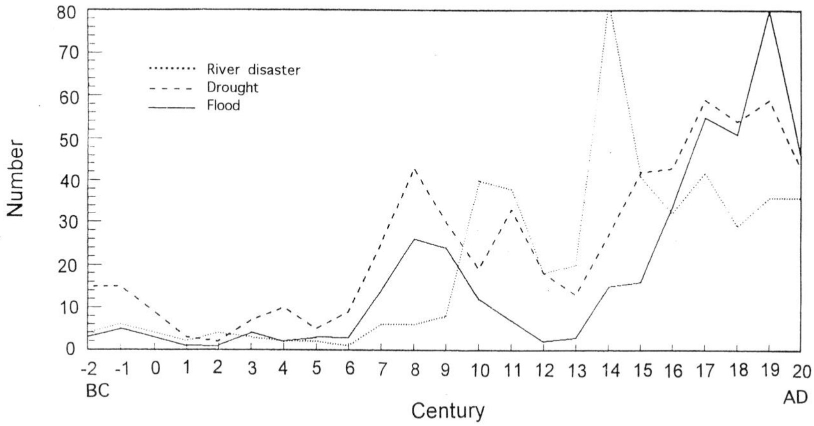

Figure 2.13 *The frequency of flood, drought and river disaster events in low reaches of the Yellow River and the Shaanxi Province has apparently increased since* AD *600. These events can be considered as a response to anthropogenic impacts (Wu* et al., *1994).*

Summary

Regional evidence has to be assembled and integrated to gain a global understanding of the climate change process. As historic documents, tree ring and ice core records show, global air temperature indicates an increasing trend since the beginning of the 20th century. The climate variation of monsoonal China in the past is characterized by the alternation of dryness and wetness or the alternating predominance of winter and summer monsoons. Precipitation or effective humidity is critical for monsoon climate study. The shift of the precipitation band in East Asia in the last 10 000 years provides temporal and spatial resolution for insights into the nature, cause and mechanism of the paleomonsoon variations.

Little Ice Age, Medieval Warm Period and Younger Dryas events in East Asia indicate both the global nature and regional characteristics of these events. The significance of regional paleoclimate and its linkage among the regions is of great importance for understanding the global climatic change process. However, there is neither one region nor a single proxy record that is representative of global paleoclimates. Dust from inland Asia contributes enormously to the dust loading in the Northern Hemisphere and strongly influences the global climate system and biogeochemical cycle.

The Asian monsoon is a very important component of the global climate system. The Tibetan Plateau plays a key role in the formation and development of the Asian monsoon system. Much greater attention should be paid to the sensitivity of the Plateau to global climate change by the scientific community. The East Asian summer and winter monsoon circulations can be viewed as the linkages between Asia and the global climate system. The study of monsoonal Asian paleoclimate is important for the prediction of future climate change in Asia where a significant fraction of the world's population resides.

Acknowledgments

We express our gratitude to Zhang De'er, Wu Xiangding, Wang Suming, Yao Tangdong and Zhang Xiaoye for providing related data and assistance. We thank Li Zhenghua, Zhou Weijiang, Liu Yu, Liu Xiaodong, Gao Wanyi, Li Xiaoqiang and M. E. Davis for their essential help. We also thank S. C. Porter, J. N. Galloway and H. Zimmerman for their useful suggestions. This work was supported by Chinese National Science Foundation and the US National Science Foundation.

References

An, Z. S. & Sun, D. H. (1995). Discussion on the monsoon variation over the loess plateau in the last glacial cycle. In *China Contribution to Global Change*, ed. D. Z. Ye, pp. 122–5. Beijing: Science Press.

An, Z. S., Liu, T. S., Lu, Y. C., Porter, S. C., Kukla, G., Wu, X. H. & Hua, Y. M. (1990). The long-term paleomonsoon variation recorded by the loess-paleosol sequence in central China. *Quaternary International*, **7/8**, 91–5.

An, Z. S., Wu, X. H., Wang, P. X., Wang, S. M. *et al.* (1991*a*). Changes in the monsoon and associated environmental changes in China since the last interglacial. In *Loess, Environment and Global Change*, ed. T. S. Liu, pp. 1–29. Beijing: Science Press.

An, Z. S., Kukla, G., Porter, S. C. & Xiao, J. L. (1991*b*). Magnetic susceptibility evidence of monsoon variation on the Loess Plateau of central China during the last 130 000 years. *Quaternary Research*, **36**, 29–36.

An, Z. S., Kukla, G., Porter, S. C. & Xiao, J. L. (1991*c*). Late Quaternary dust flux on the Chinese Loess Plateau. *Catena*, **18**, 125 - 32.

An, Z. S., Porter, S. C., Zhou, W. J., Lu, Y. C., Donahue, D. J., Head, M. J., Wu, X. H., Ren, J. Z. & Zheng, H. P. (1993). Episode of strengthened summer monsoon climate of Younger Dryas age on the Loess Plateau of central China. *Quaternary Research*, **39**, 45–54.

An, Z. S., Sun, D. H., Zhang, X. Y., Zhou, W. J., Porter, S. C., Shaw, J. & Zhang, D. Z. (1995). Accumulation sequence of Chinese loess and climate records of Greenland ice core during the last 130 Ka. *Chinese Science Bulletin*, **40**, 1272–6.

An, Z. S., Porter, S. C., Kutzbach, J. E., Wu, X. H., Wang, S. M., Liu, X. D., Li, X. Q. & Zhou, W. J. (1996). Asynchronous Holocene Optimum of the east Asian summer monsoon in central and eastern China. Submitted to *Quaternary Science Reviews.*

Berger, A. & Loutre, M. F. (1991). Insolation values for the climate of the last 10 million years. *Quaternary Science Reviews*, **10**, 297–317.

Gao, Y. X., Xu, S. Y. Guo, Q. Y. & Zhang, M. L. (1962). On some problems of Asian monsoon. In *Some Questions about the East Asian Monsoon*, ed. Y. X. Gao, pp. 1–49. Beijing: Science Press.

Gasse, F., Arnold, M., Fontes, J. C., Fort, M., Gibert, E., Huc, A., Li, B. Y., Li, Y. F., Liu, Q., Melieres, F., Van Campo, E., Wang, F. B. & Zhang, Q. S. (1991). A 13 000-year climate record from western Tibet. *Nature*, **353**, 742–5.

GESAMP, Group of Experts on Scientific Aspects of Marine Pollution (1989). *The atmospheric input of trace species to the world ocean.* Reports and Studies No.38. World Meteorological Organization, Geneva, 111 pp.

Gong, G. F. & Hameed, S. (1991). The variation on moisture conditions in China during the last 2000 years. *International Journal of Climatology*, **11**, 271–83.

Hughes, M. K., Wu, X. D., Shao, X. M. & Garfin, G. M. (1994). A preliminary reconstruction of rainfall in North-Central China since AD 1600 from tree-ring density and width. *Quaternary Research*, **42**, 88–99.

Jin, S. P., Wang, C. W. & Chen, X. H. (1990). *Water runoff and soil erosion and benefit of water and soil conservation measures in Wufundi gully experimental plot, Jungar Qi, Inner Mongolia.* The Data Bank of Northwest Institute of Water and Soil Conservation.

Kelts, K., Zao C. K., Lister, G., Hong G. Z., Niessen, F. & Bonani, G. (1989). Geological fingerprints of climate history: a cooperative study of Qinghai lake, China. *Eclogae Geologicae Helvetiae*, **82**, 167–82.

Kukla, G. & An, Z. S. (1989). Loess stratigraphy in central China. *Palaeogeography, Palaeoclimatology, Palaeoecology*, **72**, 203–25.

Kutzbach, J. E. & Gutter, P. J. (1986). The influence of changing orbital parameters and surface boundary conditions on climate simulation for the past 18 000 years. *Journal of Atmospheric Science*, **43**, 1726–59.

Li, S. J. & Shi, Y. F. (1995). The climate implication of Quaternary glacier on the Tibetan Plateau. In *Qinghai–Tibetan Plateau and Global Variation*, ed. China Society of the Qinghai–Tibetan Plateau Research, pp. 30–40. Beijing: Meteorology Press.

Li, Z. H., Liu, R. M., An, Z. S. & Wu, X. D. (1996). Climatic implications of seasonal $\delta^{13}C$ in tree rings from Huangling of Shaanxi Province. *Chinese Science Bulletin*, **41**, 326–29.

Liu, G. X., Shi, Y. F., Shen, Y. P., Hong, M. & Jiao, K. Q. (eds.) (1995). The primarily study on the environment of Tibetan Plateau during the Maximum of Holocene. In *The Formation, Evolution, and Environment Variation and Ecology System of the Tibetan Plateau*, 37 pp. Beijing, China: Science Press.

Liu, T. S. (1985). *Loess and the Environment*, pp. 171-4. Beijing: China Ocean Press.

Liu, Y., Wu, X. D., Leavitt, S. W. & Hughes, M. K. (1996). Stable carbon isotope in tree rings from Huangling, China and climatic variations. *Science in China* (Series D), **39**, 152–61.

Martinson, D. C., Pisias, N. G., Hays, J. D., Imbre, J., Moore, T. C. & Shackleton, N. J. (1987). Age dating and the original theory of the ice age: development of a high-resolution 0 to 300 000 year chronostratigraphy. *Quaternary Research*, **27**, 1–29.

Porter, S. C. & An, Z. (1995). Correlation between climatic events of the last glaciation in the North Atlantic and China. *Nature*, **375**, 305–8.

Prospero, J. M., Uematsu, M. & Savoie, D. L. (1989). Mineral aerosol transport to the Pacific Ocean. In *Chemical Oceanography*, Vol.10, ed. J. P. Riley, R. Chester & R. A. Duce, pp. 188–218. San Diego, CA: Academic Press.

Ren, M. E. & Zhu, X. M. (1994). Anthropogenic influences on changes in the sediment load of the Yellow River, China, during the Holocene. *The Holocene*, **4**, 314–20.

Suess, H. E. (1968). Climatic changes solar activity and the cosmic-ray production rate of numeral radiocarbon. *Meteorological Monographs*, pp. 30 - 47.

Tang, M. C. (1979). The average characteristics of the Plateau monsoon climate. *Acta Geographica*, **1**, 33–42.

Tang, M. C. (ed.) (1989). *The Introduction to Theoretical Climatology*, pp. 25–41. Beijing: Meteorological Press.

Thompson, L. G. (1992). Ice core evidence from Peru and China. In *Climate Since AD 1500*, ed. R. S. Bradley & P. D. Jones, pp. 517–48. London and New York: Routledge.

Thompson, L. G., Thompson, M. E., Davis, M. E., Bolzan, J. F., Dai, J., Yao, T., Gundestrup, N., Wu, X., Klein, L. & Xie, Z. (1989). Holocene-late Pleistocene climatic ice core records from the Qinghai–Tibetan Plateau. *Science*, **246**, 474–7.

Thompson, L. G., Moseley-Thompson, E., Davis, M., Lin, P.N., Yao, T., Dyurgerov, M., Dai, J., Faure, H., Faure-Denard, L. & Liu, T. (1993). Recent warming: ice core evidence from tropical ice cores with emphasis on central Asia. *Global and Planetary Change*, **7(1–3)**, 145–56.

Thompson, L. G., Thompson, E. M., Davis, M. E., Lin, P. N., Henderson, K. A., Cole-Dai, J., Bolzan, J. F. & Liu, K .B. (1995*a*). Late Glacial Stage and Holocene Tropical ice core records from Huascarán, Peru. *Science*, **269**, 46–50.

Thompson, L. G., Thompson, E. M., Davis, M. E., Lin, P. N., Dai, J. & Bolzan, J. F., Yao, T. (1995*b*). A 1000 year climate ice-core record from the Guliya ice cap, China: its relationship to global climate variability. *Annals of Glaciological*, **21**, 175–81.

Uematsu, M., Duce, R. A. & Prospero, J. M. (1985). Deposition of atmospheric mineral particles in the North Pacific Ocean. *Journal of Atmospheric Chemistry*, **3**, 123–38.

Wang, S. W. (1978). The climate change of China and global circulation during the last 100 years. In *Paper Collections on the Climate Change 4*, pp. 1–6. Beijing: Science Press.

Wang, S. W. (1981). The scale and cause on the long-range climate process. In *Paper Collections on Middle / Long-range Hydrological Meteorology Forecast* 4, ed. Project Office on the Reaches of the Yangtze River, pp. 401–10. Wuhan, China: Office on the Reaches of Yangtze River.

Wang, S. W. (1992). Climate of the Little Ice Age in China In *Proceedings of the International Symposium on the Little Ice Age Climate*, pp.

116–21. Tokyo, Japan: Tokyo Metropolitan University.

Wang, S. W. & Zhang, D. (1992). Reconstruction of the 18th century summer precipitation of Nanjing, Suzhou, and Hangzhou, China, based on the Clear and Rain Records. In *Climate Since AD 1500*, ed. R. S. Bradley & P. D. Jones, pp. 184–209. London: Routledge.

Wang, S. W. & Zhao, Z. C. (1995). A study of the trend of climatic change during the period of next 50 years. *Quaternary Journal of Applied Meteorology,* **6**, 333–42.

Wu, X. D. (1992). Dendroclimatic study in China. In *Climate Since AD 1500,* ed. R. S. Bradley & P. D. Jones, pp. 432–45. London: Routledge.

Wu, X. D. & Lin, Z. Y. (1981). Some characteristics of the climatic changes during historical times in Tibetan plateau. *Acta Meteorologica Sinica,* **39**, 90–1.

Wu, X. D., Lin, Z. Y. & Sun, L. (1988). A preliminary study on the climatic change of Hengduan Mountains area since AD 1600. *Advances in Atmospheric Sciences,* **5**, 437–43.

Wu, X. D., Niu, Z. X., Wang, S. C. (1994). Historical changes of environment and water and sediment in the Yellow River basin. In *Researches on Environment Changes of the Yellow River Basin and Law of Water and Sediment Transportation,* ed. Q. Ye, pp. 30-57. Ginan, China: Shandong Science and Technology Press.

Wu, X. D., Zhang, Z. H. & Li, J. (1996). A preliminary reconstruction of moisture index in central Qilian Mountains, China since AD 1300 from ring-width chronology. In *Tree Rings: From the Past to the Future. Proceedings of International Workshop on Asian and Pacific Dendrochronology,* ed. S. Ohta, T. Fugii, N. Okada, M. Hughes & D. Ecksten, pp. 273–280. Tokyo, Japan: Forestry and Forest Products Research Institute, Tokyo University of Agriculture.

Xiao, J., Porter, S. C., An, Z., Kumai, H., Yoshikawa, S. (1995). Grain size of quartz as an indicator of winter monsoon strength on the Loess Plateau of central China during the last 130 000 yr. *Quaternary Research,* **43**, 22–9.

Xu, X. J. & Zhang, X. G. (1983). The distribution and trend of droughts and floods in east China during the last 500 years. In *Paper Collections of Meteorology Science*, 4. Beijing: Meteorology Press.

Yao, T. D. & Thompson, L. G. (1992). Trends and features of climatic changes in past 5000 years recorded by the Dunde ice core. *Annuals of Glaciology,* **16**, 21–4.

Ye, D. Z., Gao, Y. X. (ed.) (1979). *Meteorology of Qinghai-Xizang (Tibet) Plateau.* Beijing: Science Press.

Yoshino, M. M. (1978). *Regionality of climatic change.* Climatic Change and Food Production. University of Tokyo, 31 pp.

Zhang, D. E. (1991). The Little Ice Age in China and its correlation with global change. *Quaternary Sciences,* **2**, 104–12.

Zhang, D. E. (1993). A study on the Medieval Warm Period in China. *Quaternary Sciences,* **1**, 7–15.

Zhang, D. E. (1995). High resolution records available from Chinese historical documents. *Quaternary Sciences,* **1**, 76–81.

Zhang, D. E. & Wang, P. K. (1991). A study on the reconstruction of the eighteen century Meiyu (plum rains) activity of lower Changjiang (Yangtze) region of China. *Science in China* (Series B), **34**, 1237–45.

Zhang, J. C. (1983). The droughts and floods in China during the last 500 years. In *Paper Collections of Meteorology Science* 4. Beijing: Meteorology Press.

Zhang, P. Y., Wang, Z., Liu, X. L. & Zhang, S. H. (1994). The stages of climate evolution in China during the last 2000 years. *Science in China* (Series B), **24**, 998–1008.

Zhang, X. Y., An, Z. S., Arimoto, R. Tuo, Chen, Zhang, G. Y. & Ray, B. J. (1994). Late Quaternary records of the atmospheric input of eolian dust to the center of the Chinese Loess Plateau. *Quaternary Research,* **41**, 35–43.

Zhang, X. Y., Arimoto, R. & An, Z. S. (1997). Dust emission from Chinese desert sources linked to large-scale variation in atmospheric circulation. *Journal of Geophysics Research,* in press.

Zhang, X. Y., Shen, Z., Zhang, G., Chen, T. & Liu, H. (1996). Remote mineral aerosols in Westerlies and their contributions to the

Chinese loess. *Science in China* (Series D), **39**, 67–76.

Zheng, S. Z., Zhang, F. X. & Gong, G. F. (1975). A preliminary study on the drought/flood and moisture conditions in the eastern China for the last 2000 years. In *Paper Collections of Extra Long-range Climate Forecast.*

Zhou, W. J., Donahue, D. J., An, Z. S., Porter, S. C., Jull, T.A., Li, X. Q., Stuiver, M., An, Z. S., Matsumoto, E. & Dong, G. R. (1996). Variability of monsoon climate in East Asia at the end of the last glaciation. *Quaternary Research*, **46**, 219–29.

Zhu, K. Z. (1973). A preliminary study on the climatic fluctuation during the last 5000 years in China. *Scientia Sinica* (Series B), **16**, 226–56.

Coupled biogeochemical cycles of carbon, nitrogen, phosphorus and sulfur in the land–ocean–atmosphere system

F. T. Mackenzie, L. M. Ver and A. Lerman

Introduction

During the past two decades, there has been considerable and growing attention paid to the environmental problem of global climatic change induced by industrial, agricultural, transportation, and household activities. To some degree, this problem, plus that of the complementary issue of stratospheric ozone depletion, have overshadowed a myriad of other human-induced environmental problems of worldwide extent. These include, for example, acid deposition, photochemical smog, cultural eutrophication, water use and quality, and waste disposal. All of these problems can involve the biogeochemical cycles of carbon (C), nitrogen (N), phosphorus (P), and sulfur (S).

Human inputs into the natural cycles of the elements rival or surpass the natural fluxes (cf. for example, Garrels *et al.*, 1975; Svensson & Soderlund, 1976; Likens, 1981; Schlesinger, 1991; Mackenzie *et al.*, 1993; Galloway *et al.*, 1993, 1996; Howarth *et al.*, 1996). Although the relative magnitudes of anthropogenic and natural fluxes are still debated, most scientists concede that these four element cycles are strongly influenced by fluxes from human activities. What is often overlooked is that the cycles are coupled through the biological processes of photosynthesis and respiration and decay. This cycling implies that an anthropogenic source of one of these elements, like fossil fuel use or fertilizers, is commonly the source of all four elements. Furthermore, such coupling requires consideration of the four cycles in concert when dealing with problems of global change.

The rates of global environmental change have accelerated, particularly since the Second World War. As the rates accelerated, debate about the consequences of such changes for the future of the biosphere intensified both within the scientific community and within policy-making institutions. The debate focused primarily on three realms: the atmosphere, the terrestrial biosphere, and the oceans. This choice of realms was a result of the fact that these three entities are the largest reservoirs of carbon that control, along with fossil fuel use, atmospheric CO_2 concentrations on a human time scale. Furthermore, until the recognition that tropical deforestation could be a significant source of carbon to the atmosphere, the debate was largely limited to the industrialized world. The freshwaters, estuaries, and coastal oceans, and their linkages to the land and atmosphere were conspicuously missing from the research agenda on global change until recently. Also missing was the realization that the industrializing and developing countries could follow a pathway toward development which mimics that of the two centuries of development in the industrialized world. This is the present course of many Asian countries, e.g. India, China, Indonesia, and South Korea.

In this chapter, we discuss human effects on the coupled biogeochemical cycles of carbon, nitrogen, phosphorus, and sulfur globally and with emphasis on Asia. We then employ an Earth system model, the Terrestrial Ocean aTmosphere Ecosystem Model (TOTEM) (Lerman *et al.*, 1989; Mackenzie *et al.*, 1993, 1995; Ver *et al.*, 1994*a*,*b*) to explore the behavior of the coupled biogeochemical cycles of C, N, P, and S in the land–ocean–atmosphere system during global change. In particular, we focus on the hypothesis, which has been growing in popularity recently, that the forests and soil humus are a sink of fossil fuel and land use emissions of CO_2 to the atmosphere (Keeling *et al.*, 1989; Tans *et al.*, 1990; Friedlingstein *et al.*, 1995; Galloway *et al.*, 1995; Kolchugina & Vinson, 1995; Melillo *et al.*, 1995; Schindler *et al.*, 1996; see review in Woodwell & Mackenzie, 1995). This sink has developed because of the fertilization of the land biota by increasing concentrations of CO_2 in the atmosphere and increasing nitrogen loading of terrestrial ecosystems because of fertilizer application and acid deposition of nitrogen-bearing chemical species. Another aspect of this sink is the regrowth of forests, particularly in parts of the tropics (Grace *et al.*, 1995).

We demonstrate through sensitivity analysis using TOTEM that the CO_2 fertilization effect has been maintained mainly by the release of nutrient N and P from land use activities and to a lesser extent by the application of fertilizers to croplands. Land use includes the practices of deforestation and conversion of forests to grazing and farm lands and the regrowth of forests on abandoned land. Thus, the term includes reforestation as well as deforestation (Houghton, 1995). Also, we show that coastal ocean responses to environmental changes on land occur well within a 100-year time frame. Thus, any change in industrial or agricultural practices on land in response to pollution must be considered in light of potential impacts on the coastal ocean environment. In addition, the analysis indicates that increases in carbon storage on land may be limited by nutrients other than nitrogen availability. In particular, the availability of the nutrient phosphorus may become critical, because of the fact that the ultimate source of phosphorus for bioproductivity in terrestrial ecosystems is the dissolution of phosphorus from phosphate minerals in soils and crustal rocks. Finally, we discuss qualitatively the global results in the context of Asia and conclude that this geographical region plays an important role in the dynamics of C-N-P-S biogeochemical cycling today and may play a disproportionately larger role in the future.

Coupled C, N, P and S biogeochemical cycles

Because of the importance of the coupling of the C–N–P–S biogeochemical cycles (Likens, 1981; Mackenzie *et al.*, 1993), it is worthwhile considering early the main features of this coupling. Primary biological productivity on land and in the aquatic realm forms organic materials made up of six main elements, C, H, O, N, P, S, and a dozen minor elements, such as zinc and iron, that are important to the maintenance of organic structures and the physiological functions of living organisms. Primary biological productivity and its counterpart of respiration and decay represent a coupling mechanism that joins together the individual elements and their cycles. The main elements are present in different proportions in aquatic and land plants. The following ratios denote the atomic proportions of C, N, P, and S found in terrestrial and aquatic plants representative of this coupling:

Marine plankton	C : N : S : P = 106 : 16 : 1.7 : 1 (Redfield *et al.*, 1963)
Marine benthic plants	C : N : P = 550 : 30 : 1 (Atkinson & Smith, 1983)
Land plants	C : N : S : P = 882 : 9 : 0.6 : 1 (Deevey, 1973)
(various estimates)	C : N : S : P = 510 : 4 : 0.8 : 1 (Delwiche & Likens, 1977)
	C : N : S : P = 2057 : 17 : 3 : 1 (Likens *et al.*, 1981)

The marine ratio is the "Redfield Ratio" applied to marine plankton by A. C. Redfield (Redfield, 1958; Redfield *et al.*, 1963). The ratio for marine plankton, although somewhat more variable than is often assumed when the "Redfield Ratio" is used, does not exhibit as much variability as that for land plants. It is not surprising that land plants are richer in carbon relative to marine plants because the former have a much greater proportion of carbohydrates as cellulose stored in structural tissues.

The net photosynthesis reactions that produce organic matter with the stoichiometric proportions of C, N, P, and S can be written using the mean composition of marine and terrestrial plants by balancing inorganic reactants with organic matter and free oxygen as reaction products. For example, the reaction for marine plankton can be represented as:

$$106\,CO_2 + 16\,HNO_3 + 2\,H_2SO_4 + H_3PO_4 + 120\,H_2O \Leftrightarrow C_{106}H_{263}O_{110}N_{16}S_2P + 141\,O_2$$

The reaction as written to the right represents photosynthesis of plant material, to the left, respiration and oxic decomposition. Similar reactions can be written for marine benthos and land vegetation. These reactions represent the coupling mechanisms for C, N, P, and S and their association with oxygen and hydrogen.

Thus, human additions to the environment of any one of these elements or a bio-essential trace metal will alter the mobilization and cycling rate of all the elements, particularly if the added element were limiting.

C : N : P : S ratios also exist for organic matter in soils and in sediments. These ratios along with those of marine and terrestrial plants provide the foundation for modeling the biogeochemical cycles of the bio-essential elements in an integrated and coupled fashion. This approach is discussed in detail below and in several recent articles of the authors dealing with the Terrestrial Ocean Atmosphere Ecosystem Model (Lerman *et al.*, 1989, 1993; Mackenzie *et al.*, 1993, 1995; Ver *et al.*, 1994*a*,*b*).

Framework of environmental change in the C–N–P–S bio-geochemical system: global and Asian perspective

Analogy is commonly made between the Earth's surface system, its ecosphere or exogenic system, and a giant chemical-engineering factory. This is a decidedly mechanical view of the behavior of the ecosphere. In actual fact, the Earth's surface has a natural metabolism with material circulation driven principally by energy from the Sun and much less importantly by the decay of radioactive elements in the interior of the Earth and by surficial and tidal processes. Each element within the ecosphere has a natural biogeochemical cycle. Human activities have contributed materials to these cycles at increasing rates; some of these materials are compounds, like CO_2, that have circulated for millions of years; others, like CFCs, are anthropogenic substances that are foreign to the natural environment.

In any determination of global change in the C–N–P–S biogeochemical cycles in the land–ocean–atmosphere system, it is necessary to consider the interactions between the biogeochemical system and human systems. This necessity stems from the fact that fluxes generated by the industrial, agricultural, and household sectors have become important in the cycling of these elements in the ecosphere. Economic, cultural, and social behavioral patterns give rise to human activities involving land use, fertilizer and pesticide application, fossil fuel and biomass combustion, waste disposal, and so forth. These activities have the potential to create pollution and modify climate. In general, the features of the environment that humans value aesthetically and factors that affect economic decisions are significant determinants in human perception of the negative impacts of pollution on the land–ocean–atmosphere system. On the one hand, if we did not value clean aquatic ecosystems free of algal scum, humans might not be concerned about excessive nutrient inputs into freshwater and marine environments. Because habitat and biodiversity, clean aquatic systems, and natural features of recreational areas are valued by the people of many countries, any change in the nature of these variables is viewed as an impact on the system. On the other hand, decisions concerning the effects of pollution on fisheries or the costs of pollution measures to control the impacts are often mainly economically based.

In response to the impacts and economic considerations, human systems may take several steps. Programs of remediation, mitigation, prevention, and monitoring may be developed to cope with the impacts. In turn, these programs can lead to measures that change human activities and the fluxes resulting from them. In the reconfiguration of industrial and agricultural activities to deal with environmental impacts on land due to human activities, it is necessary to consider how such reconfigurations may affect the ocean and atmosphere. This is simply a reflection of the fact that changes in terrestrial biogeochemical cycles because of land use and the response of terrestrial vegetation and the hydrologic cycle to climatic change are transferred to the coastal ocean and atmosphere on time scales of less than a year to a century (Lerman *et al.*, 1989; Holligan & Reiners, 1992; Nixon, 1992; Smith & Hollibaugh, 1993; Ver *et al.*, 1994*a*, *b*; Mackenzie *et al.*, 1995). As background for the discussion to follow, we consider briefly some of the reasons for concern with environmental degradation and human interference with the natural biogeochemical cycles of C, N, P, and S.

The global issues

There is no doubt that human activities are modifying the biogeochemical cycles of the elements on a spectrum of space and time scales, including the global to local and the century to interannual (see, for example, Detwyler, 1971; Garrels *et al.*, 1975; Likens, 1981; Smil, 1985; Turner *et al.*, 1990; Schlesinger, 1991; Oremland, 1993; Wollast *et al.*, 1993; Socolow *et al.*, 1994; Mackenzie & Mackenzie, 1995; Berner & Berner, 1996). On an annual basis, perhaps an equivalent of as much as 40% of terrestrial net primary productivity (NPP), or 24 billion tonnes of carbon, and several per cent of marine productivity of 40 billion tonnes are already affected by processes related to human activities (Vitousek *et al.*, 1986). Humanity now uses 26% of total global evapotranspiration and 54% of the *accessible* water runoff from the continents, or about 6800 cubic kilometers per year (Postel *et al.*, 1996). Total annual human mobilization of sulfur and nitrogen amounts to 0.15 billion tonnes and 0.14 billion tonnes, respectively (Galloway, 1995). Mining mobilizes approximately 12 million tonnes of phosphorus annually (Mackenzie *et al.*, 1993). Fertilizer production, fossil fuel combustion, and other human activities have more than doubled the global rate of nitrogen fixation since pre-industrial times (Kinzig & Socolow, 1994; Galloway *et al.*, 1995). During the last 40 years, fossil fuel CO_2 emissions have increased by nearly 200% (Marland *et al.*, 1994). Emissions are now equivalent to about 10% of global terrestrial net primary production. The mining of phosphate rock, principally for use in fertilizer production, has increased by 500% (Mackenzie & Mackenzie, 1995). During this period, nearly one-third of the world's arable land may have become unavailable for crop production because of erosion. Arable land continues to be lost at a rate of more than 10 million hectares per year (Pimentel *et al.*, 1995). The monetary cost of this loss in the United States alone is estimated to be $44 billion per year.

Because of human activities, the fluxes of water, sediments, organic matter and nutrients between the land and aquatic systems have been changing. Nitrogen, phosphorus, suspended sediment and organic carbon fluxes by rivers to the ocean all may have doubled during the past two centuries (Meybeck & Helmer, 1989; Wollast & Mackenzie, 1989; Milliman & Syvitski, 1992; Nixon *et al.*, 1995). Excessive river-borne nutrients and the cultural eutrophication of freshwater and coastal marine ecosystems go hand-in-hand. Meybeck and Helmer (1989) estimate that only about 20% of the world's drainage basins have pristine water quality. In turn, all these fluxes have become sensitive indicators of the broader global change issues of population growth and land use change (including water resources engineering works) in the coastal zone and upland drainage basins, climatic change, and sea level rise. Furthermore, global modeling of land–coastal ocean interactions (Lerman *et al.*, 1989; Mackenzie *et al.*, 1993; Ver *et al.*, 1994*a*; Mackenzie *et al.*, 1995) and experiential evidence (e.g. Nixon *et al.*, 1995; Dagg *et al.*, 1996; Smith, 1995; Nelsen, 1996; Howarth *et al.*, 1996) have shown that coastal ocean responses to such loadings occur on a decadal to century time scale, that of concern to human populations.

Pollution in the C–N–P–S system arises from fluxes of materials such as fuels, fertilizers, metals, and foodstuffs through the environment. To some degree, the past history of environmental pollution has been tied to the combined agents of the distribution and rates of growth of the human population and the demand for economic growth with concomitant utilization of natural resources. This has led to a plethora of human activities, the spin-off of which has been both an improvement in the human condition and an increasing potential for environmental degradation. These activities include, but are not restricted to, fossil fuel burning and biomass combustion, deforestation and other land use activities, agricultural practices, and halocarbon, pesticide and other synthetic chemical production and release to the environment.

There are several general factors whose continued future growth will undoubtedly increase problems of environmental degradation within the ecosphere, unless measures are taken to slow their growth or to alleviate their impact on the ecosphere and its biogeochemical cycles. The factors involved are principally population growth and distribution and the heavy reliance of humanity on natural resources, particularly coal, oil and gas, to fuel past, present, and future rates of industrialization. Growth in both of these factors is generally correlated to land use and disturbance.

The consequences of Asia's large and expanding population, coupled with record economic growth, have the potential to bear heavily on the very same factors that are perturbing the global processes between Earth's exogenic reservoirs: fossil fuel emissions, inorganic N and P fertilizer application, and land use activities. In addition, many socio-economic factors that contribute to the enhancement of these perturbations are also amplified in the case of Asian countries (Table 3.1). Some examples of these factors are: faster population growth and increasing population density in coastal areas; expanding human demand for food significantly exceeding the carrying capacity of the land and water resources; and accelerated conversion of cropland to other uses such as housing and other municipal requirements. Here, we discuss those factors relevant to cycling in the coupled C–N–P–S system with special reference to Asia.

Population growth

It is very likely that, in the next century, the Earth will experience further global environmental change. It certainly has naturally changed dramatically since the climax of the last glacial maximum 18 000 years ago. If left to its own recourse, the planet will experience natural climatic change in the future, including the strong likelihood of entering another glacial maximum about 20 000–30 000 years from now. On a shorter time scale, large-scale reorganization of ocean currents may modify regional climate on a century time scale (Broecker & Denton, 1989), and the El Niño Southern Oscillation phenomenon will affect climate and biogeochemical cycles on an interannual to decadal scale (Diaz & Pulwarty, 1994; Karl *et al.*, 1995). However, the flywheel of population growth is turning rapidly and will be difficult to slow in the next century. The growing human population of the Earth is now so large that, through its activities, it is placing a severe strain on natural resources and causing pollution problems that could alter the climate of the Earth and the productivity of natural and agricultural ecosystems. In 1995 the population of the world was growing at a rate of 1.54% per year, a doubling time of 45 years, which implies a population of about 10 billion by 2050 (Fig. 3.1). The rate of growth of the world population has been slowing in recent decades. However, most future population growth will be in the developing world, despite recent decreases in fertility rates in this region (United Nations Population Division, 1995).

In 1995, Asia had 60% of the world's population of 5.72 billion people (World Resources Institute, 1996). Its anticipated average annual change in population early in the next century is projected at about 1.4%, second only to the continent of Africa (Table 3.1). The Asian population could reach 5 billion people, or 60% of the world population in the year 2025 (Table 3.1; Lutz, 1994). China is anticipated to experience a relative decline in the rate of future population growth, but due to

Table 3.1 *Comparison of selected socio-economic indicators for the world and for each region*

(a) Population

Socio-economic indicator	Population, millions				Average annual % change in population			Population density per 1000 hectares	Urban population % of total		Average annual % change in urban population
	1950	1990	1995	2025	1980–1985	1990–1995	2000–2005	1995	1965	1995	1965–1995
World	2520	5285	5716	8294	1.7	1.6	1.4	436	35.5	45.2	2.7
Asia	1403	3186	3458	4960	1.9	1.6	1.4	1119	22.2	34.0	3.5
Africa	224	633	728	1496	2.9	2.8	2.6	246	20.6	34.7	4.7
N./C. America	220	424	454	616	1.3	1.4	1.1	209	67.4	74.0	1.8
S. America	112	293	320	463	2.1	1.7	1.4	182	55.9	78.0	3.3
Europe	549	722	727	718	0.4	0.2	0.0	320	63.8	75.0	1.0
Oceania	13	26	29	41	1.5	1.5	1.3	34	68.6	71.0	1.8

(b) Energy

Socio-economic indicator	Energy consumption (10^{15} Btu)			Energy consumption Average annual (% change)		Carbon dioxide emissions Industrial (10^6 tonnes)			Carbon dioxide emissions Average annual (% change)		Land use (% of global)
	1990	1995	2015	1995–1990	1995–2015	1990	1995	2015	1995–1990	1995–2015	1991
World	344.9	364.9	561.9	1.2	2.7	6012	6241	9704	0.8	2.8	100
Asia	80.6	105.1	207.7	6.1	4.9	1603	2090	4068	6.1	4.7	32
Africa	10.2	12.1	17.8	3.7	2.4	205	248	352	4.2	2.1	18
N. America	99.9	108.3	140.2	1.7	1.5	1561	1663	2170	1.3	1.5	n.a.
C./S. America	13.8	16.8	30.2	4.3	4.0	189	220	452	3.3	5.3	47
Europe	77	76.7	101.9	−0.1	1.6	1325	1254	1597	−1.1	1.4	0.003
Former USSR	58.5	40	56.8	−6.3	2.1	1029	653	933	−7.3	2.1	n.a.
Oceania	5.1	5.8	7.3	2.7	1.3	100	112	133	2.4	0.9	0.009

(c) Land use and agriculture

Socio-economic indicator	Cropland: Land area, 1000 hectares 1991–1993	Cropland: % change since 1981–1983	Permanent pasture: Land area, 1000 hectares 1991–1993	Permanent pasture: % change since 1981–1983	Cropland and permanent pasture (% of land area) 1991–1993	Per capita index of agricultural production 1979–81=100 1992–1994	Average annual fertilizer use (kg / ha cropland) 1993
World	1 450 834	1.3	3 364 537	3.6	36.8	104	83
Asia	470 322	2.9	799 881	12.9	41.1	127	118
Africa	187 357	5.8	853 049	1.2	35.1	94	21
N./C. America	271 300	−0.9	362 033	0.1	29.1	97	95
S. America	104 567	1.6	495 884	3.9	34.2	105	59
Europe	136 757	−2.7	80 794	−5.6	9.6	101	116
Oceania	51 619	4.3	430 738	−3.7	57.1	97	41

(d) Water resources

Socio-economic indicator	Freshwater resources: Per capita annual renewable water resources (1000 m^3) 1995	Freshwater resources: Per capita annual withdrawals (1000 m^3) 1987	Coastal areas: Population in coastal urban centers (millions) 1980	Coastal areas: Population in coastal urban centers (millions) 2000	Coastal areas: % change 1980–2000
World	7.18	0.64	617	997	61.5
Asia	3.82	0.54	282	487	72.8
Africa	5.49	0.20	43	112	158.4
N./C. America	15.37	1.45	89	121	36.6
S. America	29.79	0.33	60	105	75.7
Europe	8.58	0.63	130	154	18.3
Oceania	56.54	0.59	13	18	35.1

Sources: Energy Information Administration (1997); World Resources Institute (1996).

Notes: n.a. = not available.

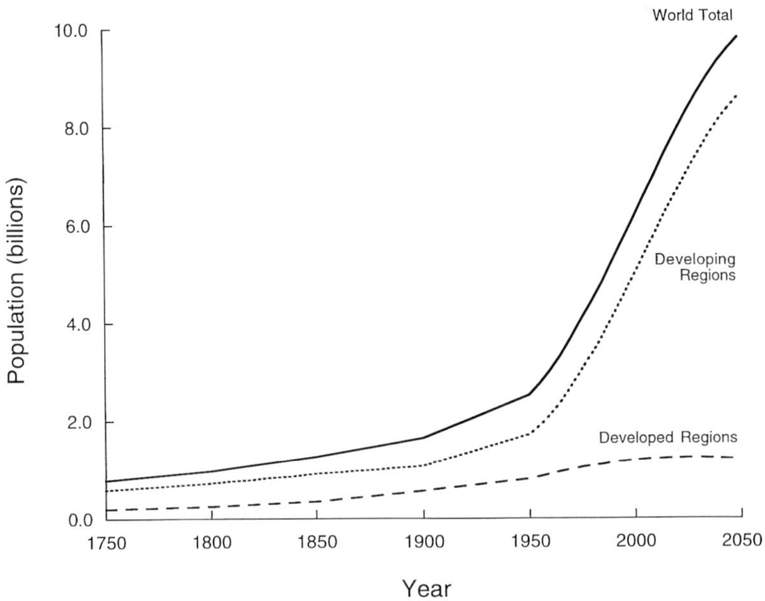

Figure 3.1 *The growth of the human population in the developed and developing regions of the world from 1750 to* AD *2100 (Joyce, 1975; United Nations Population Division, 1995).*

its present large population, will remain a major region of world population growth well into the next century.

The implications of population growth for the world and Asia are thought-provoking. For example, it is estimated that 1.7 billion people in the developing world today, or about 40% of the world population, do not have sanitary means to dispose of sewage. Early in the 21st century, it is likely that Asia will produce 50% of global CO_2 emissions, and China alone will exceed the United States in fossil fuel emissions of CO_2 to the atmosphere (Energy Information Administration, 1997). Asian emissions of CO_2 to the atmosphere from industrial processes alone currently account for about 33% of global emissions (Table 3.1). For Asia, sulfur and nitrogen emissions from human activities to the atmosphere possibly will increase by 650% and 530%, respectively, and amount to 52% and 51% of world emissions, respectively, by the year 2020 (Galloway, 1989, 1995). About half of the suspended sediment entering the oceans today comes from Asian countries (Milliman & Syvitski, 1992). With extensive land use activities and development of dams and reservoirs and other modifications of waterways, this flux will obviously change. The direction of change is difficult to predict and will vary regionally. However, it is likely that on a global scale, the future construction of dams and waterway modification will lead to a decrease in the suspended load discharge to the oceans (Halim, 1991).

Currently about 60% of the global population lives within 100 kilometers of the coastline and about 20% resides in coastal urban centers (World Resources Institute, 1996). More than 50% of the United States population in 1990 lived within 70 km of the strandline (World Resources Institute, 1996). Population density will increase in the coastal region in the 21st century, particularly in the developing countries of the world, including those of Asia. By the year 2000, Asia will have experienced a 73% increase in the population of its coastal regions in a span of two decades. This percentage change in coastal population is less than that of Africa and South America but more than the world average (Table 3.1).

During the 20th century, most of the consumption of natural resources and generation of wastes occurred in temperate-latitude, industrialized nations. During the first quarter of the 21st century, more than 80% of the human population of the world is projected to reside in the developing and industrializing nations, and more than half the world population increase will be in nations of the tropical and subtropical climatic zones (World Resources Institute, 1994). A large area of Asian countries is found in these climatic zones. Much less is known about the behavior of the biogeochemical cycles of C, N, P, and S in these climatic regimes than in the temperate environment from which most of our information has been derived (Schlesinger, 1991; Wollast *et al.*, 1993; Tiessen, 1995; Howarth, 1996). Thus, there is a strong need to obtain a clear understanding of the behavior of these cycles in natural and disturbed ecosystems of tropical and subtropical latitudes. This understanding is important to both the countries of Asia and to the world at large. Particular attention needs to be paid to the land–coastal ocean ecosystem which will be heavily impacted by growing world and regional human populations and their activities. Table 3.2 shows some examples of human forcings, other than the pandemic issue of climatic change, that impact the land–ocean–atmosphere system and their time scales of change. These forcings are strongly dependent on population density and rates of population growth.

Fossil fuel burning

Perhaps the major single impact of humans on the coupled biogeochemical cycles of C, N, P, and S is the burning of the fossil fuels, coal, oil, and gas. Combustion of these fuels releases radiatively active gases of carbon, nitrogen, and sulfur to the atmosphere. In 1994 about 6 billion tonnes of carbon were transferred from the fossil fuel reservoir to the atmosphere (Houghton *et al.*, 1995), along with approximately 30 million tonnes of nitrogen as nitrogen oxides and 80 million tonnes of sulfur as sulfur oxides (Galloway, 1995). In addition, about 50 million tonnes and 10 million tonnes of N as NH_x and as NO_x, respectively, are emitted to the atmosphere because of agricultural activities annually (cf. Mackenzie *et al.*, 1993; Galloway, 1995). In 1991 Asia contributed about 27% and 32% of global total CO_2 emissions from industrial processes and land use change, respectively (Table 3.1).

Total world consumption of commercial energy, mostly fossil fuel, is approaching 9 billion metric tonnes of oil equivalent (MTOE) (Energy Information Administration, 1997). The consumption of commercial energy in the developing world, about 90% of which is coal, oil, and gas, reached 2.7 billion MTOE in 1995, rivaling that of North America, which consumed 2.6 billion MTOE. Total energy consumption in Asia and Oceania (a minor contributor) exceeded that of Europe in the mid-1980s (Fig. 3.2). Asia's total energy consumption has increased about 350% since 1970, about twice the rate of increase for the world. Most scenarios of future global energy use project a continuous heavy reliance on fossil fuel into the 21st century (Fig. 3.3). Asia is poised for large future increases in fossil fuel usage, unless the historical path of industrialization of the presently developed nations is foregone.

Strong reliance on coal, oil, and gas will result in continuous emissions of the greenhouse trace gases CO_2, CH_4 (which also, by oxidation, increases the concentration of stratospheric H_2O, a greenhouse gas), and nitrous oxide (N_2O) to the atmosphere, along with trace metals, NMHCs (non-methane hydrocarbons), SO_x, and NO_x. These latter three substances react with other chemical components of the climate system, particularly OH radical. Also, SO_x and NO_x are the principal constituents in acid deposition, and NO_x and NMHCs are involved in the formation of tropospheric O_3, another greenhouse gas, and photochemical smog. All of the greenhouse gases are biogenic in origin. Their production or consumption at Earth's surface is mediated by biological reactions. In the case of O_3, natural emissions of precursor biogenic trace gases lead to formation of this atmospheric gas and thus indirectly affect the climate system (Moore & Schimel, 1992; Mackenzie, 1995). As a result of historical anthropogenic releases of these gases, the composition of the atmosphere has changed substantially during the past 200 years. This change in composition is the underlying cause of the environmental problem of an enhanced greenhouse effect and global warming.

Any global temperature change will influence the biogeochemical dynamics of cycling of C, N, P, and S. For example, denitrification and both production and respiration of terrestrial plant material, are influenced by temperature. This dependence is expressed in the Q_{10} function which represents the change in the rate of a process for each 10 °C change in temperature. The global mean temperature change of 0.3 to 0.6 °C (Houghton *et al.*, 1990) during the last century most probably had an effect on the rates of these processes, which in turn further affected the element cycles on land and in the ocean (Ver *et al.*, 1994*a*, *b*; this chapter). The eruption of Mt Pinatubo in 1991 and the injection of sulfate aerosols into the stratosphere led to a cooling of the planet of about 0.5 °C (Houghton *et al.*, 1995). The cooling was responsible for at least some of the slow-down in the rate of growth of the atmospheric greenhouse gases CO_2, CH_4, and N_2O in the early 1990s. The content of water vapor in the atmosphere is primarily a function

Table 3.2 *Human forcing functions that bear directly on the behavior of C-N-P-S cycling in the land–ocean–atmosphere system and the time period over which each of these forcings has changed significantly. Several of these forcings are discussed in the text in terms of Asia*

Human forcing	Time period of maximum change	Some examples of degree of change
Increase in population size and density	The past 100 years	World population has grown by 1.6 to 2% per year in the last 40 years. The developing world and global urban populations are currently growing at a more rapid rate than the populations of the developed world and rural areas, respectively.
Land deforestation and changes in carbon residence time on land	The Northern Hemisphere temperate forests were extensively deforested in the 18th and early 19th centuries. Currently, the tropical forests are being heavily deforested.	Tropical rain forests are currently being deforested at a rate of about 1% per year, resulting in the conversion of forest to grazing land, farm land and urban centers, and the release of 1.6 ± 1 billion tonnes carbon per year to the atmosphere.
Increased use of fertilizers of varying chemical composition	Post-Second World War	World fertilizer consumption has increased at a rate of 2.7 million tonnes per year since 1960, with a decline of 16% between 1989 and 1995.
Changes in water balances of large regions, owing to consumption of groundwater by agriculture and urban needs, at rates exceeding the present rates of recharge in certain regions	The past 50 years	Between 1950 and the year 2000, water withdrawal will have risen 280% and water consumption will have risen 220%.

Urbanization leading to the concentration of sources of human-derived pollutant materials (e.g. from landfills, wastewater treatment plants, radioactive waste disposal, etc.)	Primarily the past 100 years	In the industrialized world, the annual rate of urban growth was 2.8% per year in the late 1880s. Urban growth is now centered in the developing world at a rate of 4% per year between 1965 and 1995.
Chemical contamination by industrial wastes, pesticides and herbicides	Primarily the past 50 years	World pesticide consumption has increased about 240% since 1970. In 1991 the US alone had 2.2 million tonnes of toxic materials emissions.
Industrial and commercial activities, including storage of human-derived pollutant materials on land, increased shipping in international commerce and accidental (or sometimes intentional) discharges and spillage	Primarily the past 50 years	In the mid-1980s, global municipal wastes were generated at a rate of about 240 kg per person. In 1990, the total global accumulation of used nuclear fuel was 84 million tonnes.

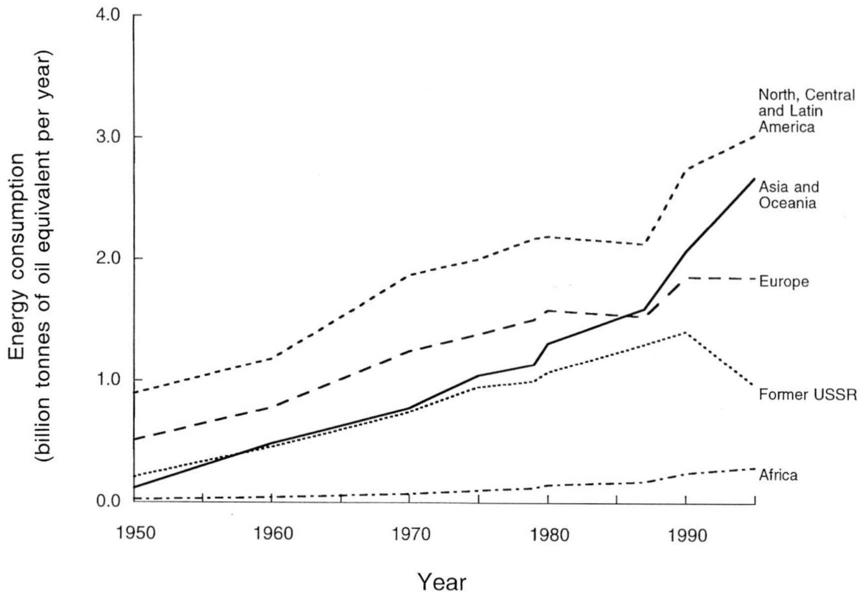

Figure 3.2 *Commercial energy consumption by region (United Nations, various years).*

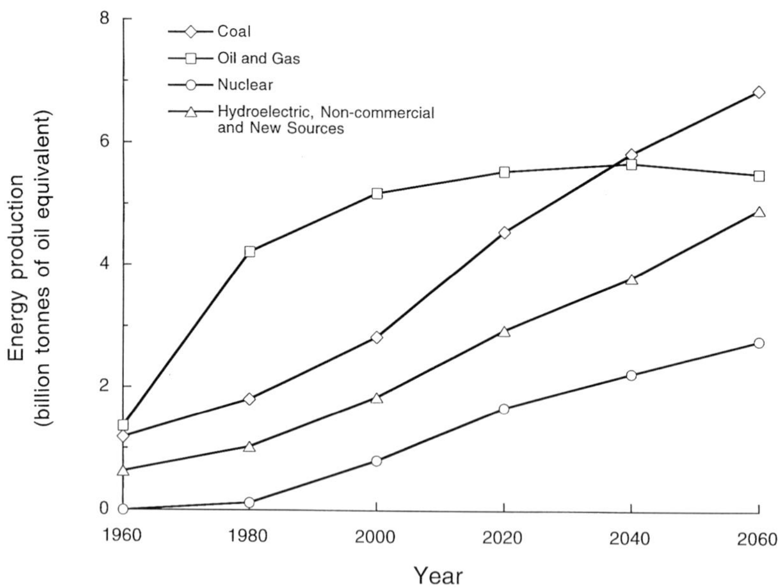

Figure 3.3 *Evolution of world energy supplies from 1960 to 2060 (Frisch, 1986).*

of the planetary mean temperature. The amount of water vapor, in turn, influences cloudiness and the magnitude and distribution of precipitation. Water is the substance that drives other biogeochemical cycles. Thus, the possibility of a future global warming accompanied by changes in the water cycle portends shifts in fluxes involved with the cycles of C, N, P, and S.

As mentioned above, the burning of fossil fuels releases sulfur and nitrogen gases to the atmosphere. The net flux of non-seasalt sulfur between the continental and the ocean atmosphere has been reversed with transport now in the direction of the ocean. Anthropogenic sulfur in the atmosphere after reaction to sulfate aerosol may result in atmospheric cooling on a regional and subcontinental scale because of adsorption and scattering of solar radiation (Charlson *et al.*, 1992; Charlson, 1995). This phenomenon is probably partly responsible for the difference between the global temperature record of the past 100 years and the global warming record for that same period as predicted from general circulation models of the enhanced greenhouse effect (Houghton *et al.*, 1995).

Nitrous oxide has a dual environmental role in the atmosphere. In the troposphere, it is a greenhouse gas and accounts for about 9% of the predicted greenhouse warming of the past 100 years. In the stratosphere, it reacts with ozone and naturally helps in the regulation of stratospheric ozone levels. With increasing emissions of N_2O from anthropogenic sources and its long residence time in the atmosphere, it may contribute to human-induced stratospheric ozone depletion (Crutzen & Andreae, 1990; Matson & Vitousek, 1990; McElroy, 1994).

Emissions of chemical species of sulfur and nitrogen to the atmosphere due to the combustion of fossil fuels, their subsequent reaction in the atmosphere, and the deposition of products on the continents is the chain of events that leads to the acidification of soils and freshwater aquatic systems. It is difficult to prove quantitatively but it is likely that both chemical species are being stored in continental reservoirs. If this is the case, and as long as the concentrations of these species remain elevated, over natural levels, acidification of ecosystems will continue (Galloway, 1995). Weathering rates are increased by these additional acid fluxes reaching the land surface. Bischoff *et al.* (1984), using a material balance approach for evaluation of the S and N components of acid deposition in the eastern United States, were able to show that the rate of chemical denudation of the region has increased about 15% during the 20th century because of acid deposition. Furthermore, the accumulation of nitrogen (and phosphorus) in terrestrial and aquatic systems can fertilize ecosystems and lead to the storage of organic carbon (see below). Also, nitrogen may contaminate groundwater leading to direct adverse effects on the health of humans and animals.

Land use activities

Land use activities, including deforestation and biomass burning, certainly constitute an important human impact on the coupled C–N–P–S biogeochemical cycle system. Land use activities may lead to the degradation of land. There are currently about 1.9 billion hectares of degraded vegetated land, or about 17% of total vegetated land (Mackenzie & Mackenzie, 1995). Land is degraded by water and wind erosion, inputs of chemicals, and physical compaction due to the grazing activities of domesticated animals. In order of importance, overgrazing, deforestation, agricultural practices, fuelwood gathering, and industrial waste pollution are the principal causes of soil degradation on a world-wide basis. Asian countries are important contributors to global land degradation, accounting for 38% of the global human-induced land degradation from 1945 to 1990 (Oldeman *et al.*, 1991). Currently, cropland and permanent pasture constitute about 41% of Asian land area, or about 1.3 billion hectares. Between 1983 and 1993, cropland area in Asia has grown by 2.9%, more than twice the global average but only half of that of Africa, the leading region in cropland area expansion for the past three decades (Table 3.1; World Resources Institute, 1996). Asia has led all regions in areal expansion of permanent pasture at 12.9% since the 1980s.

Deforestation

The largest single human influence of land use on the C–N–P–S system is that of progressive deforestation, first of Northern Hemisphere forests in the 18th, 19th, and early 20th centuries and then of tropical forests in the latter half of the 20th century (Fig. 3.4). The latter process currently is responsible for a net amount of carbon release to the atmosphere of 1.1 ± 0.5 billion tonnes annually (Houghton *et al.*, 1995). Deforestation has slowed dramatically in the developed countries of the world during this century but in the last half of the century, the deforestation of the world's tropical forests poses a threat to this ecosystem. Table 3.3 shows the total change in the area of various tropical forest regions of the world between 1980 and 1995. The world average decline in tropical forest area for the first half of this decade has been about 0.7%. For Asia and Oceania, the estimated loss in forested area from 1980–1990 was about 1.2% per year (Table 3.1). In Thailand and the Philippines, the deforestation rate exceeds 3% per year of initial forest area and if sustained would result in the elimination of their forests in 2010.

Table 3.3 shows the total change in the area of various tropical forest regions of the world between 1980 and 1995. The world average decline in tropical forest area for the first half of this decade has been about 0.7% per year. For Asia and Oceania, the estimated loss in forested area from 1990–1995 was about 0.95% per year. In Thailand and the Philippines, the deforestation rate exceeds 2.5% per year which, if sustained, would result in the elimination of their forests by 2010 (World Resources Institute, 1996). Recent analysis of Landsat satellite data confirms that

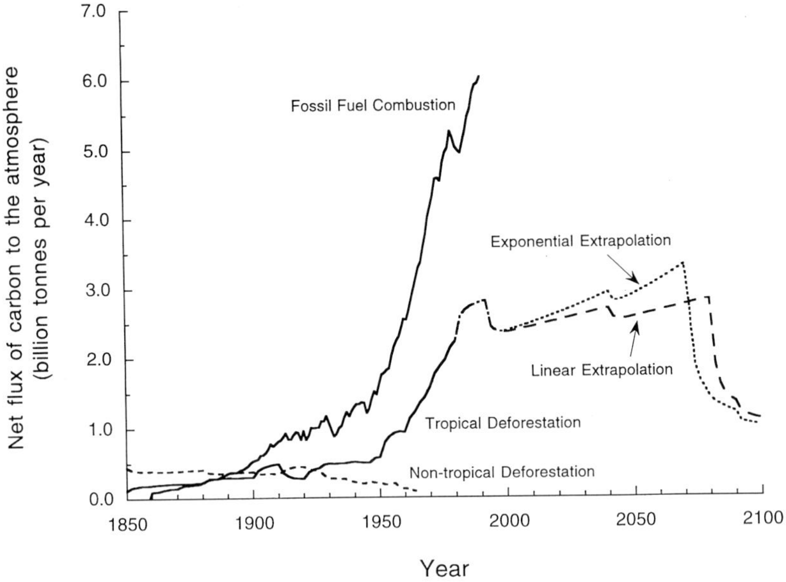

Figure 3.4 *Emissions of carbon to the atmosphere from non-tropical deforestation, tropical deforestation, and fossil fuel burning. Future emissions from tropical deforestation are model projections based on linear and exponential extrapolations of deforestation (Houghton, 1990).*

Table 3.3 *Total change in tropical forest area for various regions of the world between 1980 and 1995*

Region	Total forests, million ha			Annual change			
	1980	1990	1995	1980-1990		1990-1995	
				(million ha)	(% of 1980)	(million ha)	(% of 1990)
Total	1924.93	1796.93	1733.96	−12.80	−0.66	−12.59	−0.70
Asia and Oceania	362.60	337.70	321.67	−2.49	−0.69	−3.21	−0.95
Caribbean and Latin America	997.95	935.85	907.39	−6.21	−0.62	−5.69	−0.61
Africa	564.38	523.38	504.90	−4.10	−0.73	−3.69	−0.71

Source: United Nations FAO (1997); *World Resources Institute* (1996).

the tropical deforestation rates in Southeast Asia have generally declined since the late 1980s (see Skole *et al.*, this volume).

Asia is second to Central and South America as the leading source region for CO_2 emissions to the atmosphere from land use activities. In 1991, total CO_2 emissions attributed to land use activities in Asia, particularly deforestation practices, were about 32% of global emissions, or roughly 1.3 billion tonnes of carbon annually (Table 3.1). Note that these global and regional estimates of land use-derived CO_2 emissions are based on decadal mean values and therefore do not reflect the large interannual variations in deforestation rates as indicated by satellite data of Southeast Asian and Amazonian tropical forests (see Skole *et al.*, this volume).

Changing river suspended sediment flux

Land use activities are also responsible for changing fluxes of suspended sediments, organic carbon, nitrogen, and phosphorus to the ocean. Milliman and Syvitski (1992) concluded that the mid-20th century discharge of sediment to the ocean may have been approximately 20 billion tonnes annually, with about half of this amount coming from Oceania and another third from southern Asia. In comparison, the annual sediment discharge to the ocean 2500 years ago was of the order of 10 billion tonnes or less (Garrels & Mackenzie, 1971; Saunders & Young, 1983; Berner & Berner, 1987). The high sediment loads from Asia are a result of the influence of human activities, climate, and geology on the rivers draining these regions. Extensive farming and deforestation activities in Oceania and southern Asia probably account for disproportionately high sediment discharges to the ocean from these regions (Milliman & Syvitski, 1992).

More than 36 000 dams are now located on the world's rivers (Halim, 1991; Mackenzie & Mackenzie, 1995). Perhaps as much as 66% of the world's total river discharge to the ocean will be controlled by dams in the early part of the next century. There is presently slightly more than 2000 billion cubic meters of reservoir volume world wide, sufficient to account for nearly 300 times the total amount of sediment transported annually to the ocean. This storage, with an average organic carbon content of 0.8 weight %, is equivalent to 50 billion tonnes of buried organic carbon, or about ten times the annual release of CO_2 to the atmosphere by human activities in 1994.

Thirteen per cent of all fluvial sediment discharge, or 2.6 billion tonnes annually, are presently trapped by dams (Pearce, 1991). Asian rivers may fill their dammed reservoirs quickly because of their high sediment yields and relatively high sediment loads (Milliman & Syvitski, 1992). For example, the Yellow River (Huanghe) is the second most important river in the world in terms of its sediment discharge rate of 1.5 billion tonnes per year in the mid-1960s (see Hu *et al.*, this volume; Milliman & Syvitski, 1992). The total amount of sediment stored behind dams on the Yellow River over the past 30 years amounts to approximately 7.5 billion cubic meters of material, equivalent to an annual decrease in the sediment discharge of the river by 0.3 billion tonnes of sediment (see Hu *et al.*, this volume; Saito *et al.*, 1995). For the United States, Meade and Parker (1985) concluded that post-dam sediment discharges may not be too different from those prior to European colonization because of the offsetting factors of dam construction and anthropogenically enhanced erosion.

During the past 20 years, it is estimated that 480 billion tonnes of topsoil have been lost from farmland worldwide by erosion (Mackenzie & Mackenzie, 1995). If the topsoil erosion figures are correct, the rate of loss exceeds the worldwide transport of sediment to the ocean by 24 times! Apparently much of this sediment is being stored in dammed reservoirs and on other non-agricultural areas on land. Whatever the case, the construction of dams on river drainages disrupts the C–N–P–S biogeochemical system. For example, the emissions of CO_2 and CH_4 from flooding of forested areas because of damming may be substantial. Shortly after dam construction, a large pulse of CO_2 is released from the above-water aerobic decay of trees left standing. However, methane emissions tend to be continuous and long lasting because of the slow decay of biomass underwater due to anaerobic processes. Fearnside (1995) calculated that planned hydroelectric development in Brazilian Amazonia, which would flood 20 times the area flooded in 1990, could result in the emission of 5.2 million tonnes of methane annually. This emission is equivalent to a CH_4 flux of 52 tonnes per km^2 of flooded area per year, roughly equivalent to the global flux from natural wetlands. One might anticipate that the construction of the Three Gorges dam on the Yangtze River (Changjiang) in China could result in similar effects (see Hu *et al.*, this volume).

The global picture of changing nutrient and organic carbon fluxes

On a global scale, the river fluxes of the nutrients nitrogen and phosphorus and that of organic carbon to the ocean may have doubled over pristine values (Meybeck, 1993). Prior to human influence on the system, there was a net flux of CO_2 out of the ocean owing to organic metabolism (net heterotrophy; Smith & Mackenzie, 1987; Wollast & Mackenzie, 1989; Smith & Hollibaugh, 1993). This flux was supported by organic matter transported by rivers to the oceans and subsequent decay of the organic matter in the oceans. Carbon dioxide was also released to the atmosphere due to the precipitation of carbonate minerals in the oceans. Despite the net heterotrophy of the ocean and the release of CO_2 to the ocean–atmosphere upon formation of calcium carbonate in the ocean, the modern ocean is currently a net sink of atmospheric CO_2 because of the burning of fossil fuels and the practice of deforestation. Over the past 200 years, as CO_2 built up in the atmosphere, the gradient of CO_2 concentration across the atmosphere–ocean interface changed, favoring uptake of anthropogenic CO_2 in the ocean (Smith &

Mackenzie, 1987; Sarmiento & Sundquist, 1992). The net average oceanic carbon uptake during the last decade was 2 ± 1 billion tonnes annually (cf. Keeling *et al.*, 1984; Siegenthaler & Sarmiento, 1993; Tans *et al.*, 1995).

The present-day burial rate of organic carbon in the ocean may be about double that of the late Holocene flux (cf., for example, Walsh *et al.*, 1981; Mackenzie, 1981, 1995; Berner, 1982; Smith & Mackenzie, 1987; Wollast & Mackenzie, 1989; Sarmiento & Sundquist, 1992; Galloway *et al.*, 1995), supported by increased fluxes of N, P, and organic carbon to the ocean via rivers and groundwater flows, and of N deposition to the atmosphere. This organic carbon flux into sediments may constitute a sink of anthropogenic CO_2 and a minor negative feedback on accumulation of CO_2 in the atmosphere.

The increased flux of land-derived organic carbon delivered to the ocean by rivers may accumulate there or be respired back to the atmosphere. If respired, then CO_2 is released to the atmosphere. This release flux may be great enough to offset the increased burial flux of organic carbon to the seafloor due to enhanced fertilization of the ocean by nutrients derived from human activities. The magnitude of the CO_2 exchange is a poorly constrained flux today (Smith, 1995). One region for which there is a substantial lack of knowledge is the Asia Pacific. This is an area of several large seas, a region of important riverine inputs to the ocean of N, P, organic carbon, and sediments from land, and a region of important CO_2 exchange between the ocean and the atmosphere (see Chen & Tsunogai, this volume).

N and P fertilizers and aquatic systems

Chemical fertilizers of N and P applied to croplands are the most important anthropogenic inputs of these elements to aquatic systems via leaching from the land surface. Other major human sources of excess nitrogen in aquatic systems in decreasing importance are sewage, runoff from feedlots, atmospheric deposition, and sediments. Other important sources of phosphorus are synthetic laundry detergents and water-treatment chemicals. Deforestation, conversion of forest to grazing land or to agriculture, urbanization, and chemical inputs from factories are the major anthropogenic sources of excess organic matter in river waters.

In 1994 the total world commercial fertilizer consumption was approximately 121 million tonnes, or 22 kg per person per year, of which about 60%, or 73 million tonnes was in the form of nitrogen fertilizer and about 24%, or 29 million tonnes, was in the form of phosphate fertilizer (P_2O_5) (Fig. 3.5; United Nations FAO, various years). The average annual rate of application of fertilizer to cropland in Asia was the highest in the world at 118 kg ha^{-1} in 1993 (Table 3.1).

Fig. 3.5 shows the past 45 years of global and Asian N and P_2O_5 commercial fertilizer consumption. The application of nitrogen fertilizer to cropland in Asia in 1995 was equivalent to about 55% of the world total commercial nitrogen fertilizer

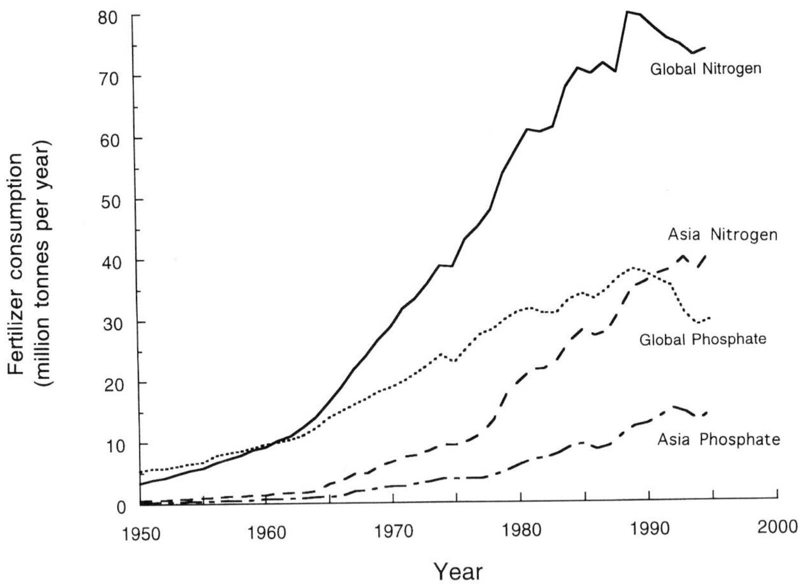

Figure 3.5 *World and Asia commercial nitrogen and phosphate fertilizer consumption for the period 1950–1994 (United Nations FAO, various years). About 60% of the consumption of fertilizer in Asia comes from commercial fertilizers.*

consumption. Application of commercial P_2O_5 fertilizer in Asia was about 50% of the total P_2O_5 fertilizer applied to the world's croplands in the same year (Fig. 3.5). The generous application of fertilizer in Asia was, in part, responsible for the close to 30% increase in per capita agricultural production from 1979 to 1994 (Table 3.1).

It is essentially since the Second World War that the use of N and P fertilizers has increased dramatically. If fertilizer consumption were given as a loading factor in units of tonnes of N or P per square kilometre of arable and permanent cropland per year, a *nearly linear* growth in both N and P loadings after the Second World War up until the late 1980s would be seen, at which time the loadings began to decrease. The reasons for the slowing are complicated, but probably entail the worldwide economic conditions of the late 1980s and 1990s, the return, for example, in the United States, to the use of organic fertilizers, better crop management, and perhaps a sign that additional N and P do little to increase further the productivity of crops presently managed.

To gain further insight into the fertilizer loadings for the world and for Asia, the relationship between population density and commercial synthetic fertilizer consumption per unit area of arable and permanent cropland for the world and Asia are compared in Fig. 3.6. The data in the figure are for the years 1960–1994. In general, both variables increase in value over this span of time. Many regions and countries of the world show a similar relationship between fertilizer use per

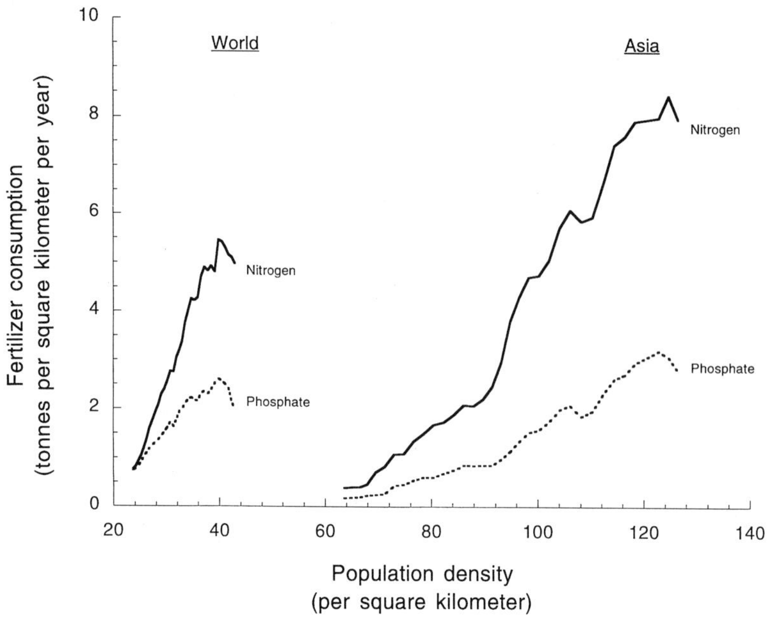

Figure 3.6 *Trends in world and Asia commercial nitrogen and phosphate fertilizer loading vs. population density for the years 1960–1994.*

unit area and population density as that seen for the world. As the population density of these regions grew through time, fertilizer loading of the land surface also increased to provide food for the growing world and regional populations. It is interesting to note that Asia as a whole, and China in particular, have followed a pattern of commercial fertilizer consumption similar to that of the world, despite their much higher population densities.

Traditional Asian agriculture has been labor intensive, relying strongly on the use of draft animals, crop rotation, and organic fertilizers. In recent decades, the situation has changed. In China, the consumption of commercial fertilizers tripled between 1978 and 1988 (World Resources Institute, 1994). Organic manure, once the principal source of nutrients for crops, now accounts for less than 40% of the fertilizer consumed in China. The area fertilized with green manure has shrunk from 9.9 million hectares in 1975 to 4.2 million hectares in 1990 (Dazhong & Pimentel, 1990). Other countries of Asia have followed the same general pattern in shifting from organic fertilizers to commercial sources. In India, for example, where rain-fed agriculture accounts for 68% of the total cultivated land area, the application of commercial fertilizer increased from 0.55 kilograms per hectare in 1950–1951 to 72 kilograms per hectare in 1990 (World Resources Institute, 1994).

The global organic nitrogen production in the form of urine and feces from domesticated animals can be calculated using a base estimate of 2.2×10^{15} grams dry matter consumed by livestock annually (Vitousek *et al.*, 1986) with an efficiency factor of 20%. With an average N content of 2.5 weight per cent, nitrogen production in urine and feces from domesticated animals is calculated to be 12.5 million tonnes per year. This value compares well with Smil's (1991) estimates of 15 million tonnes of N applied annually to croplands in the form of animal wastes. Thus, world production of N from domesticated animals was about 20% of commercial nitrogen consumption in the early 1990s.

The global mobilization of nitrogen due to human activities includes N mobilized from agricultural and farming use of synthetic and organic fertilizers (95 million tonnes in the early 1990s) and N fixation from legume and rice cultivation (40 million tonnes) (Galloway *et al.*, 1995; Mackenzie, 1995) for a total of 135 million tonnes (95 + 40 million tonnes N). This estimate is about 35% greater than that of Smil (1991), whose estimate for total N added to global croplands from synthetic fertilizers, animal wastes, and biotic fixation was 90–100 million tonnes.

The correlation between population density and nitrogen fertilizer use leads to a moderately strong correlation between the population densities of the watersheds of major rivers and the nutrient nitrate exported to the ocean by these rivers (Fig. 3.7). The latter relationship further documents the conclusion that many major rivers of the world are significantly polluted with nitrogen derived from fertilizer use and other sources. Fig. 3.7 also shows that Asian rivers are no exception.

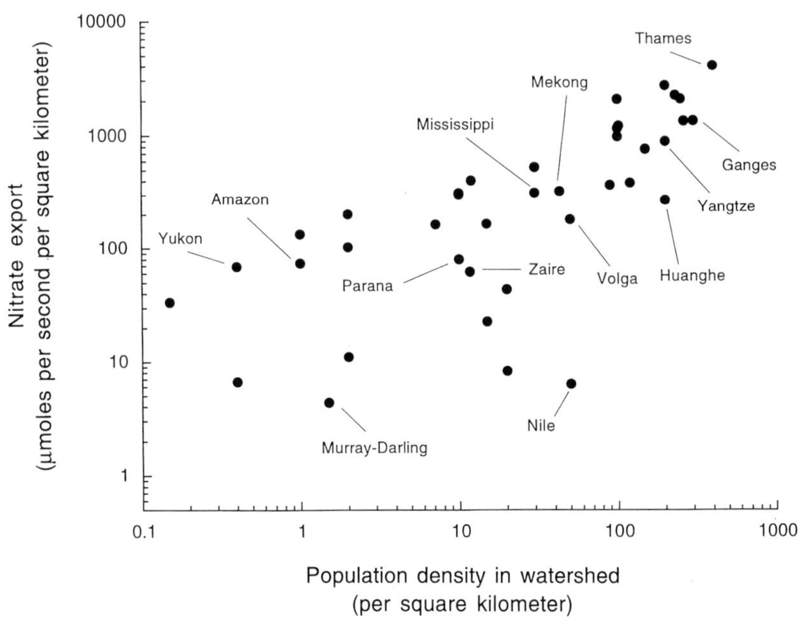

Figure 3.7 *Nutrient nitrogen export of some major rivers of the world vs. the population density of their watersheds. Note the moderately good correlation between these variables (Cole* et al., *1993).*

Small streams draining deforested, agricultural, and urbanized areas in temperate regions may contain as much as 30 times more nutrient and organic carbon than streams draining pristine forests (Billen, 1993). The total daily per capita discharge of carbon, nitrogen, and phosphorus from domestic activity (excrement, household wastewater, washing powder) into aquatic systems in industrialized countries was 65 grams by weight of the elements in the early 1990s (Billen, 1993). For a population of 1.35 billion people in the industrialized world, this discharge amounts to 32 million tonnes annually, of which 78% is organic carbon, 16% is nitrogen, and 6% is phosphorus. For some western European countries, less than 10% of the total nitrogen and phosphorus found in surface waters of these countries is of natural origin (Isermann, 1993). The rest is derived from anthropogenic sources, mainly as discharges from municipal sewage plants and agriculture. In the Federal Republic of Germany in 1989, only 2% of the total phosphorus and 6.5% of the total nitrogen entering surface waters were of natural origin. The domestic activity discharges are very rich in phosphorus relative to living matter. The excess phosphorus and other nutrients in water can result in the set of processes collectively termed cultural eutrophication.

If the pattern of historical industrialization of the developed countries holds into the future, similar changes in organic carbon, N, and P loadings for Asian countries may be anticipated, unless strong control measures are implemented. For example, in the year 2025, with a population of five billion people and a per capita daily discharge from domestic sources of organic carbon, N, and P equivalent to that of the industrialized nations today (65 grams), Asia's discharge of these substances to the environment would amount to 0.12 billion tonnes of C annually. This flux is nearly four times that from the industrialized countries in the early 1990s.

There are two major coupled processes in the phosphorus cycle that are of importance to global environmental change. One involves enhanced euthrophication of aquatic systems because of inputs of fertilizer and sewage phosphorus to these systems. The global dissolved phosphorus flux of rivers to coastal marine margins has doubled because of these human activities (Wollast & Mackenzie, 1989; Meybeck, 1993). This has resulted in eutrophication of coastal marine systems and a potential accumulation of organic carbon in these systems of about 0.1–0.2 billion tonnes of carbon per year (Mackenzie, 1995), provided the nutrient limiting biomass accumulation is phosphorus. This enhanced flux may be a small negative feedback on accumulation of anthropogenic CO_2 in the atmosphere today.

The land appears to be accumulating phosphorus today because of mining activities and its utilization as a fertilizer and because of sewage inputs to the land surface (Mackenzie *et al.*, 1993; Mackenzie, 1995). With an anticipated increase in most mineral chemical weathering and biological decomposition rates with warming temperature, this terrestrial P reservoir may be more easily leached into aquatic systems, enhancing eutrophication processes in these systems and leading to enhanced accumulation of organic matter. This flux is a small negative feedback on CO_2 accumulation in the atmosphere and global warming. Provided co-limiting nutrients in coastal marine systems are available, the total phosphorus presently stored on land, if leached and transported by rivers to those systems, would amount to a maximum of 0.8 billion tonnes of organic carbon storage in marine sediments (Mackenzie, 1995).

Greening of the Earth

There is no doubt that there has been an increase in the availability of phosphorus- and nitrogen-bearing nutrients in the terrestrial and oceanic environments. Also, there has been an unequivocal increase in atmospheric CO_2 concentrations during the past two centuries owing to fossil fuel combustion, cement manufacturing, and land use activities. We shall discuss these increases in the context of storage of carbon in ecosystems and the possibility that this storage is a present negative feedback on increasing CO_2 concentrations in the atmosphere. Later we show results from a global Earth System Model related to this hypothesis and their application to Asia.

Excess nutrients enter the environment from the human activities of application of fertilizers to the land surface, burning of fossil fuels and biomass, and discharges of N- and P-laden sewage, and in the case of N, fixation of N in rice paddies. The nutrients are available to stimulate plant growth in the soil and aquatic environments. This cultural eutrophication of both land and marine environments acts as a negative biofeedback on warming of the Earth in that atmospheric CO_2 can be sequestered in plant production. Kauppi *et al.* (1992) estimated from studies of western European forests that NO_3^- stimulation of plant growth was sequestering 85–120 million tonnes of carbon per year in the late 1970s and 1980s. The effect of eutrophication and storage of organic matter in both coastal and open ocean marine systems may be of the order of 0.5 billion tonnes of carbon per year (Garrels *et al.*, 1975; Wollast & Mackenzie, 1989; Sabine & Mackenzie, 1991). Melillo (1983) estimated that the effect of eutrophication, both on land and in the ocean, could be as large as one billion tonnes of carbon per year. Unless human-induced inputs of nutrients to the environment are reduced, it is likely that the cultural eutrophication sink for CO_2 will be enhanced in the future and act as a small negative feedback on increasing CO_2 concentrations in the atmosphere and global warming (Mackenzie *et al.*, 1993; Mackenzie, 1995). This statement must be conditioned by the assumption that increasing levels of acid deposition, air pollution, and degradation and erosion of soils caused by human activities will not prevent increased terrestrial plant growth.

One of the more controversial of potential negative feedbacks in terrestrial ecosystems is that of CO_2 fertilization. It is known from experimental simulations that certain agricultural crop plants and a few perennials when subjected to increased CO_2 levels will exhibit an increase in their photosynthetic and growth rates (e.g. Lemon, 1977; Idso, 1982; Strain & Cure, 1985; Bazzaz, 1990; Norby *et al.*, 1992; and reviews in Woodwell & Mackenzie, 1995). If this enhancement occurred in large-scale terrestrial ecosystems, more carbon could be stored in vegetation or soil organic matter. Thus, CO_2 stimulation would lead to withdrawal of CO_2 from the atmosphere and sequestration in plant organic matter, a potentially strong negative feedback on atmospheric CO_2 accumulation and hence global warming.

Although not a demonstrable observational fact, CO_2 stimulation has been invoked as a means of balancing the carbon cycle (e.g. Tans *et al.*, 1990, 1995; Friedlingstein *et al.*, 1995; Keeling *et al.*, 1996), part of the argument of several geochemists for a global greening of the Earth (Broecker & Severinghaus, 1992). It is difficult for some ecologists and others to accept this hypothesis of increased carbon storage in terrestrial ecosystems on a global scale in recent decades because of the uncontroversial evidence of degradation of terrestrial ecosystems both from satellite images and ground-truth data. Here we demonstrate from the global modeling results that the hypothesis is not incompatible with the observations of degradation of terrestrial ecosystems.

The quantitative role of terrestrial ecosystems today and in the future, in terms of the carbon balance, awaits further resolution; however, the potential for strong biotic feedbacks on atmospheric CO_2 concentrations and hence global warming is large. Terrestrial ecosystem net primary production is about 60 billion tonnes of carbon per year, or 60% of terrestrial plus marine net primary production. Total terrestrial ecosystem (living plants, litter, soil organic matter) carbon biomass today is about 2400 billion tonnes, 30 times all the organic matter presently stored in the ocean as living, particulate, and dissolved organic carbon (Mackenzie *et al.*, 1993). Obviously, biotic feedbacks involving CO_2 and the terrestrial ecosystem, whether positive (enhanced respiration) or negative (CO_2 stimulation and eutrophication), can represent large mass transfers and substantially affect the course of CO_2 accumulation in the atmosphere.

The future behavior of Asia's terrestrial ecosystems is important in this context. For example, China has 0.8–1.2% of the world's vegetational terrestrial carbon pool and 5.9–11.9% of the soil carbon pool (Fang *et al.*, 1995). Pan and colleagues (1995) predict a warmer and wetter country in the future with a doubling in the atmospheric concentration of CO_2. Along with the redistribution of vegetation, these climatic changes may lead to an increase of 20–29% in NPP for the whole country. The forests of Russia, the greatest expanse of northern forests in the world, may have sequestered in the mid- and late-1980s, 0.4 billion tonnes of carbon annually (Kolchugina & Vinson, 1995). The future of this forest ecosystem in response to climate change and faced with severe pressure for timber extraction will be critical to the course of future atmospheric CO_2 change. Furthermore, tropical forest losses are continuing at a rapid rate in Southeast Asian countries (United Nations FAO, 1997). Thus, the future carbon balance for Asian terrestrial ecosystems taken as a whole is equivocal.

Terrestrial Ocean aTmosphere Ecosystem Model

The system of C–N–P–S coupled biogeochemical cycles represents the main interactions between the biosphere and the inorganic world of the planetary surface environment. The *conceptual* model of the C–N–P–S biogeochemical system is shown in Fig. 3.8. This model is known as the Terrestrial Ocean aTmosphere Ecosystem Model (TOTEM). Details of the application of this model to environmental systems are given in several recent publications (Lerman *et al.*, 1989; Mackenzie *et al.*, 1993; Ver *et al.*, 1994*a*, *b*; Mackenzie *et al.*, 1995). Here we briefly discuss the general framework of the model.

Fig. 3.8 shows the major global reservoirs of C, N, P, and S in the system. The land part of the system is represented by the reservoirs of vegetation (land biota), humus, and mineral soil. Humus and mineral soils, respectively, are the dead organic matter and mineral components of the soils and weathering bedrock. Continental aquatic systems are represented by the soil water reservoir (soil water and very shallow groundwater to a depth of 10 meters, the depth of communication of root systems with nutrients dissolved in groundwater), shallow groundwater from 10 to 750 meters (Berner & Berner, 1987), and lakes.

Water is the principal agent responsible for the delivery of materials to the coastal zone. The residence time of water in rivers with respect to net precipitation on land is less than a year and that of water in lakes and shallow groundwater is about 5 and 100 years, respectively.

The atmosphere is an important part of the model for C, N, and S but not for P. The atmospheric burden of P is small and there is no important gas phase of P. However, release of P from biomass burning and atmospheric P inputs to some ecosystems may be important. Atmospheric transport of materials from land to the coastal environment may take only a matter of days or less. Thus, environmental changes on land have the potential to initiate change in the coastal zone world wide via rivers and the atmosphere on a decadal to century time scale, a time scale of concern to human society.

The ocean part of the model is subdivided into coastal margin and open ocean environments. The coastal margin, which includes the estuaries, lagoons, embayments, and the continental shelf to the shelf-slope break is represented by

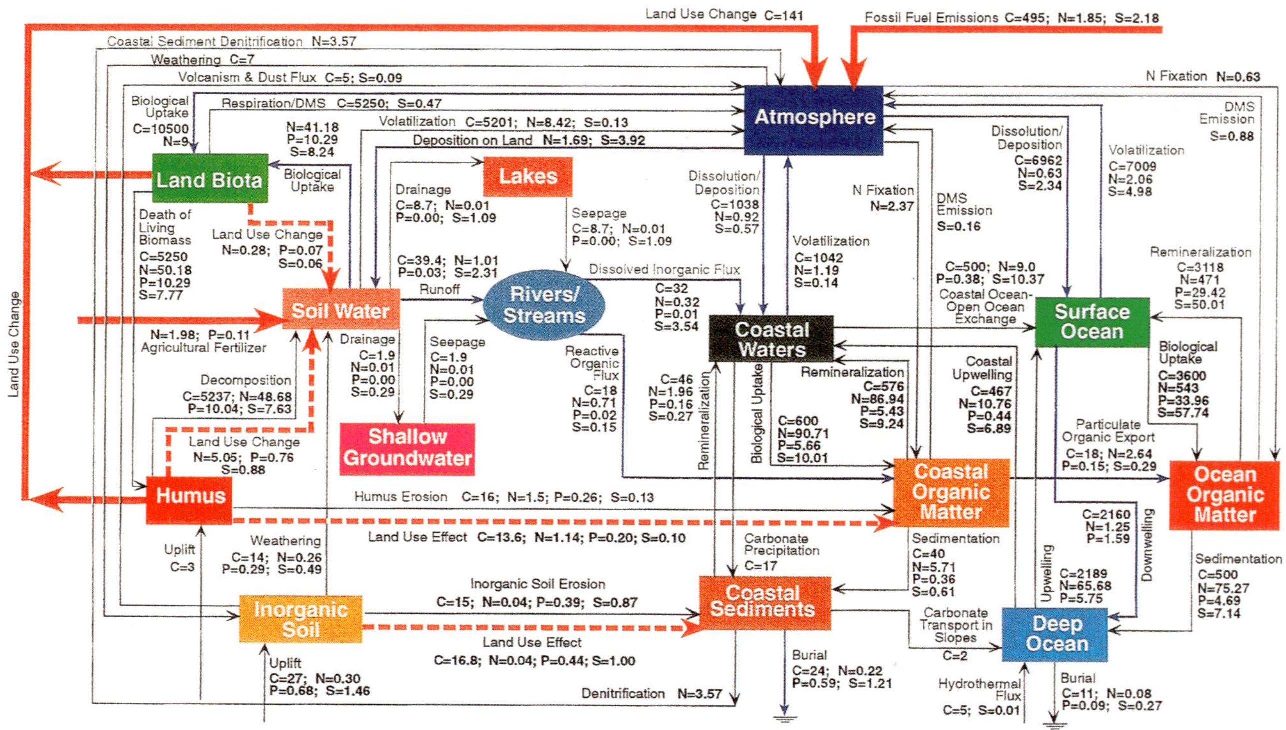

Figure 3.8 *TOTEM, a model of the C-N-P-S coupled biogeochemical system, representing main interactions between the biosphere and the inorganic world of the planetary surface environment. Heavy red arrows are fluxes of elements owing to perturbations from fossil fuel combustion, land use change, and agricultural fertilizer application. Heavy dashed red arrows are effects of land use activities on fluxes of elements from terrestrial organic and inorganic reservoirs. Medium-sized blue arrows denote natural fluxes significantly modified by human activities; thin black arrows represent fluxes less affected by perturbations. Flux estimates for blue and black arrows are at initial condition; those for red arrows are for 1990. Flux units are 10^{12} moles yr^{-1} of the element shown. (To recalculate fluxes in units of billion T yr^{-1}, multiply C flux by 0.012, N by 0.014, P by 0.031, and S by 0.032.) Significant figures are retained to preserve material balance at steady state.*

the reservoirs of coastal ocean water, coastal organic matter (phytoplankton, macrophytes and organic particulates), and sediments. The boundary between the coastal ocean and the open ocean is taken as the shelf-slope break at an average depth of 140 meters. Exchange between the open ocean and the coastal zone occurs by input to coastal margins from the upwelling of intermediate and deep ocean waters and output from coastal waters to the surface ocean.

Adjacent to the coastal margin lies the open ocean, with its reservoirs of ocean organic matter, and surface (thickness = 100 meters), and deep waters. Vertical mixing and sedimentation of organic and inorganic matter represent processes leading to exchange of materials between the shallow and deep ocean.

The coupling of the four biogeochemical cycles in the land and ocean biotic reservoirs and the organic matter reservoirs of soils and sediments is realized through the C : N : P : S ratios of organic matter (the "Redfield Ratios") that characteristically vary within certain and known limits (Peterson & Melillo, 1985). The photosynthetic uptake by the terrestrial biota is regulated by the mass of CO_2 in the atmospheric reservoir, the mass of carbon in terrestrial biota, the nutrient elements nitrogen and phosphorus in the continental soil water, and global average temperature. The colimiting roles of each of these five variables with respect to biological uptake is mathematically represented by a Michaelis–Menten kinetic formulation for each of the three nutrient terms (atmospheric CO_2 and soil water N and P), a first-order linear function with respect to terrestrial biotic C, and a temperature limiting function (as described in the next section). This approach involves non-linear functions and definition of mean nutrient concentrations in the continental soil water reservoir. Errors associated with this approach are discussed by Rastetter (1996).

It is important to note that the biochemical uptake of C, N, P, and S by the terrestrial biota is also dependent on other environmental variables, such as water stress (see Schulze & Heimann, this volume) and trace element concentrations that are not presently included in this model. Although our current model includes parametrizations for only the five variables mentioned above, a planned future version of the model may include one or two other significant environmental variables.

Fig. 3.8 shows the many processes and associated fluxes moving materials about in the system. The initial values for both the initial fluxes and the reservoir sizes (Table 3.4) of the elements are for a quasi-steady state system prior to extensive human interference in the coupled global biogeochemical cycles of C, N, P, and S. These estimates are derived from a vast literature on the subject and from geological constraints on the system. Also shown for comparison in Fig 3.8 are the current (1990) fluxes of the elements due to fossil fuel combustion, land use change, and application of agricultural fertilizers. The medium-sized arrows

Table 3.4 *Reservoir sizes and residence times for the various exogenic reservoirs of carbon, nitrogen, phosphorus and sulfur*

	Carbon		Nitrogen		Phosphorus		Sulfur	
	Reservoir size (10^9 tonnes)	Residence time (years)	Reservoir size (10^9 tonnes)	Residence time (years)	Reservoir size (10^9 tonnes)	Residence time (years)	Reservoir size (10^9 tonnes)	Residence time (years)
Atmosphere	588	10	3 920 000	1.8×10^7			0.003	0.01
Terrestrial biota	598	9.5	5	9.5	3	9.5	2.4	9.5
Humus	246	4	14	20	5	14	5	22
Inorganic soil	718	1800	1.7	500	36	4000	2 675 300	6.0×10^7
Continental soilwater	3	0.05	0.07	0.09	0.005	0.02	1.4	4
Lakes	1.2	12	0.002	12	0.0004	12	8	200
Shallow groundwater	41	1800	0.3	1800	0.012	1800	48 050	5.4×10^6
Rivers	0.04		1.6		0.9		0.014	
Coastal organic matter	0.5	0.07	0.09	0.07	0.012	0.07	0.02	0.07
Coastal waters	72	200	1.5	1	0.14	0.75	2623	4000
Coastal sediments	2484	3000	32	400	285	12 000	375	8000
Ocean organic matter	2.7	0.07	0.5	0.07	0.06	0.07	0.11	0.07
Surface ocean	788	30	2.9	0.4	8	7.5	28 737	1.5×10^4
Deep ocean	36 480	1400	715	700	85	400	1 178 000	5.0×10^4

denote natural processes that have been significantly modified by the human activities of fossil fuel combustion, land use activities, and application of fertilizers to the land surface. The thin arrows represent fluxes that have been less affected. All of the flux estimates are in units of 10^{12} moles per year of the elements shown. To convert to 10^{12} grams (Tg) element per year, multiply the C flux by 12, the N flux by 14, the P flux by 31, and the S flux by 32 grams per mole.

The large number of parameters in the model reflect the complex nature of the interacting C–N–P–S biogeochemical cycles. It would be misleading to show a diagram any less complex than Fig. 3.8, if we seek to portray the reality of the interactive cycles. The parameters in the model are not usually what are termed "adjustable parameters" in numerical models, because the full range of their values is constrained by observational evidence. This applies, in particular, to the coupling of the main reservoirs of land plants, soil and sediment organic matter, and aquatic biota. The stoichiometric proportions of C, N, P, and S may vary in response to environmental (and model) conditions, but these variations are not permitted to run freely in order to satisfy the mathematical solutions of the model. The variations in all the essential parameters of the model are constrained by bounds based on historical and geologic data and on present-day data from process studies.

We constructed and executed the model using STELLA®, a computer simulation package developed by High Performance Systems, Inc. Standard numerical recipes are used to solve the set of simultaneous differential equations describing the model. Rate mechanisms used to describe the various biogeochemical process are presented in Table 3.5.

The global TOTEM can be forced by one or a combination of parameters, e.g. changes in population density, fertilizer consumption, land use, vegetation, soil erosion rates, and climate. In the next section, we present the results of model scenarios forced by human activities on land, paying particular attention to the coupled response of C, N, P, and S in the land–atmosphere–ocean system. These forcings include CO_2, phosphorus, and nitrogen fertilization of the land biota (the "greening of the Earth"), and land use changes. Because of space limitations, we cannot describe the responses of all the components of the C–N–P–S system in detail and thus we limit our discussion to the more important and interesting changes that are applicable to Asia.

Anthropogenic and climatic perturbations on the C–N–P–S system

Several authors have suggested that, at least for the 1980s, forest and humus ecosystems have been a sink of fossil fuel and land use emissions of CO_2 to the atmosphere (Tans *et al.*, 1990; Keeling *et al.*, 1989, 1996; Galloway, 1995; for review see Woodwell & Mackenzie, 1995). We attempt to establish some answers to this discussion with simulations of our model. Results of the model simulations

Table 3.5 *Main flux mechanisms for components of land-ocean-atmosphere CNPS-ESM model. See footnotes for definition of variables*

Fluxes	Flux mechanism	Element cycle	Comments
1. Uptake by land biota	Michaelis–Menten kinetics with respect to atmospheric CO_2, terrestrial biotic C, continental soil water N and P, and global temperature. We formulated an equation of the type $F_{Uptake,t} = F_{Uptake,t=0} \times f_{C_{Atm}} \times f_{N_{sw}} \times f_{P_{sw}} \times f_{C_{LBio}} \times f_T$ where $f_{C_{Atm}} = \frac{C_{Atm,t} / (k_C + C_{Atm,t})}{C_{Atm,0} / (k_C + C_{Atm,0})}$ $f_{N_{sw}} = \frac{N_{sw,t} / (k_N + N_{sw,t})}{N_{sw,0} / (k_N + N_{sw,0})}$; $f_{P_{sw}} = \frac{P_{sw,t} / (k_P + P_{sw,t})}{P_{sw,0} / (k_P + P_{sw,0})}$ and $f_{C_{LBio}} = \frac{C_{LBio,t}}{C_{LBio,0}}$; $f_T = Q_{10}^{[(T - T_{1700})/10]}$ The last four factors are dimensionless and scaled to vary from –1 to 1.	C, N, P, S	Constrained by the "Redfield" ratios and limiting threshold concentrations.
2. Land biota to humus		C, N, P, S	
3. Humus to water[a]		C, N, P, S	
4. Water[a] to atmosphere		C, N, S	
5. Mineral to water[a]	First-order equations of the form $F_{ij} = kM_i = \frac{M_i}{\tau} = Uc$	C, P, S	
6. Riverine input to coastal water		C, N, P, S	
7. Coastal biota to coastal waters		C, N, P, S	
8. Coastal sedimentation		C, N, P, S	
9. Coastal waters to open ocean[b]		C, N, P, S	
10. Atmospheric deposition		N, S	

11. Atmosphere to coastal water/open ocean	Gas-solution equilibrium and diffusion controlled, based on the equation type $R(C_{Atm}) = \frac{(C_{Atm} - C_{Atm,t=0})/C_{Atm,t=0}}{(C_{SurfaceOcean} - C_{SurfaceOcean,t=0})/C_{SurfaceOcean,t=0}}$ where $R(C_{Atm}) \approx R_0 + d\frac{(C_{Atm} - C_{Atm,t=0})}{C_{Atm,t=0}}$; *(Revelle factor)*	C, N, S	Comment as in (1)
12. Uptake by coastal biota and ocean biota	Michaelis–Menten kinetics with respect to N, uptake of the form $F_{Uptake,t} = \frac{R_{max\,N} \times N_{CoastalWater}}{(k_N + N_{CoastalWater})}$	C, N, P, S	
13. Downwelling to deep ocean waters	Equilibrium equation of the type $F_{exchange} = \frac{(C_{SurfaceOcean,t} - C_{SurfaceOcean,t=0})}{h_{SurfaceOcean} \times \tau_{Deep}} - \frac{(C_{Deep,t} - C_{Deep,t=0})}{h_{Deep} \times \tau_{Deep}}$	C	Constrained by C : N : S : P ratios in downwelling water
14. Regeneration from sediments	Zero-order equations of the form $F = k$	C, N, P, S	

[a]Surface water, soil water and shallow groundwater that are in contact with humus and minerals and that feed into the land-to-ocean runoff. [b]Water reservoir distal of the shelf-slope break. F = flux, in units of moles/time; C_{atm}, N_{sw}, P_{sw}, C_{LBio}, $C_{SurfaceOcean}$, $C_{DeepOcean}$, $N_{CoastalWater}$ = reservoir mass in moles of atmospheric carbon, N/P in soil water, terrestrial biotic reservoir, carbon in surface ocean and deep ocean reservoir, and N in coastal water reservoir respectively; k = rate parameter (for first-order flux, 1/time); M = a generic reservoir; t, residence time; U = volumetric flow rate, volume/time; c = concentration, moles/volume; and $h_{SurfaceOcean}$, $h_{DeepOcean}$, depth of surface and deep ocean reservoirs. In Michaelis–Menten equation, the parameters are: R_{max} in units of 1/time, k_x in units of mass.

will allow us to (1) describe the response of the Earth system to the simultaneous human-derived perturbations on the atmosphere, terrestrial organic, and soil water reservoirs; (2) determine the potential terrestrial fertilization response and how this fertilization response affects the coastal margin environment of the world's ocean; and (3) discuss qualitatively the results in the context of the Asian region.

We begin with an Earth system at quasi-steady state in the year 1700. Through the year 1990, the behavior of the elements C, N, P, and S in Earth's exogenic reservoirs are modeled in response to perturbations imposed on the system due to three human activities on land: fossil fuel emissions, land use change, and application of inorganic N and P fertilizers, and due to changes in the climatic variable of temperature.

We use data and calculations by Kammen and Marino (1993) for the early land use perturbation flux (pre-industrial, anthropogenic non-fossil fuel CO_2 emissions from 1700–1850) and estimates by Houghton (1991) for the more recent land use CO_2 emissions (1850–1990). We model the land use perturbation to affect directly the remineralization of terrestrial biota and humus, resulting in the release of CO_2 into the atmosphere and inorganic N, P, and S into the continental soil water system. Additionally, land use activities directly affect the erosion of humus and mineral soil, enhancing the transport of particulate materials to the coastal margin.

In TOTEM, we assume that 75% of the land use CO_2 emissions is derived from perturbations on the humus reservoir while the rest is derived from perturbations on the terrestrial biota. Typically, 20–40% of the soil carbon is lost upon conversion of forest virgin land to agriculture (Detwiler, 1986; Houghton *et al.*, 1987; Mann, 1986; Schlesinger, 1986). Model results show that 14% of the initial soil organic carbon is lost between 1700 and 1990, just outside the range of estimates published in the literature.

Fossil fuel emissions (1860–1990) are based on data collated by Marland and colleagues (1994). We compiled data on global inorganic nitrogen and phosphorus fertilizer consumption from the United Nations Food and Agricultural Organization (FAO) (United Nations FAO, various years) for the years 1950–1990. For data pre-dating FAO's collection (1917–1950), we used Smil's (1991) estimates. The modeled fertilizer N and P perturbation inputs do not include the total amount of N and P applied to croplands. We assume that only 10% of the applied fertilizer N and P is leached into soils, following the global estimate of Smil (1991).

The temperature history of the Earth for the period 1700–1990 was derived from data presented in UCAR/OIES (1991) and Houghton *et al.* (1990). We model the response to changes in temperature with the Q_{10} function which describes a unit change in the rate process for every 10 °C change in temperature. For the TOTEM, we assume that $Q_{10} = 2$, i.e. a biological process rate doubles for a 10 °C increase in temperature relative to initial conditions or is halved for a 10 °C decrease.

Discussion of results of model scenarios

We shall first discuss the results of several model scenarios aimed at defining the controls on the global historical record of CO_2 concentrations during the past 300 years. We show the time course responses of the exogenic reservoirs resulting from human perturbations and temperature change and describe the partitioning of anthropogenic CO_2 into the sinks of atmosphere, terrestrial biota and humus, and ocean. In summary, we discuss the implications of the model results with respect to Asia.

Model run 1: simultaneous perturbations from four forcing functions

The simultaneous imposition of four forcing functions on the model (three human-derived and a climatic perturbation) results in a complex interaction of responses, including positive and negative feedbacks, involving Earth's exogenic reservoirs. The three major reservoirs directly affected by these perturbations are the terrestrial organic reservoir (humus + biota), the oceanic reservoir (coastal ocean and sediments, surface and deep oceans), and the atmosphere.

When the model simulation begins, the Earth is in the middle of the Little Ice Age. A minor warming and subsequent cooling occur between 1700 and 1800 followed by a general but erratic warming trend spanning the rest of the model simulation time into 1990. Temperature increases very rapidly during the later part of the simulation when the highest recorded temperature change relative to 1700 also occurs.

An increase in global temperature relative to the temperature in the year 1700 stimulates plant and soil respiration, denitrification, and terrestrial biotic uptake of carbon and nutrients governed by the "Redfield Ratio" of terrestrial plant matter. A decrease in relative temperature has an opposite effect. Note that, for this work, we chose to allow temperature to affect only the biochemical processes mentioned above and not any other Earth process normally affected by changes in global temperature, such as water stress, climatological processes, and oceanic biological processes. This constraint on the temperature effect was imposed in order to analyze and understand specifically the direct effect of temperature on terrestrial biochemical processes. Also, it appears from our global modeling results discussed below that the processes whose temperature dependence have been ignored here may be less important in terms of the historical atmospheric record of CO_2 during the past 300 years.

During early pre-industrial time in the 1700s, land use activities annually contributed about 0.3 billion tonnes C to the atmosphere, while 14 million tonnes N and 5 million tonnes P were released to the continental soil water reservoir. Kammen and Marino (1993) attribute this pre-industrial, non-fossil fuel flux to the combustion of carbon for agricultural land management, domestic needs, and small-scale industrial and craft activities. The linear rate of increase of this flux is small and remains fairly constant through 1950. Houghton (1991) estimates that land use activities accelerated rapidly from 1950 to the present resulting in large increases in the land use-derived CO_2 flux, to a maximum (although controversial) of about 1.7 billion tonnes C per year in the early 1990s. We calculate the inorganic nutrient flux to the continental soil water reservoir owing to land use activities in 1990 to be 74 million tonnes N and 26 million tonnes P per year.

We calculate the inorganic nutrient flux to the continental soil water reservoir owing to land use activities in 1990 to be 74 million tonnes N and 26 million tonnes P per year.

Fossil fuel and industrial emissions commence in 1860 with 0.1 billion tonnes of C, 3 million tonnes of N, and 4 million tonnes of S per year. As with land use activities, the rates of emissions accelerate rapidly after 1950 to 5.9 billion tonnes of C, 26 million tonnes of N, and 70 million tonnes of S per year in 1990. Application of inorganic fertilizer N and P began in 1920 at the rate of 1 million tonnes of N and 2 million tonnes of P per year, 10% of which we assume leaches into the continental soil water reservoir. In 1990, 79 million tonnes N and 16 million tonnes P were applied to croplands.

Simulation results from TOTEM (Fig. 3.9) show that the model atmospheric CO_2 reservoir behaves in very close agreement with the only firm constraint on the presently accepted CO_2 budget, the combined observed carbon dioxide data derived from ice cores taken at East Antartica, Siple, and South Pole stations

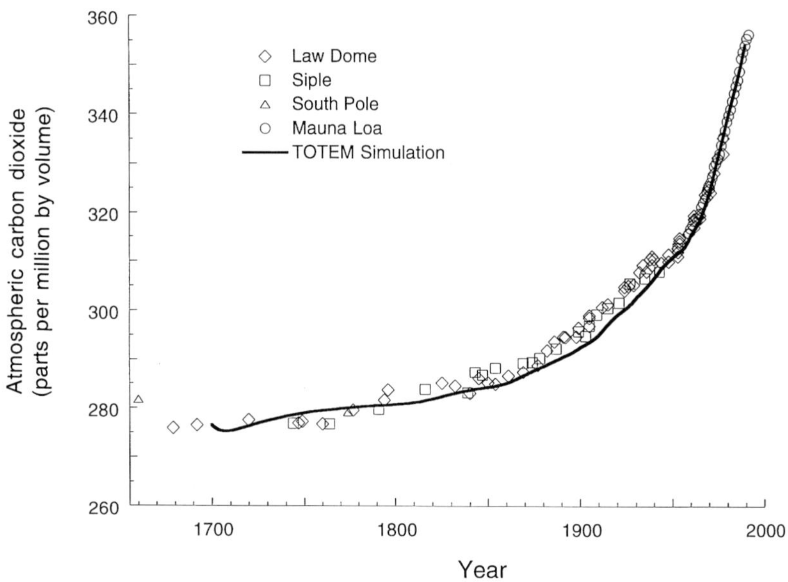

Figure 3.9 *TOTEM simulation of the behavior of atmospheric CO_2 compared with the observational data derived from ice cores taken at East Antarctica, Siple, and South Pole stations, and measurements made at the Mauna Loa Observatory. Note the very good correspondence between TOTEM results and the proxy and direct atmospheric measurements for the past 290 years.*

(Neftel *et al.*, 1985; Friedli *et al.*, 1986; Barnola *et al.*, 1995; Etheridge *et al.*, 1996) and from measurements made at the Mauna Loa Observatory (Keeling & Whorf, 1994). In 1990, of the total of 7.6 billion tonnes C added to the atmosphere annually due to fossil fuel and land use perturbations, our model estimates that 3.9 billion tonnes C accumulate in the atmosphere, 2.1 billion tonnes C are taken up by the oceans, and 1.6 billion tonnes C are taken up by the terrestrial organic reservoir. During the decade of the 1980s (1980–1989), fossil fuel (*c.* 5.3 billion tonnes C per year) and land use (*c.* 1.6 billion tonnes C per year) emissions are balanced by atmospheric accumulation (*c.* 3.2 billion tonnes C per year), oceanic uptake and terrestrial accumulation (*c.* 1.9 billion tonnes C per year each). The oceanic uptake rate compares very closely with the calculations of Tans *et al.* (1990).

TOTEM results parallel the findings obtained from other models (such as Siegenthaler & Sarmiento, 1993; Keeling *et al.*, 1989; see Fig. 3.10) in spite of fundamental differences in modeling approach, structure, and components between the TOTEM and these other models. Using the coupled global biogeochemical cycles of C, N, P and S as a basis, our model results confirm the assertion of several authors (Tans *et al.*, 1990; Watson *et al.*, 1990; Friedlingstein *et al.*, 1995; Houghton, 1995; Keeling *et al.*, 1996) that fossil fuel and non-fossil fuel sources of CO_2 are balanced by atmospheric accumulation, oceanic uptake, and terrestrial sinks. Our model-calculated cumulative terrestrial biospheric sink of 85 billion tonnes C over the period 1850–1990 is conceptually equivalent to Siegenthaler and Sarmiento's (1993) "residual" or "missing" sink (the unknown sink in Fig. 3.10*b*). From their modelling results, Siegenthaler and Sarmiento estimate an integrated residual of 97 billion tonnes C for the period 1850–1990, a difference of only 12% between their ocean–atmosphere model results and that of TOTEM.

Over the entire period of simulation, changes in net terrestrial biospheric metabolism are mainly controlled by the availability of soilwater P in the continental soil water reservoir (Fig. 3.11). Through the first 200 years of simulation, nitrogen in the soil water reservoir remains at a fairly constant mass while P exhibits a general steady decline in mass, largely by removal to the coastal zone. During the period of 1900–1990, N and P from the soil water are more rapidly assimilated into the terrestrial biota, and eventually into the humus reservoir. Some of the terrestrial and humus mass is not recycled directly back into the soil water but rather is transported to the coastal zone. Phosphorus is lost from the terrestrial realm without an available atmospheric transport pathway. Contrary to the fate of P, nitrogen accumulates in the soil water reservoir because of increased atmospheric deposition and nitrogen fixation.

TOTEM results suggest that atmospheric CO_2 fertilization of the terrestrial biota is thus primarily sustained by the recycling and remineralization of nutrients

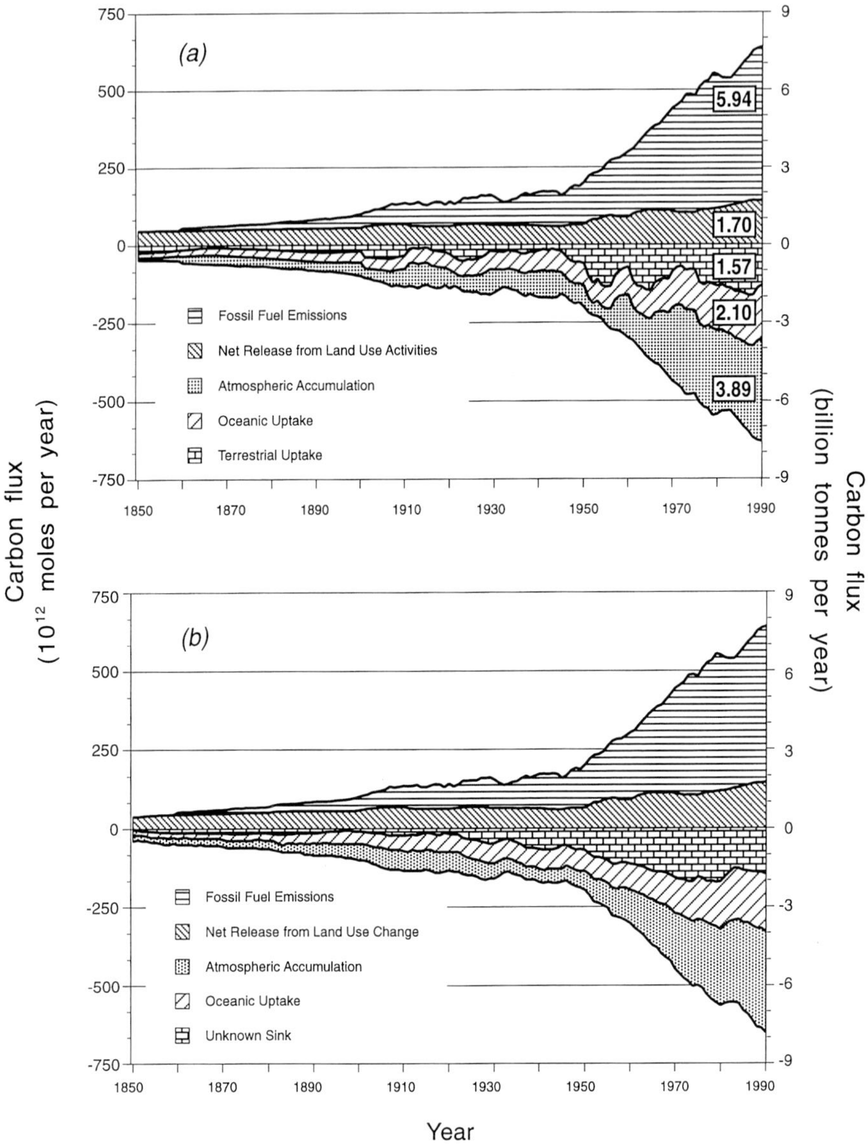

Figure 3.10 *(a) The model-calculated annual carbon perturbation fluxes to the atmosphere and the major sinks of excess atmospheric* CO_2 *for the period 1850 – 1990. The system is perturbed by a combination of three human-derived forcings (fossil fuel burning, land use activities, and application of agricultural fertilizer-derived inorganic N and P) and global temperature change. Simulation results from TOTEM shown in (a) parallel the findings of the deconvolution analysis by Siegenthaler and Sarmiento (1993) shown in (b).*

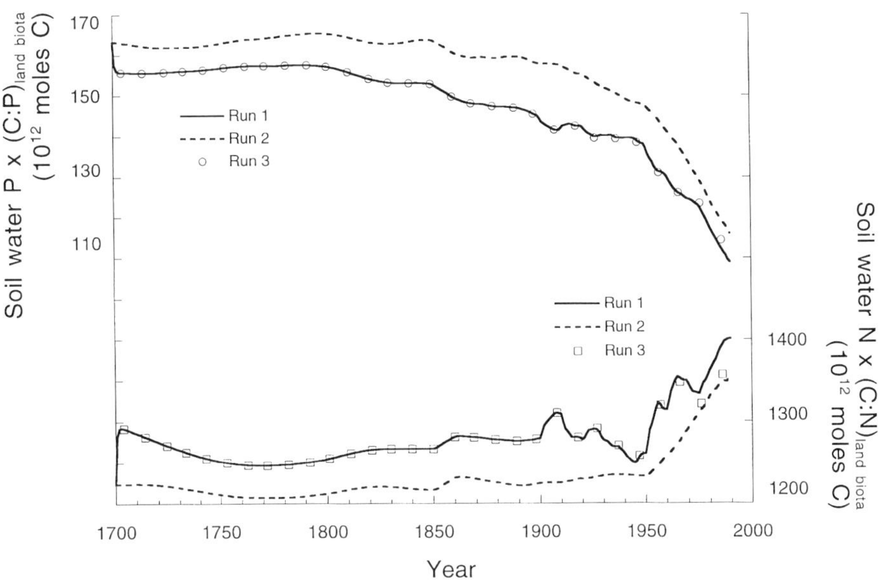

Figure 3.11 *Soilwater P and N vs. time for model simulations. Phosphorus and nitrogen are shown in equivalent carbon units, calculated by multiplying the P and N values by the $(C:P)_{land\ biota}$ and $(C:N)_{land\ biota}$ ratios, respectively. Note that, in all the simulations, there is abundant N to fuel bioproductivity on land. In contrast, the soil water P reservoir continuously shrinks in mass since about 1850, and P becomes limiting to plant growth. Indeed, after 1950, the soil water N reservoir actually grows because of land use, the application of N fertilizer to croplands, and the rainout from the atmosphere of anthropogenic nitrogen.*

from the humus reservoir over time periods of decades to centuries. Relative to the humus reservoir, land use remobilized inorganic N and P constitute a minor source of nutrients, while leached inorganic fertilizer nutrients and atmospheric sources of N are of even less importance (see below). Note that land use activities not only cause the remineralization and remobilization of organic matter but also enhance the transport of particulate matter to the coastal zone. The ultimate fate of the terrestrial mass is a net transfer of material to the atmosphere and to the oceans.

During the first half of the simulation (1700–1850), prior to extensive anthropogenic activities on land, the average net terrestrial accumulation of about 0.3 billion tonnes C per year is exceeded by an estimated loss of 0.5 billion tonnes C per year owing to land use activities. The atmosphere accumulates 80 million tonnes C per year while the oceans accumulate 0.1 billion tonnes C per year. Over the span of 150 years, the terrestrial realm has lost only 3% of its biomass or a total of 25 billion tonnes to the oceans and atmosphere.

Over the period 1850–1990, the terrestrial biosphere accumulates 85 billion tonnes C but emits 121 billion tonnes C owing to land use activities, roughly 4%

of the terrestrial organic reservoirs. Fossil fuel emissions dominate the atmospheric input with a cumulative flux of 215 billion tonnes. In the transition period between 1950 and 1980, the time history of terrestrial accumulation exhibits some erratic fluctuations which we attribute to the rapid increase in fossil fuel emissions and fluctuations in the land use emission flux. It is uncertain whether this noise represents real fluctuations or uncertainties in the data. These uncertainties in the land use activities perturbation were also invoked as part of the reason for differences between model results of other investigators and observed atmospheric CO_2 data (Friedlingstein *et al.*, 1995).

In the following sections, we explore separately the sensitivity of the model to the forcings of N and P fertilizer application to croplands and to land use-derived CO_2 emissions to the atmosphere. In the following sections, we explore separately the sensitivity of the model to the forcings of N and P fertilizer application to croplands and to land use-derived CO_2 emissions to the atmosphere. The perturbations from fossil fuel emissions and from global temperature changes were allowed to operate in both scenarios. Here we discuss the response of the various Earth reservoirs and the partitioning of anthropogenic CO_2 in each scenario.

Model run 2: contribution of agriculture fertilizer-derived inorganic N and P to CO_2 fertilization effect

Results from Run 2 suggest that the additional source of continental soilwater inorganic N and P from leaching of agricultural fertilizers accounts for very little of the increase in terrestrial biotic uptake over quasi-steady state values (Fig. 3.12*a*). Recall that the effects on biotic processes owing to temperature change and fossil fuel emissions are already accounted for in the above calculations. Thus, the terrestrial biotic uptake determined by the model for this scenario is mainly due to the process of inorganic fertilizer-enhanced atmospheric CO_2 fertilization. In 1990, the terrestrial organic reservoirs accumulated only 0.14 billion tonnes C per year. The cumulative increase in the organic reservoir mass over the entire history of fertilizer use is 8 billion tonnes C, about 1% of the initial organic mass or 3.5% of the fossil fuel input.

A critical parameter in this model scenario is the assumed fraction of applied fertilizer that is retained in the continental soil water reservoir. We adapted Smil's (1991) estimate of 10% retention even though lysimeter experiments conducted by Vormel and Ewert (1981) using ^{15}N as a tracer indicate that the loss of fertilizer N by leaching ranges between 0.4 and 3.4%. If we assume that the percentage retention of N is doubled to 20% and that the response to changes in the inorganic fertilizer perturbation is linear, the cumulative terrestrial uptake would only increase to about 7% of the fossil fuel perturbation. Clearly, applied inorganic N and P cannot be relied upon as a viable nutrient source to support the terrestrial drawdown of excess atmospheric CO_2.

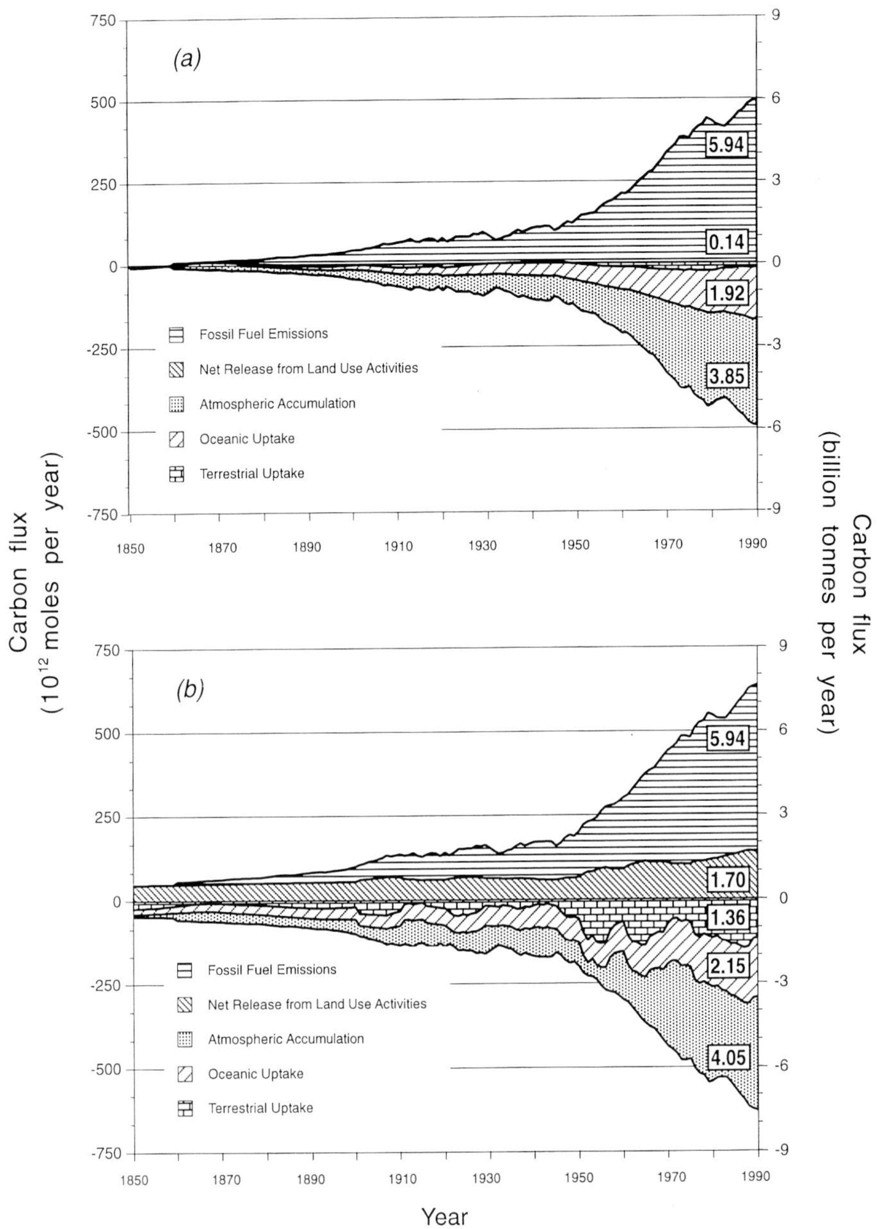

Figure 3.12 *A comparison of Earth system responses to two human-derived activities on land: (a) Increasing flux of inorganic N and P to continental soilwater from agricultural fertilizer application; and (b) Land use activities increasing the remobilization of organic and inorganic C, N, P, and S into the continental soilwater and sediments of the coastal zone. Each of these scenarios are in combination with forcings from fossil fuel burning and global temperature change. The figure shows the annual carbon perturbation fluxes to the atmosphere and the major sinks for excess atmospheric CO_2 for the period 1850–1990.*

Also for this simulation, it was assumed that 70–100% of the retained P becomes readily available for terrestrial biotic uptake. If, as discussed previously, most of the fertilizer P is immediately sequestered into insoluble inorganic phosphate forms, then the small model-calculated contribution to the fertilization effect from this perturbation may even be overestimated.

Model run 3: response to land use activities perturbation
The "land use activities" emission of CO_2 to the atmosphere includes the harvest of timber, establishment of tree plantations, and conversion of natural ecosystems to permanent cropland, shifting cultivation, pasture, and urban land. This perturbation is represented in our model by the flux of carbon and nutrients from the terrestrial organic reservoir to the atmosphere, continental soilwater reservoir, and coastal sediments.

Results (Fig. 3.12*b*) from this modeling experiment show that land use-remobilized inorganic N, P, and S can be a fairly important source of nutrients contributing to the atmospheric CO_2 fertilization of the terrestrial biota. In 1990, inorganic nutrients made available by this human-influenced perturbation, coupled with the warming temperature effect, supported 1.4 billion tonnes C per year accumulating in the terrestrial biosphere, close to 90% of that under the all perturbations scenario (Run 1) and an order of magnitude larger than that under the inorganic fertilizer perturbation (Run 2). This result should not come as a surprise. The ratio of both N and P to C is much higher in humus than in living land plants. Thus, remineralization of one mole of N from humus, for example, will support plant growth equivalent to about 100 moles of C. This implies that, with an average C:N ratio in humus of 12 : 1, roughly 88 moles of this "new" C must come from the atmosphere.

A significant impact of the land use activities perturbation is the increased transport of particulate organic and inorganic C, N, P, and S to the nearshore coastal sediments where a portion of these materials is buried and sequestered from the short-term cycling of carbon. This enhanced flux slightly increases the capacity of the oceans as a sink for anthropogenic CO_2 emissions to the atmosphere.

It is imperative to understand at this point that land use activities play a dual and complex role in the perturbation of, and response to, the perturbation of the Earth system. In their role as a source of perturbation, land use activities contribute to the flux of anthropogenic CO_2 into the atmosphere. During pre-industrial time, this flux was a small but constant perturbation on the atmosphere. With the advent of industrialization, the perturbation from land use activities increased roughly exponentially, enhancing the remobilization and transport of terrestrial organic material.

In their role as a potentially strong biofeedback, land use activities increase the supply of inorganic N, P, and S to the continental soil water, thus enhancing the biotic sink. Land use activities also contribute to an increase in the burial of particulate organic materials in the coastal sediments (and to a lesser extent in lacustrine sediments), enhancing the oceanic sink. Results from the model demonstrate that the role of land use activities as a strong biofeedback via the enhanced remobilization of inorganic N, P, and S to the continental soil water reservoir and increased transport of particulate matter to the coastal zone may be more critical than its role as a source of perturbation.

Effect of perturbations on net ecosystem productivity of terrestrial organic reservoirs and coastal zone

It is interesting to explore the effects on net ecosystem productivity (NEP) of the terrestrial and coastal margin reservoirs to the environmental perturbations in our model runs. For the scenario Runs 1 and 3 discussed in the previous sections, the model predicts, through the entire period of the simulations until about 1950, a decrease in NEP of the terrestrial organic reservoirs (Fig. 3.13). This simply means that terrestrial organic carbon is removed to the atmosphere and the coastal zone via rivers, principally due to land use practices. This is not surprising considering the extensive deforestation, changes in land use patterns and urbanization that have gone on for much of the past three centuries. However, at about 1950, this trend reverses itself, and there is an increase in the NEP of the terrestrial organic reservoirs. Relative to 1950 conditions, organic carbon is being stored on land, perhaps a result of the CO_2 fertilization effect, fed mainly by nutrients made available by land use activities. This model observation is consistent with recent works (Keeling *et al.*, 1989; Tans *et al.*, 1990, 1995; Friedlingstein *et al.*, 1995; Schimel *et al.*, 1995; Keeling *et al.*, 1996) proposing that, during the 1980s and the early 1990s, the terrestrial biota and soil humus were storing anthropogenic carbon. It is interesting to note, however, that the tendency for storage was not the general tendency for 200 years, when according to our model calculations, the terrestrial organic reservoirs were shrinking. This latter trend is consistent with much observational and experiential evidence pointing to a general degradation of the terrestrial environment for much of the 19th and 20th centuries.

In general, for the scenarios in Runs 1 and 3, the global coastal zone NEP is the opposite of that for the terrestrial environment. Of particular interest is the strong tendency for increased heterotrophy in the coastal zone after 1950 (Fig. 3.13). This trend is driven by the increase in NEP of the terrestrial organic reservoirs, owing to increasing land use activities, with a resultant general tendency for an increase in the riverine discharge of particulate and dissolved organic matter to the coastal margin. These model results are in agreement with Smith's (1995) conclusion based on process-oriented studies that the NEP of the coastal margin system is controlled by the delivery of terrestrial organic matter to the coastal zone rather than by temperature. These results mean that, due to perturbations on land, the

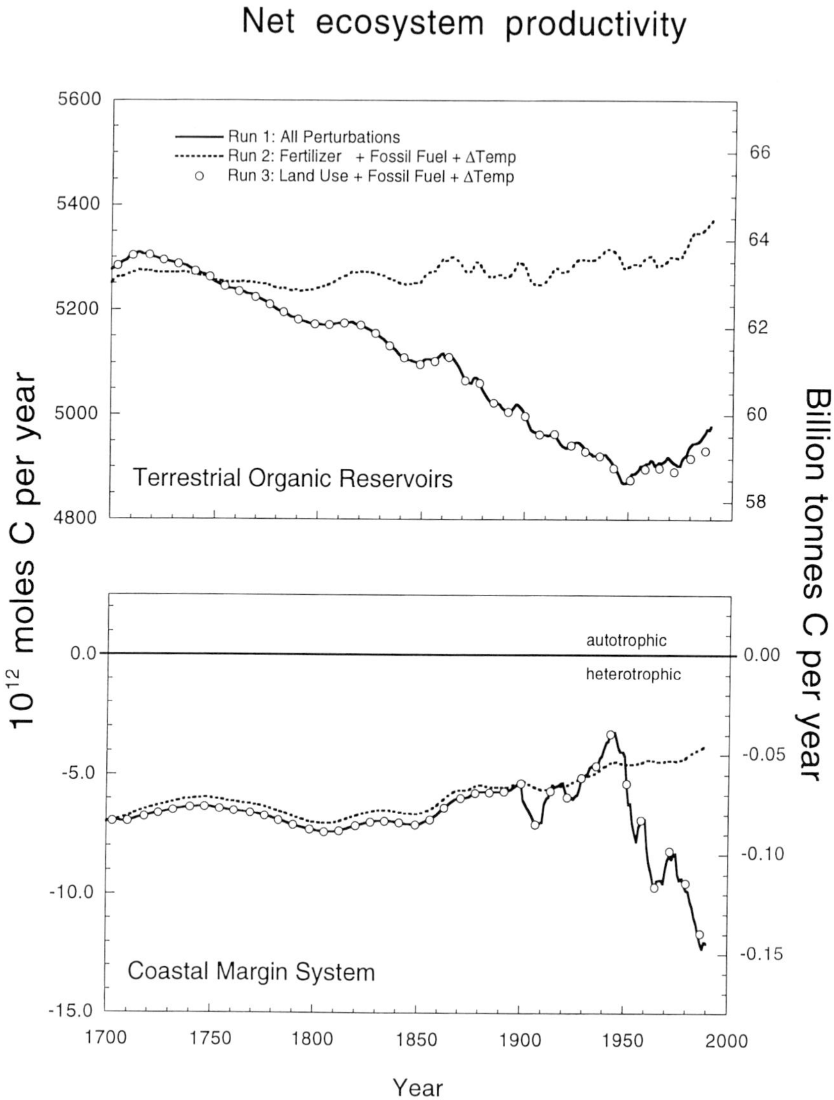

Figure 3.13 *A comparison of the response of the net ecosystem productivity (NEP) in the terrestrial realm and the coastal margin system under three model scenarios. NEP is expressed in units of 10^{12} moles carbon per year and billion tonnes carbon per year. The NEP of a system describes its net trophic state and is equivalent to the balance between organic material inputs from outside the system and outputs. A positive NEP indicates that the system is a net producer of organic matter (autotrophic), while a negative NEP indicates net consumption (heterotrophy).*

coastal zone becomes a greater net source of carbon dioxide to the atmosphere from the *in situ* consumption of organic matter. Model calculations indicate a change in NEP of the coastal zone of about 200% from 1950 to 1990.

Run 2 scenario differs from the other two scenarios in that the application of inorganic fertilizer N and P fuels terrestrial bioproductivity and increases the loading of inorganic nutrients to the coastal zone. These "new" inorganic nutrients lead to an increase in the NEP of the terrestrial organic reservoirs and reverse the trophic state of the global coastal margin system from heterotrophy towards autotrophy. Our modeling results suggest that, were it not for land use activities, the continued application of fertilizer on land would eventually cause the coastal zone to become net autotrophic, a net sink for CO_2 involved with organic metabolism. This modeling result is in agreement with the observations that regional-scale coastal environments with principally inorganic excess nutrient inputs have a tendency to develop a more autotrophic state (Smith & Mackenzie, 1987; Smith & Hollibaugh, 1993).

It should be pointed out that the changes in the net trophic state (autotrophy vs. heterotrophy) of the coastal margin in all of the above scenarios are due to increased loading of particulate and dissolved forms of organic and inorganic C, N, P, and S. At steady state, the coastal zone is net heterotrophic at the rate of 80 million tonnes C per year (Smith & Hollibaugh, 1993). An increase in the flux of inorganic nutrients owing to perturbations on land processes drives the net system trophic state towards autotrophy. An increase in the flux of organic material from land enhances coastal system heterotrophy. Previous studies conducted with TOTEM (Mackenzie *et al.*, 1993; Ver *et al.*, 1994*a*, *b*) have determined that the resultant effect of increased loading of organic material on NEP of the global coastal zone is far stronger than that of increased loading of inorganic nutrients.

Implications of modeling for Asia

In the preceding sections, we analyzed the behavior of the coupled biogeochemical cycles of C, N, P, and S in the land and coastal environment during environmental change in the global context. From model results, we are able to demonstrate that the net effect of land use activities is reduction in carbon storage in the terrestrial biosphere and increasing storage in the sediments of the coastal zone, owing to the enhanced remobilization of inorganic N, P, and S to the continental soil water and increased transport of particulate terrestrial materials to the coastal sediments. The implications and applicability of model-derived findings are of particular interest to modelers in the context of larger or finer scale geographical areas. The Asian region in general, and vast and populous China in particular, are interesting cases in point.

Thus, following the template for the behavior of the global environment in response to perturbations on land, we anticipate that Asia would experience similar, possibly even greater, loss of storage of carbon on land and increased storage in coastal marine sediments per unit area, when compared to the global scenario. However, accelerated increases in the sources of the perturbation from Asia may be anticipated to compete with the potential sink. For example, total carbon emissions from the developing countries of Asia, particularly China and India, are expected to increase by 1.8 billion tonnes over current levels by 2015, growing at an average rate of 4.0% compared to a global rate of 2.2% (Energy Information Administration, 1997). As mentioned before, because of the region's heavy reliance on fossil fuels as an energy source, SO_x and NO_x emissions to the atmosphere will increase substantially in the next century. These emissions are also responsible for acid deposition and, along with emissions of hydrocarbons, NO_x emissions participate in the processes leading to the formation of tropospheric ozone and photochemical smog. If the historical pattern in the development of photochemical smog in the US (Malm, 1989) is followed in Asian countries, one may anticipate an increase in the area of Asia covered by such smog in the 21st century. Also, SO_x emissions increase subcontinental-scale atmospheric burdens of sulfate aerosol and act radiatively to cool the continental troposphere (Charlson *et al.*, 1992; Houghton *et al.*, 1995). Thus, it is likely that the combination of acid deposition, increasing areas of photochemical smog and tropospheric ozone, and moderation of any global warming due to the cooling effect of regional and subcontinental-scale atmospheric burdens of sulfate aerosol in Asia will compromise to some degree the ability of terrestrial ecosystems in this region to store carbon in the 21st century. The direct effects of the destruction of Asian forests will have the same result.

As a generalized summary of the preceding paragraphs, we see Asia as a continent of potentially increasing contributions to the loading of the environment, owing to a combination of such factors as its increasing population, increasing industrialization dependent on fossil fuels, concentration of its population in the major river drainage basins and in the coastal area, and expansion of land use practices. The more detailed data and a quantitative picture of these processes were discussed in earlier sections, in the context of the background information and results of TOTEM. Here, we shall further discuss the potentially more important consequences for the Asian environment, and the uncertainties inherent in our current reasoning. A parallel discussion on the implications of anthropogenic change on the terrestrial water cycle and drainage basins with respect to China is presented by Vorosmarty and colleagues (see Vorosmarty *et al.*, this volume).

The relatively rapid growth of the population in the oceanic coastal zone indicates that higher inputs of both dissolved and particulate organic nutrients may be expected to enter coastal waters. A similar trend of population concentration in agricultural areas inland, in the drainage basins of the main rivers flowing into the ocean, is also expected to produce growth in the dissolved and particulate organic nutrient loads that may eventually reach the ocean. Inputs from inland regions to the ocean would be relatively more important if no entrapment or depositional storage occurred *en route*, such as in the dammed sections of rivers or in soils. In the case of many of China's rivers, the decline in sediment discharge from large rivers such as the Yangtze and the Yellow Rivers is expected to continue due to the increased construction of dams. The average decadal sediment discharge from the Yangtze River, for example, has decreased by 50% from the 1950s to the 1980s (see Hu *et al.*, this volume). If the evidence proposed by Milliman and Meade (1983) for the continental United States applies to Asia, the damming of major rivers would not effectively reduce the suspended material flow to the ocean because of the changes in the erosional patterns on land that accompany river damming and more intensive land use practices. These flows on land and into coastal ocean waters are contributing factors to the relative importance of autotrophic and heterotrophic processes, competition between the two, and the consequences for carbon exchange between the atmosphere and land, and the atmosphere and ocean water.

The change from the practices of land fertilization by manure to the more recent usage of chemical fertilizers in Asia suggest a shift away from solid organic nutrients and therefore a reduced flow of materials promoting heterotrophy. There is, however, a large uncertainty associated with this expectation, based on observations of smaller systems, such as lakes and rivers. In small lakes surrounded by pasture land, substantial amounts of manure from grazing herbivores are transported into the lakes where dissolved nutrients (nitrogen and phosphorus) can reach very high levels not observable in other natural bodies of water, and where accumulation of organic matter on the lake bottom indicates net storage of organic carbon, at least on the time scales of years to decades. In rivers, situated similarly where land is heavily used pastures and input of organic manure from the herbivore populations is significant, heterotrophy is likely to be an important process for recycling of organic carbon. This is suggested by observations of shallow rivers draining pastures, where benthic macroflora decreases and water becomes generally clearer downstream from the areas of organic matter input. To relate such observations to geographically much larger areas of land and coastal ocean, in order to draw conclusions on the relative importance of autotrophic and heterotrophic processes, is obviously difficult. It is conceivable that net storage of organic carbon may predominate in some depositional environments that are in relatively deeper water, and heterotrophic processes may predominate in shallow, well oxygenated waters. In this context, a wide continental shelf or a rising sea level that creates a shallow coastal zone may favor a heterotrophic mode of nutrient recycling, thereby returning CO_2 to the atmosphere. Conversely, an environment of deeper near-shore basins or estuaries may promote storage of a large fraction of the land-derived organic matter and overcome the effects of the heterotrophic processes.

One of the prominent changes in land use in Asia and elsewhere is deforestation. This process attracts considerably more attention than its reverse, the natural reinvasion of the forest on to agricultural lands in disuse or, more generally, the sum of the "greening" processes. In humid tropics and subtropics, recolonization of land by plants is a relatively fast process. New growth on land where flora was devastated by volcanic eruptions and lava flows, and incursions of forests on to abandoned fields and human habitats, are phenomena of short duration. On a continental or a global scale, however, the extent of such regreening processes and their contribution to the nutrient element and carbon budgets are matters of continuous study and debate (meaning their estimates are uncertain). In Asia, we do not expect the population growth and the increasing population density in cities and oceanic coastal areas to reduce or even stabilize the current level of land use practices. An expectation of reduced land use may be valid for relatively small areas which import most of their agricultural produce from elsewhere, but it is not a likely development for the geographically big and populous geopolitical units in Asia. Thus, the environmentally favorable conditions for natural regreening that exist in major parts of Asia are not likely to overcome the effects of the human land-use practices.

Continuation of the current trends of land use, as discussed earlier, would lead at its limit to the following effects for Asia.

- Burning of forests and grasslands, studied by numerous investigators, is a well-established direct source of CO_2 emissions to the atmosphere, and a process that reduces the mass of the carbon reservoir on land.
- Conversion of forests to grassland and agricultural land has been reported to increase the rates of denitrification in soils and emissions of the greenhouse gas, nitrous oxide (N_2O).
- Despite the increasing amounts of chemical fertilizers used on agricultural land, the magnitude of this regreening effect is not expected to be comparable to the forests because of the much shorter lifetime of most agricultural crops and their rapid recycling by heterotrophic consumption on land.
- Stripping of the soil layer down to agriculturally unsuitable regolith or rock is another potential factor in the disruption of nutrient cycles on land in Asia. Whether caused by natural processes or human activity, replacement of the soil layer by weathering, bacterial and micro-floral activity is a process of relatively long duration. Reduction of the agricultural soil area, which may occur despite the social needs for more crops, would reduce further the magnitude of the carbon sink on land and probably contribute to the heterotrophic recycling of carbon.

In conclusion, the outlook for Asia in the 21st century is that it has the potential to affect its own environment significantly in terms of C–N–P–S cycling as well as that of the rest of the world. In turn, global climatic change will overprint the internal perturbations of C–N–P–S cycling in Asia and lead to both positive and negative feedbacks to change.

Conclusions

It is very difficult, if not impossible, from experimental field programs alone to determine the responses of the coupled C–N–P–S biogeochemical cycles in the land–ocean–atmosphere system during global environmental change. Models need to be used as tools to organize system data and to provide a framework for mounting field expeditions. Careful attention needs to be paid to interactions between events on land and the coastal zone and its atmosphere.

Using TOTEM, we were able to model very closely the historical rise in atmospheric CO_2 concentrations during the past 300 years, and the partitioning of anthropogenic CO_2 emissions during this period among the sinks of atmosphere, ocean, and terrestrial ecosystems. It appears that the magnitude and maintenance of any CO_2 fertilization effect on the terrestrial biosphere are partially due to nutrients mobilized and translocated during land use activities, which in themselves can be sources of CO_2 to the atmosphere. Thus, modifications of land use activities could potentially decrease or amplify the sink strength associated with the CO_2 fertilization effect. For example, the world wide pressure for decreasing the rates of land use activities coupled with the strong possibility of continuously increasing global temperatures leading to increasing rates of terrestrial gross respiration (e.g. Woodwell *et al.*, 1995) could lead to a considerable weakening of the proposed terrestrial sink of anthropogenic CO_2.

Model calculations using TOTEM also demonstrated that environmental changes on land are rapidly conveyed downstream to the coastal zone, not a surprising conclusion because of the relatively short residence times of fresh water on land. Thus, any reconfiguration of agricultural and industrial practices on land has the potential to affect downstream reservoirs like groundwaters and the coastal ocean on the decadal to century time scale. Of particular interest is the fact that, if terrestrial ecosystems have been undergoing fertilization by anthropogenic CO_2, nitrogen, and phosphorus, then we would expect changes in the organic metabolism of coastal ecosystems downstream from the regions affected by fertilization processes. If terrestrial fertilization were the only process affecting organic metabolism in coastal ecosystems, we would anticipate a global trend toward increasing heterotrophy of such systems during the 21st century. Obviously, coastal ecosystems are influenced by a myriad of natural and human-induced forcings, and their behavior in the future depends on these interactive forcings.

Finally, the model development discussed in this paper and applied to several environmental scenarios affecting the global system is applicable to smaller space scales, including regional, subcontinental, and continental. It can be considered as one possible template for an analysis of environmental change affecting C–N–P–S cycling in the Asian region. Demographic and socio-economic factors indicate that the current practices of rapid development, industrialization, and land use in this region will continue on into the 21st century. These activities will modify the biogeochemical cycles of C–N–P–S in Asia significantly and contribute strongly to modification of the global cycles of these elements.

Acknowledgments

This research was supported by NSF Grant EAR93-16133 and by a grant from the NOAA Office of Global Programs NA37RJ0199. We thank Dr Steve Smith and three anonymous reviewers for their critical comments. During writing of this paper, FTM was a Fellow at the Wissenschaftskolleg zu Berlin and wishes to express his thanks to the Rector, Prof Dr Wolf Lepenies, for providing space and atmosphere for unbridled intellectual thought. This is UH School of Ocean and Earth Sciences and Technology Contribution No. 4511.

References

Atkinson, M. J. & Smith, S. V. (1983). C : N : P ratios of benthic marine plants. *Limnology Oceanography*, **28**, 568–75.

Barnola, J. M., Anklinm, M., Procheron, J., Reynaud, D., Schwander, J. & Stauffer, B. (1995). CO_2 evolution during the last millennium as recorded by Antarctic and Greenland ice. *Tellus*, **47B**, 264–72.

Bazzaz, F. A. (1990). The response of natural ecosystems to the rising global CO_2 levels. *Annual Review of Ecology Systematics*, **21**, 167–96.

Berner, R. A. (1982). Burial of organic carbon and pyrite sulfur in the modern ocean: its geochemical and environmental significance, *American Journal of Science*, **282**, 451–73.

Berner, E. K. & Berner, R. A. (1987). *The Global Water Cycle*, 397 pp. Englewood Cliffs, NJ: Prentice-Hall, Inc.

Berner, E. A. & Berner, R. A. (1996). *Global Environment: Water, Air and Geochemical Cycles*, 376 pp. Upper Saddle River, NJ: Prentice Hall.

Billen, G. (1993). The PHISON river system: a conceptual model of C, N and P transformations in the aquatic continuum from land to sea. In *Interactions of C, N, P and S Biogeochemical Cycles and Global Change*, ed. R. Wollast, F. T. Mackenzie & L. Chou, pp. 141–61. Berlin: Springer-Verlag.

Bischoff, W. D., Paterson, V. L. & Mackenzie, F. T. (1984). Geochemical mass balance for sulfur-and nitrogen-bearing acid components: Eastern United States. In *Geological Aspects of Acid Deposition*, ed. O. P. Bricker, pp. 1–21. Boston: Butterworth Publishing.

Broecker, W. S. & Denton, G. H. (1989). The role of ocean-atmosphere reorganization in glacial cycles. *Geochimica et Cosmochimica Acta*, **53(10)**, 2465–501.

Broecker, W. & Severinghaus, J. P. (1992). Diminishing oxygen. *Nature*, **358**, 710–11.

Charlson, R. J. (1995). The vanishing climatic role of dimethyl sulphide. In *Biotic Feedbacks in the Global Climatic System: Will the Warming Feed the Warming?*, ed. G. M. Woodwell & F. T. Mackenzie, pp. 251–62. Oxford: Oxford University Press.

Charlson, R. J., Schwartz, S. E., Hales, J. M., Cess, R. D., Coakley Jr., J. A., Hansen, J. E. & Hofmann, D. J. (1992). Climate forcing by anthropogenic aerosols. *Science*, **255**, 423–30.

Cole, J. J., Peierls, B. L., Caraco, N. F. & Pace, M. L. Nitrogen loading of rivers as a human-driven process. (1993). In *Humans as Components of Ecosystems: The Ecology of Subtle Human Effects and Populated Areas*, ed. M. J. McDonnell & S. T. A. Pickett, pp. 141–157. New York: Springer-Verlag.

Crutzen, P. J. & Andreae, M. O. (1990). Biomass burning in the tropics: impact on atmospheric chemistry and biogeochemical cycles. *Science*, **250**, 1669–78.

Dagg, M. J., Green, E. P., McKee, B. A. & Ortner, P. B. (1996). Biological removal of fine-grained lithogenic particles from a large river plume. *Journal of Marine Research*, **54(1)**, 149–59.

Dazhong, W. & Pimentel, D. (1990). Ecological resource management for a productive, sustainable agriculture in Northeast China. In *Agricultural Reform and Development in China*, ed. T. Tso, 302 pp. Beltsville, MD: IDEALS.

Deevy, E. S. (1973). Sulfur, nitrogen and carbon in the atmosphere. In *Carbon and the Biosphere*, ed. G. M. Woodwell & E. V. Peacan, pp. 182–90. Washington, DC: USAEC.

Delwiche, C. C. & Likens, G. E. (1977). Biological response to fossil fuel combustion products. In *Global Chemical Cycles and Their Alterations by Man, Dahlem Konferenzen*, ed. W. Stumm, pp. 73–88. Berlin.

Detwiler, R. P. (1986). Land use change and the global carbon cycle: The role of tropical soils. *Biogeochemistry*, **2**, 67–93.

Detwyler, T. R. (1971). *Man's Impact on Environment*, 731 pp. New York: McGraw-Hill.

Diaz, H. F. & Pulwarty, R. S. (1994). An analysis of the time scales of variability in centuries-long ENSO-sensitive records in the last 1000 years. *Climatic Change*, **26(2-3)**, 317–42.

Energy Information Administration, International Energy Outlook 1997, US Department of Energy, Washington, DC, 1997.

Etheridge, D. M., Steele, L. P., Langenfelds, R. L., Francey, R. J., Barnola, J.-M. & Morgan, V. I. (1996). Natural and anthropogenic changes in atmospheric CO_2 over the last 1000 years from air in Antarctic ice and firn. *Journal of Geophysical Research*, **101**, 4115–28.

Fang, J. Y., Su, S. L. & Liu, G. H. (1995). Carbon pools on terrestrial ecosystems of China and their global significance. In *Natural and Anthropogenic Changes: Impacts on Global Biogeochemical Cycles*. Beijing, China: IGBP.

Fearnside, P. (1995). Hydroelectric dams are significant CO_2 source. *Environmental Conservative*, **22**, 7–19.

Friedli, H., Lotscher, H., Oeschger, H., Siegenthaler, U. & Stauffer, B. (1986). Ice core record of the $^{13}C/^{12}C$ ratio of atmospheric carbon dioxide in the past two centuries. *Nature*, **324**, 237–8.

Friedlingstein, P., Fung, I., Holland, E., John, J., Brasseur, G., Erickson, D., & Schimel, D. (1995). On the contribution of CO_2 fertilization to the missing biospheric sink. *Global Biogeochemical Cycles*, **9**, 541–56.

Frisch, J-R. (1986). *Future Stresses for Energy Resources: Energy Abundance, Myth or Reality?*, 226 pp. London: Graham & Trotman.

Galloway, J. N. (1989). Atmospheric acidification: Projections for the future. *Ambio*, **18**, 161–6.

Galloway, J. N. (1995). Acid deposition: Perspectives in time and space. *Water, Air, and Soil Pollution*, **85(1)**, 15–24.

Galloway, J. N., Savoie, D. L. & Keene, W. C. (1993). The temporal and spatial variability of scavenging ratios for nss sulfate, nitrate, methanesulfonate and sodium in the atmosphere over the north Atlantic Ocean. *Atmospheric Environment*, **27(2)**, 235–50.

Galloway, J. N., Schlesinger, W. H., A. Michaels,

H. L. II & Schnoor, J. L. (1995). Nitrogen fixation: Anthropogenic enhancement-environmental response. *Global Biogeochemical Cycles*, **9**, 235–52.

Galloway, J. N., Howarth, R. W., Michaels, A. F., Nixon, S. W., Prospero J. M. & Dentener, F. J. (1996). Nitrogen and phosphorus budgets of the North Atlantic Ocean and its watershed. *Biogeochemistry*, **35(1)**, 3–25.

Garrels, R. M. & Mackenzie, F. T. (1971). *Evolution of Sedimentary Rocks*, 397 pp. New York: W. W. Norton.

Garrels, R. M., Mackenzie, F. T. & Hunt, C. (1975). *Chemical Cycles and the Global Environment*, 206 pp. Los Altos, CA: W. Kaufmann, Inc.

Grace, J., Lloyd, J., McIntyre, J., Miranda, A. C., Meir, P., Miranda, H. S., Nobre, C., Monroe, J., Massheder, J., Malhi, Y., Wright, I. & Gash, J. (1995). Carbon dioxide uptake by an undisturbed tropical rain forest in southwest Amazonia, 1992 to 1993. *Science*, **270(5237)**, 778–80.

Halim, Y. (1991). The impact of human alterations of the hydrologic cycle on ocean margins. In *Ocean Margin Processes on Global Change*, ed. R. F. C. Mantoura, J. M. Martin, and R. Wollast, pp. 301-28. Chichester: Wiley-Interscience Publishers.

Holligan, P. M. & Reiners, W. A. (1992). Predicting the responses of the coastal zone to global change. *Advances in Ecological Research*, **22**, 211–55.

Houghton, J., Jenkins, G. & Ephraums, J. (1990). *Climate Change: The IPCC Scientific Assessment*, 365 pp. Cambridge: Cambridge University Press.

Houghton, J. T., Filho, L. G. M., Bruce, J., Lee, H., Callander, B. A., Haites, E., Harris, N. & Maskell, K. (1995). *Climate Change 1994. Radiative Forcing of Climate Change and an Evaluation of the IPCC IS92 Emission Scenarios*, 339 pp. Cambridge: Cambridge University Press..

Houghton, R. A. (1991). Tropical deforestation and atmospheric carbon dioxide. *Climatic Change*, **19**, 99–118.

Houghton, R. A. (1995). Effects of land-use change, surface temperature and CO_2 concentration on terrestrial stores of carbon. In *Biotic Feedbacks in the Global Climatic System: Will the Warming Feed the Warming?*, ed. G. M. Woodwell & F. T. Mackenzie, pp. 333–350. Oxford: Oxford University Press.

Houghton, R. A., Boone, R. D., Fruci, J. R., Hobbie, J. E., Melillo, J. M., Palm, C. A., Peterson, B. J., Shaver, G. R. & Woodwell, G. M. (1987). The flux of carbon from terrestrial ecosystems to the atmosphere in 1980 due to changes in land use: geographic distribution of the global flux. *Tellus*, **39B**, 122–39.

Howarth, R. W. (1996). Nitrogen Cycling in the North Atlantic Ocean and its Watersheds. In *Biogeochemistry*, pp. 304. Dordrecht, The Netherlands: Kluwer Academic Publishers.

Howarth, R. W., Billen, G., Swaney, D., Townsend, A., Jaworski, N., Lajtha, K., Downing, J. A., Elmgren, R., Caraco, N., Jordan, T., Berendse, F., Freney, J., Kudeyarov, V., Murdoch, P. & Zhao-liang, Z. (1996). Regional nitrogen budgets and riverine N & P fluxes for the drainages to the North Atlantic Ocean: natural and human influences. *Biogeochemistry*, **35(1)**, 75–139.

Idso, S. B. (1982). *Carbon Dioxide: Friend or Foe*, 92 pp. Tempe, AZ: Institute for Carbon Dioxide Research Press.

Isermann, K. (1993). Territorial, continental and global aspects of C, N, P and S in soils of terrestrial ecosystems. In *Interactions of C, N, P and S Biogeochemical Cycles and Global Change*, ed. R. Wollast, F. T. Mackenzie & L. Chou, pp. 79–121. Berlin: Springer-Verlag.

Joyce, J. A. (1975). *World population: Basic Documents*, 432 pp. Dobbs Ferry, New York: Oceana Publications, Inc.

Kammen, D. & Marino, B. (1993). On the origin and magnitude of pre-industrial anthropogenic CO_2 and CH_4 emissions. *Chemosphere*, **26**, 69–86.

Karl, D. M., Letelier, R., Hebel, D., Tupas, L., Dore, J., Winn, C. & Christian, J. (1995). Ecosystem changes in the North Pacific subtropical gyre attributed to the 1991–92 El Niño. *Nature*, **373(6511)**, 230–4.

Kauppi, P. E., Miellikainen, K. & Kuusela, K. (1992). Biomass and carbon budgets of European forests, 1971–1990. *Science*, **256**, 70–4.

Keeling, C. D. & Whorf, T. P. (1994). Atmospheric CO_2 records from sites in the SIO air sampling network. In *Trends '93: A Compendium of Data on Global Change*, ed. T. A. Boden, D. P. Kaiser, R. J. Sepanski & F. W. Stoss, pp. 16–26. Oak Ridge National Laboratory, Oak Ridge, Tennessee, USA: Carbon Dioxide Information Analysis Center.

Keeling, C. D., Carter, A. F. & Mook, W. G. (1984). Seasonal, latitudinal and secular variations in the abundance and isotopic ratios of atmospheric CO_2. 2. Results from Oceanographic cruises in the tropical Pacific Ocean, *Journal of Geophysical Research*, **89**, 4615–28.

Keeling, C. D., Bacastow, R. B., Carter, A. F., Piper, S. C., Whorf, T. P., Heimann, M., Mook, W. G. & Roeloffzen, H. (1989). A three-dimensional model of atmospheric CO_2 transport based on observed winds: 1. Analysis of observational data. In *Aspects of Climate Variability in the Pacific and the Western Americas. Geophysical Monograph 55*, ed. D. H. Peterson, pp. 165–235.

Keeling, R. F., Piper, S. C. & Heimann, M. (1996). Global and hemispheric CO_2 sinks deduced from changes in atmospheric O_2 concentration. *Nature*, **381**, 218–21.

Kinzig, A. P. & Socolow, R. H. (1994). Human impacts on the nitrogen cycle. *Physics Today*, 24–31.

Kolchugina, T. P. & Vinson, T. S. (1995). Role of Russian forests in the global carbon balance. *Ambio*, **24**, 258–64.

Lemon, E. (1977).The land's response to more carbon dioxide. In *The Fate of Fossil Fuel CO_2 in the Oceans*, ed. N. R. Anderson & A. Malahoff, pp. 97–130. New York: Plenum Press.

Lerman, A., Mackenzie, F. T. & Geiger, R. J. (1989). Environmental chemical stress effects associated with the carbon and phosphorus biogeochemical cycles. In *Ecotoxicology: Problems and Approaches*, ed. S. A. Levin, M. A. Harwell, J. R. Kelly & K. D. Kimball, pp. 315–350. New York: Springer-Verlag.

Lerman, A., Mackenzie, F. T. & Ver, L. M. (1993). Global nitrogen cycle within the coastal C-N-P system. *Chemical Geology*, **107**, 389–92.

Likens, G. E. (1981). Some perspectives of the major biogeochemical cycles, SCOPE 17, 175 pp. New York: John Wiley.

Likens, G. E., Bormann, H. F. & Johnson, N. M. (1981). Interactions between major biogeochemical cycles in terrestrial ecosystems. In *Some Perspectives of the Major Biogeochemical Cycles, SCOPE 17*, ed. G. E. Likens, pp. 93–112. New York: John Wiley.

Lutz, W. (1994). *The Future Population of the World. What Can We Assume Today?*, pp. 484. London: Earthscan Publications Ltd.

McElroy, M. B. (1994). Climate of the earth: an overview. *Environmental Pollution*, **83**, 3–21.

Mackenzie, F. T. (1981). Global carbon cycle: some minor sinks for CO_2. In *Flux of Organic Carbon by Rivers to the Sea, USDOE Conference Report 8009140*, ed. G. E. Likens, F. T. Mackenzie, J. Richey, J. R. Sedwell & K. K. Turekian, pp. 360–384, Washington, DC.

Mackenzie, F. T. (1995). Global climatic change: Climatically important biogenic gases and feedbacks. In *Biotic Feedbacks in the Global Climatic System: Will the Warming Feed the Warming?*, ed. G. M. Woodwell & F. T. Mackenzie, pp. 22–46. New York: Oxford University Press.

Mackenzie, F. T. & Mackenzie, J. A. (1995). *Our Changing Planet: An Introduction to Earth System Science and Global Environmental Change*, 387 pp. Englewood Cliffs, NJ: Prentice Hall.

Mackenzie, F. T., Ver, L. M., Sabine, C., Lane, M. & Lerman, A. (1993). C, N, P, S global biogeochemical cycles and modeling of global change. In *Interactions of C, N, P and S Biogeochemical Cycles and Global Change*, ed. R. Wollast, F. T. Mackenzie & L. Chou, pp. 1–62. Berlin: Springer-Verlag.

Mackenzie, F. T., Ver, L. M. & Lerman, A. (1995). Modeling the behavior of bio-geochemical cycles in the land-coastal margin system during global change. In *International Symposium on Earth Environment*, pp. 149–200. Pusan, South Korea: Korea Inter-University Institute of Ocean Science.

Malm, W. C. (1989). Atmospheric haze, its sources and effects on visibility in rural areas of the Continental United States. *Environmental Monitoring and Assessment*, **12**, 203–25.

Mann, L. K. (1986). Changes in soil carbon storage after cultivation. *Soil Science*, **142**,

279–88.

Marland, G., Andres, R. J. & Boden, T. A. (1994). Global, regional and national CO_2 emissions. In *Trends '93: A Compendium of Data on Global Change*, ed. T. A. Boden, D. P. Kaiser, R. J. Sepanski & F. W. Stoss, pp. 505–84. Oak Ridge National Laboratory, Oak Ridge, Tennessee, USA: Carbon Dioxide Information Analysis Center.

Matson, P. A. & Vitousek, P. M. (1990). Ecosystem approach to a global nitrous oxide budget: processes that regulate gas emissions vary in predictable ways. *BioScience*, **40(9)**, 667.

Meade, R. H. & Parker, R. S. (1985). Sediments in rivers of the United States, US Geological Survey.

Melillo, J. M. (1983). Will increases in atmospheric CO_2 concentrations affect decay processes?, Annual Report, The Ecosystems Center, Marine Biological Laboratory.

Melillo, J. M., Borchers, J., Chaney, J., Fisher, H. & Fox, S. (1995). Vegetation/ecosystem modeling and analysis project: comparing biogeography and biogeochemistry models in a continental-scale study of terrestrial ecosystem responses to climate change and CO_2 doubling. *Global Biogeochemical Cycles*, 4, 407–37.

Meybeck, M. (1993). C, N, P and S in rivers: From sources to global inputs. In *Interactions of C, N, P and S Biogeochemical Cycles and Global Change*, ed. R. Wollast, F. T. Mackenzie & L. Chou, pp. 163–193. Berlin: Springer-Verlag.

Meybeck, M. & Helmer, R. (1989). The quality of rivers: from pristine stage to global pollution. *Palaeogeography, Palaeoclimatology, Palaeoecology (Global Planetary Change)*, **1**, 283–309.

Milliman, J. D. & Meade, R. H. (1983). Worldwide delivery of river sediment to the oceans. *Journal of Geology*, **91**, 1–21.

Milliman, J. D. & Syvitski, J. P. M. (1992). Geomorphic/tectonic control of sediment discharge to the ocean: the importance of small mountainous rivers. *Journal of Geology*, **100**, 525–44.

Moore III, B. & Schimel, D. (1992). *Trace Gases and the Biosphere*, 236 pp. Boulder, CO: UCAR/Office for Interdisciplinary Earth Studies.

Neftel, A., Moor, E., Oeschger H., Turekian, K. K. & Dodge, R. E. (1985). Evidence from polar ice cores for the increase in atmospheric CO_2 in the past two centuries. *Nature*, **315**, 45–7.

Nelsen, T. A. (1996). Coastal ocean water resources: Linkages with terrestrial freshwater resources, anthropogenic influences and climate change. In *Diachronic Climatic Impacts on Water Resources*, ed. A. N. Angelakis & A. S. Issar, pp. 319–47. Berlin: Springer-Verlag.

Nixon, S. W. (1992). Quantifying the relationship between nitrogen input and the productivity of marine ecosystems. *Advances in Marine Technology*, **5**, 57–83.

Nixon, S. W., Granger, S. L. & Nowicki, B. L. (1995). An assessment of the annual mass balance of carbon, nitrogen, and phosphorus in Narragansett Bay. *Biogeochemistry*, **31**, 15–61.

Norby, R. J., Gunderson, C. A., Wullschleger, S. D., O'Neill, E. G. & McCracken, M. K. (1992). Productivity and compensatory responses of yellow poplar trees in elevated CO_2. *Nature*, **357**, 322–4.

Oldeman, L. R., van Engelen, V. W. P. & Pulles, J. H. M. (1991). The extent of human-induced soil degradation. In *World Map of the Status of Human-Induced Soil Degradation: Annex 5*. Wageningen, Netherlands: United Nations Environment Programme and International Soil Reference and Information Center.

Oremland, R. S. (1993). *Biogeochemistry of Global Change: Radiatively Active Trace Gases*, pp. 879. New York: Chapman and Hall.

Pan, Y., Melillo, J. M., Kicklighter, D. W., Xiao, X. & McGuire, A. D. (1995). Potential changes in net primary productivity in terrestrial ecosystems of China with projected climate changes: a simulation study of the terrestrial ecosystem model coupling with vegetation redistribution. In *Natural and Anthropogenic Changes: Impacts on Global Biogeochemical Cycles*. Beijing, China: IGBP.

Pearce, F. (1991). A dammed fine mess. *New Scientist*, **130(1767)**, 36–9.

Peterson, B. J. & Melillo, J. M. (1985). The potential storage of carbon caused by the eutrophication of the biosphere, *Tellus*, **37B**, 117–27.

Pimentel, D., Harvey, C., Resosudarmo, P., Sinclair, K., Kurz, F., McNair, M., Crist, S., Shpritz, L., Fitton, L., Saffouri, R. & Blair, R. (1995). Environmental and economic costs of soil erosion and conservation benefits. *Science*, **267**, 1117–23.

Postel, S. L., Daily, G. C. & Ehrlich, P. R. (1996). Human appropriation of renewable fresh water. *Science*, **271**, 785–8.

Rastetter, E. B. (1996). Validating models of ecosystem response to global change. *BioScience*, **46(3)**, 190–8.

Redfield, A. C. (1958). The biological control of chemical factors in the environment. *American Scientist*, **46**, 205–21.

Redfield, A. C., Ketchum, B. H. & Richard, F. A. (1963). The influence of organisms on the composition of seawater. In *The Sea, Vol. 2*, ed. M. N. Hill, pp. 26–77. New York: John Wiley.

Sabine, C. L. & Mackenzie, F. T. (1991). Oceanic sinks for anthropogenic CO_2. *Journal of Energy, Environment and Economics*, **1**, 119–27.

Saito, Y., Katayama, H., Ikehara, K., Matsumoto, E. & Kato, Y. (1995). Cultivation history of the loess plateau in China recorded in sediments of the East China Sea. In *Natural and Anthropogenic Changes: Impacts on Global Biogeochemical Cycles*. Beijing, China: IGBP.

Sarmiento, J. L. & Sundquist, E. T. (1992). Revised budget for the oceanic uptake of anthropogenic carbon dioxide. *Nature*, **356**, 589–93.

Saunders, I. & Young, A. (1983). Rates of surface processes on slopes, slope retreat, and denudation. *Earth Surface Processes and Landforms*, **8**, 473–501.

Schimel, D., Enting, I. G., Heimann, M., Wigley, T. M. L., Raynaud, D., Alves, D. & Siegenthaler, U. (1995). CO_2 and the carbon cycle. In *Climate Change 1994. Radiative Forcing of Climate Change and an Evaluation of the IPCC IS92 Emission Scenarios*, ed. J. T. Houghton, L. G. M. Filho, J. Bruce, H. Lee, B. A. Callander, E. Haites, N. Harris & K. Maskell, pp. 35–71. Cambridge: Cambridge University Press.

Schindler, D. W., Curtis, P. J., Parker, B. R. & Stainton, M. P. (1996). Consequences of climate warming and lake acidification for UV-B penetration in North American boreal lakes. *Nature*, **379(6567)**, 705–8.

Schlesinger, W. H. (1986). Changes in soil carbon storage and associated properties with disturbance and recovery. In *The Changing Carbon Cycle: A Global Analysis*, ed. J. R. Trabalka & D. E. Reichle, pp. 194–220. New York: Springer-Verlag.

Schlesinger, W. H. (1991). *Biogeochemistry: An Analysis of Global Change*, 443 pp. San Diego, CA: Academic Press, Inc.

Siegenthaler, U. & Sarmiento, J. L. (1993). Atmospheric carbon dioxide and the ocean. *Nature*, **365**, 119–25.

Smil, V. (1985). *Carbon Nitrogen and Sulfur: Human Interference in Grand Biospheric Cycles*, 459 pp. New York: Plenum Press.

Smil, V. (1991). Population growth and nitrogen: An exploration of a critical existential link. *Population and Development Review*, **17**, 569–601.

Smith, S. V. (1995). Net carbon metabolism of oceanic margins and estuaries. In *Biotic Feedbacks in the Global Climatic System: Will the Warming Feed the Warming?*, ed. G. M. Woodwell & F. T. Mackenzie, pp. 251–262. Oxford: Oxford University Press.

Smith, S. V. & Hollibaugh, J. T. (1993). Coastal metabolism and the oceanic organic carbon balance. *Reviews of Geophysics*, **31(1)**, 75–89.

Smith, S. V. & Mackenzie, F. T. (1987). The ocean as a net heterotrophic system: Implications from the carbon biogeochemical cycle. *Global Biogeochemical Cycles*, **1**, 187–98.

Socolow, R., Andrews, C., Berkhout, F. & Thomas, V. (1994). *Industrial Ecology and Global Change*, 500 pp. Cambridge: Cambridge University Press.

Strain, B. R. & Cure, J. D. (1985). Direct effects of increasing carbon dioxide on vegetation, US Department of Energy, DOE/ER-0.238.

Svensson, B. H. & Soderlund, R., (1976). Nitrogen, phosphorus and sulphur – global cycles : SCOPE 7. In *Ecological bulletins 22*, 192 pp. Stockholm: Swedish SCOPE

Committee of the Royal Swedish Academy of Sciences.

Tans, P. P., Fung, I. Y. & Takahashi, T. (1990). Observational constraints on the global atmospheric CO_2 budget. *Science*, **247**, 1431–38.

Tans, P. P., Fung, I. Y. & Enting, I. G. (1995). Storage versus flux budgets: The terrestrial uptake of CO_2 during the 1980s. In *Biotic Feedbacks in the Global Climatic System: Will the Warming Feed the Warming?*, ed. G. M. Woodwell & F. T. Mackenzie, pp. 351–66. Oxford: Oxford University Press.

Tiessen, H. (1995). Phosphorus in the Global Environment: Transfers, Cycles and Management. SCOPE 54, pp. 462 Chichester: John Wiley.

Turner, B. L., Clark, W. C., Kates, R. W., Richards, J. F., Mathews, J. T. & Meyer, W. B. (1990). *The Earth as Transformed by Human Action*, 713 pp. Cambridge: Cambridge University Press.

UCAR/OIES (1991). *Changes in Time in the Temperature of the Earth.* Boulder, CO: UCAR/OIES.

United Nations (various years). *Statistical Yearbook.* New York: United Nations.

United Nations FAO (1997). State of the World's Forests, Oxford, UK.

United Nations FAO (various years). *Annual Fertilizer Review.* Rome, Italy: Food and Agriculture Organization of the United Nations.

United Nations FAO (1961–1995). *Production Yearbook.* Rome, Italy: Food and Agriculture Organization of the United Nations.

United Nations Population Division (1995). *World Population Prospects: The 1994 Revision*, 884 pp. New York: United Nations.

Ver, L. M., Mackenzie, F. T. & Lerman, A. (1994*a*). Modeling pre-industrial C–N–P–S biogeochemical cycling in the land coastal margin system. *Chemosphere*, **29**, 855–87.

Ver, L. M., Mackenzie, F. T. & Lerman, A. (1994*b*). Biogeochemical controls of nutrient trace-metal fluxes from land to ocean margins. In *1994 Annual Meeting, Geological Society of America*, Geological Society of America, Seattle, WA.

Vitousek, P. M., Ehrlich, P. R., Ehrlich, A. H. & Matson, P. A. (1986). Human appropriation of the products of photosynthesis. *BioScience*, **36**, 368–73.

Vormel, A. & Ewert, U. (1981). Stickstoffdungung und stickstoffverluste auf ackerbaulich genutzten Boden. *Berichte uber Landwirtschaft Sonderheft*, **197**, 205–22.

Walsh, J. J., Rowe, G. T., Iverson, R. L. & McRoy, P. (1981). Biological export of shelf carbon is a sink of the global CO_2 cycle. *Nature*, **291**, 196–201.

Watson, R. T., Rodhe, H., Oeschger, H. & Siegenthaler, U. (1990). Greenhouse gasses and aerosols. In *Climate Change: The IPCC Scientific Assessment*, ed. J. Houghton, G. Jenkins & J. Ephraums, pp. 1–40. Cambridge: Cambridge University Press.

Wollast, R. & Mackenzie, F. T. (1989). Global biogeochemical cycles and climate. In *Climate and Geo-Sciences*, ed. A. Berger, S. Schneider & J. C. Duplessy, pp. 453–73. Dordrecht: Kluwer Academic Publishers.

Wollast, R., Mackenzie, F. T. & Chou, L. (1993). *Interactions of C, N, P and S Biogeochemical Cycles and Global Change*, pp. 521. Berlin: Springer-Verlag.

Woodwell, G. M. & Mackenzie, F. T. (1995). *Biotic Feedbacks in the Global Climatic System: Will the Warming Feed the Warming?*, 416 pp. New York: Oxford University Press.

Woodwell, G. M., Mackenzie, F. T., Houghton, R. A., Apps, M. J., Gorham, E. & Davidson, E. A. (1995). Will the warming speed the warming?. In *Biotic Feedbacks in the Global Climatic System: Will the Warming Feed the Warming?*, ed. G. M. Woodwell & F. T. Mackenzie, pp. 393–411. New York: Oxford University Press.

World Resources Institute (1994). *UN Environment Programme, and UN Development Programme, World Resources 1994–95: A Guide to the Global Environment*, 400 pp. New York: Oxford University Press.

World Resources Institute (1996). *UN Environment Programme, and UN Development Programme, World Resources 1996–1997: A Guide to the Global Environment*, 384 pp. New York: Oxford University Press.

Biomass burning: an ecosystem process of global significance

J-P. Malingreau and Y. H. Zhuang

Introduction

The global significance of biomass burning in natural and managed ecosystems is now widely recognized (Crutzen & Andreae, 1990) and the need to better characterize the impact of fire upon vegetation and the atmosphere has been identified as a priority in global change studies (IGAC, 1994; Steffen & Shvidenko, 1996). The link between vegetation fires and both atmospheric chemistry and aerosols is especially relevant to the climate change debate and the issue is understandably a driving force in biomass burning research. Biomass burning is a major source of greenhouse gases second only to industrial emissions. It is important to remember, however, that fires are closely associated with human activities and that their pattern of distribution is largely, although not entirely, related to land use practices. This means that vegetation fires will not only follow relatively fixed patterns that are directly related to vegetation conditions, climate and established land use but that patterns may also change under new and prescribed land management practices, consequently giving to the issue a policy dimension. In Asia particularly, biomass burning is very closely related to human activities since most of the fire-prone ecosystems of the region are under intense management. Predicting future state and conditions of the land cover represents a major objective of global change research because of its multiple environmental implications (Turner *et al.*, 1991) and it is in this context that the role of fire deserves to be more widely investigated.

The need for a global analysis of biomass burning is thus justified outside a single reference to atmospheric chemistry impacts. Non-climate-related issues are usually more regional or local, yet shifts in land use policies or trends can involve very large areas and become global in their impact through cumulative effect. In addition to being a process of global significance, vegetation burning is increasingly taken into consideration at the regional level and, as reviewed further in this chapter, Southeast Asia has, for example, been much concerned by problems associated with regional air pollution due to smoke. Locally, fire is of very immediate concern as it is widely used as a land management tool or as representing a direct threat to life and property.

The purpose of a global fire monitoring system is to better assess the role of biomass burning as a key component of a largely human-controlled chain of events leading to large scale changes in the Earth's system. Although fire is a generic process, data collection, interpretation and analysis will proceed differently whether one is concerned with its impact on global atmospheric chemistry, its role in controlling vegetation distribution (as, for example, in deforestation), or as a consequence of land use policies. Such diversity of requirements must be taken into account in the design and operation of monitoring systems. Comprehensive fire assessments must take into consideration the complex web of relationships between fire, climate, human activities and vegetation at the scale at which the various controlling processes happen to occur. In this context a range of scales must obviously be investigated. The production of satellite-derived global data sets of fire occurrences has significantly expanded the scope of biomass burning analyses and it has opened new avenues of investigations in the study of global processes (Kaufman *et al.*, 1988; Malingreau & Grégoire, 1996; Dwyer *et al.*, 1996).

The present chapter proposes a general framework of analysis of biomass burning in the terrestrial biosphere as driven mainly but not exclusively by its impact on atmospheric chemistry. Novel contributions of satellite observations will be described as one of the major sources of information on the extent and dynamics of vegetation fires at the global level. The significance of biomass burning events of Asia in this global picture will be highlighted.

Global biomass burning

While fire in ecosystems is a well-known process which has been intensively studied for decades (for reviews see Mueller-Dombois, 1981; Goldammer, 1990; Levine, 1991), new perspectives on the phenomenon have recently been brought to the foreground. They relate to the need to assess the role of vegetation fire in changing the characteristics of the atmosphere. Biomass burning has also acquired a critical status in the context of assessing the impact of deforestation and land use changes. As a consequence, questions related to biomass burning are now addressed at scales hitherto little explored.

Atmospheric chemistry is usually the point of entry for an examination of *global* biomass burning issues. Indeed, through gas and particle emissions, all fires contribute to a varying extent and for a certain period of time to the chemical composition of the atmosphere (see a review of those impacts in the next section). Uncertainties in the quantification of those emission sources remain a major issue in global fire assessment. Basic questions are related to how much of the Earth's vegetation is burning, where and when? A simplified biomass burning equation can guide the requirements for data and, by extension, observation systems:

$$Q = B \xi k \xi A$$

where Q is the burned biomass (kg year^{-1})
B is the total above ground combustible biomass (kg ha^{-1} year^{-1})
A is the area burned (ha)
k is the combustion efficiency.

Since combustion efficiency can be highly variable in space, the above equation can be rewritten as: $Q = S\ (B\ \xi k\ \xi A)_i$

where i = 1 to n represents the collection of areas characterized by a given type of vegetation and combustion efficiency.

Using emission factors experimentally derived for different vegetation types and fire temperature, the burned biomass identified above is transformed into emission rates (i.e. grams year^{-1}) for constituent gases (Seiler & Crutzen, 1980; Ward & Hardy, 1991).

Without entering in the details of the above budget equation, the following implications for global assessments can be noted.

- The above-ground combustible biomass is known in general terms for a range of ecosystems (see Fearnside, 1990; Houghton *et al.* 1995; Menaut *et al.*, 1991; Hao & Liu, 1994). However, the spatial variability in ecosystems and, more important perhaps, the climate dependence of biomass accumulation in fire-prone ecosystems (e.g. savanna, dry seasonal forests, etc.) make questionable the use of simplified look-up tables of biomass values. Furthermore, the combustion process will concern a more or less important amount of fuel depending upon the structure of the canopy, timing of the fire during the season and local weather conditions (Albini, 1984) and thus it determines the combustion efficiency. Vegetation productivity models, coupled with remote sensing data, could provide improved databases for estimating the fuel loading in terms of the ratio between live and senescent vegetation and fuel size distribution.
- The estimation of the area burned can proceed in two ways: the extent of the burn scars can be directly measured, or observations may focus on the number of "fire events". In this latter case a relationship between a "typical" fire in a given ecosystem or under a given management practice and the associated burned area must be established. Emissions can then be estimated through the use of propagation models (Moula, 1996). The measurement of the burned area itself is difficult when large regions are concerned because it requires detailed surveys, which must be organized as close as possible to the burning period itself. Attempts have been made to use satellite data at various resolutions to map burns in many different ecosystems (Vickos, 1986; Kasischke *et al.*, 1994). Field reports usually cover the areas of concern

(usually those most economically important) and are themselves based on sampling. The feasibility of producing global burned area assessments is currently under investigation.

- Various approaches to the evaluation of emissions from burned vegetation have been proposed (Ward & Hardy, 1991; Delmas *et al.*, 1991). Many experiments have been conducted leading to the *in-situ* measurements of emission. Each study provides the measurements reflecting site, meteorological or instrumental conditions, and it is recognized that generalization is difficult; it must be noted, in addition, that emission is largely controlled by the respective contribution of smouldering and flaming phases at the time of sampling. The need to establish a global emission factor database which can serve the objectives of global atmospheric modeling has now been considered in the framework of IGAC-GEIA activity.

Attempts at developing a global picture of biomass burning should not hide the fact that, as a rule, interest in fire events usually concentrates on regional and local scales. *Regional* issues are somewhat more tractable and a regional perspective can, among other things, give rise to recognition of international problems associated with biomass burning. The degradation of air quality in Southeast Asia is typically a regional feature since it is the burning in Sumatra and Kalimantan (Indonesia) which, almost every year, affects the air quality in Singapore and Malaysia to the point of creating health problems and impeding air traffic. As described by Koffi *et al.* (1995) regional transport mechanisms determine the deposition in remote areas of chemical components produced by intensive fire activity (as in the case of Central Africa where savanna fires to the north influence the characteristics of rain and deposition in the rainforest biome to the south). As described later in the text, regional ozone patterns have been linked to biomass burning as in the case of Africa (Fishman *et al.*, 1990) and South America (Kirchoff & Rasmussen, 1990). Increasingly, evaluations of the national contribution to atmospheric biogenic gases are also being integrated in a regional framework which is then used in international treaty negotiations on global warming. Biome-wide analyses have classically been conducted when fire plays a particular function in ecosystem processes, such as nutrient cycling, forest regeneration, pseudo-climax maintenance; this is the case of the boreal forest, savannas and other biomes which cover extensive "regions" of the world.

Finally, transnational boundary biomass burning situations will undoubtedly assume a growing importance as more pressure to control the phenomena is being felt. Clear differences between adjacent countries in terms of biomass burning practices, calendars and patterns have been recognized on satellite data sets (see further section). This is the case, for example, between Laos and Vietnam, Burma and China, India and Bangladesh as described in Malingreau & Grégoire, 1996. Such "border effect" has also been noted in some parts of Africa (Malingreau *et al.*, 1990). In many ways, regional perspectives on biomass burning patterns and calendars may thus dominate the range of interests linked to the impacts of burning in vegetation.

Local monitoring and assessments of fire activity and impact are, of course, those which are of more immediate concern in terms of fire prediction, fire fighting and overall land cover management. Local land use practices are the immediate cause of most vegetation burning; yet, practices associating fire in land management systems are usually better known in the anthropology or resources and environmental communities than in the scientific community concerned with global issues. There is, therefore, a need to more fully understand these local level processes which determine the occurrence, dynamics and impacts of individual fires, and to integrate that knowledge in the regional and global pictures.

A comprehensive framework for biomass burning analysis could be structured around three possible points of entry: the fire regime, objectives and driving forces or fire impacts.

Fire regime has often been used for characterizing the general dynamics of fire in ecosystems. It is related to the frequency of fire (1/n years), the date of occurrence, the area affected and the interannual variability in those characteristics. This probability of occurrence is essentially associated with the vegetation conditions (type and climate) and the availability of ignition sources, and there is currently enough general knowledge on those items to propose a practical stratification. Goldammer (1991) uses the present distribution of vegetation formations such as deciduous and monsoon forests, tropical pine forest and tropical savannas to spatially characterize fire regimes (for example, "regular short-term interval fire area"). Every region at the surface of the Earth is thus attributed an empirically determined probability of burning. This has naturally led to the priority examination of African savannas, boreal forests and, more locally, the Mediterranean basin, Northeast and Eastern Australian bush, etc. Menaut *et al.* (1991) established the now often cited figures of annual burning of 5–15% of the savanna in the Sahel zone, 25–50% in the Sudan zone and 60–80% in the Guinean zone of West Africa. Hao and Liu (1994) use generalized fire regime approaches per land cover and land use type in their estimate of global emissions from biomass burning. However useful it is in a first pass analysis, a frequency or intensity classification must be considered nominal since (1) it is often based on a partial sampling of the ecosystems and (2) it does not take into consideration the interannual variability in both the magnitude and calendar of burning (Cooke *et al.*, 1996). In such context, a generalized fire regime alone will not adequately describe a highly dynamic situation, and research should be more concerned with the identification of changes in fire regimes in a changing environment.

Analyses geared at supporting policy formulation in resource management and environmental protection will require a correct identification of the immediate *objective* pursued in burning and, by implication, of the *driving forces* which lead to changing biomass burning situations. This analysis can be guided by a classification based upon the actual use of fire on land. This intended (or unintended) use then determines in a first pass the relevance of a biomass burning situation for global analysis. For example, if grassland management foresees the yearly use of fire as a means of regenerating a palatable grass cover, and if this practice is stable in terms of area concerned and calendar, then burning has relatively little interest for global change *per se*. A national energy policy favoring the use of non-renewable energy sources in rural areas may increase the burning of crop residues in the open fields. The following stratification is proposed to guide this type of analysis (Table 4.1). The apportioning of biomass burning situations in a given region or continent to one of those categories is a prerequisite to the evaluation of the role of fire in land use change models.

Using the *impact* of biomass burning as the point of entry in the stratification is the most direct way to proceed to establish a global framework. In this case, fire events which significantly affect one of the relevant Earth's system attributes will receive priority. Impacts on atmospheric chemistry, on local and regional economy or on vegetation distribution can be considered.

Table 4.1 *Types of biomass burning situations*

(*a*) Human-induced fires

1. land management
   (a) land clearing: shifting cultivation, agricultural expansion, deforestation
   (b) land maintenance, bush control, weed burning
   (c) agricultural residue burning: straw, stubble

2. forest management
   prescribed fires

3. grassland management (early and late fires can be separated)
   (a) grazing (bush removal and grass regeneration)
   (b) hunting
   (c) others

4. accidental, arson fires (in relation to population density, political situation, poor management practices, etc.).

(*b*) Natural fires (see Fire frequency)

1. in grassland
2. in forest

Impact of biomass burning on atmospheric chemistry

Atmospheric chemistry, including both the gas and aerosol aspects, is the most direct link between biomass burning and the global Earth system. The main mechanisms whereby emissions from vegetation fires do contribute to changes in atmospheric composition are summarized in Fig. 4.1. We present here a summary of the types of burning and their contribution to the atmosphere, and a quantitative assessment of gaseous and particulate emissions from global biomass burning. Finally, a few highlights regarding the spatial and temporal distribution of emissions are given. Case studies and examples from Asia have been used in priority, when available.

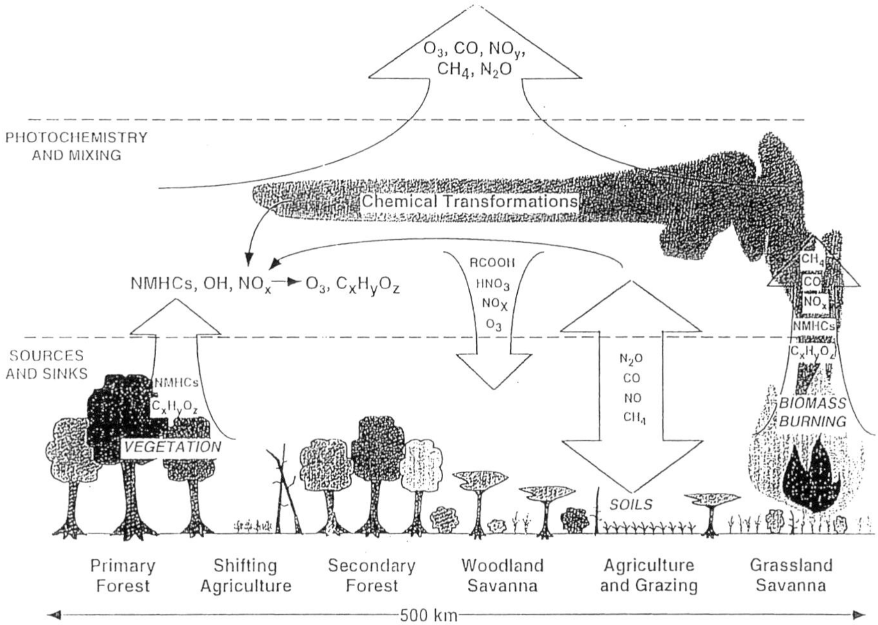

Figure 4.1 *Summary diagram of biosphere–atmosphere interactions.*

Types of burning and amount of biomass burned globally

Table 4.2 summarizes recent estimates of the amount of biomass burned annually. The species and quantity of gaseous and particulate emissions resulting from burning depend on the type of ecosystem, the moisture content of the vegetation, and the different practices and stages of combustion, i.e. open vs. enclosed burning and flaming vs. smouldering stages. A distinction is made between tropical and temperate-boreal regions. In tropical regions, the dominant practices are savanna and forest burning, use of fuel wood and agricultural burns, whereas in temperate and boreal regions (of the Northern Hemisphere) forest fires and agricultural waste burning dominate.

Table 4.2 *Estimates of yearly biomass burning for various ecosystems*

Source of biomass	Biomass burned (Tg dry matter year^{-1})					Carbon released (Tg carbon year^{-1})
	Levine (1991)	Andreae (1991)	Hao & Liu (1994)	Hao & Ward (1993)	Seiler & Crutzen (1980)	
Savannas	3690	3690	2670	2600	1190	1660
Agricultural waste	2020	2020	280	290	1900	910
Tropical forests	1260	1260	1620	1770	2430	570
Temperate and boreal forests	280	280	1620	280	280	130
Fuel wood	1430	1430	620	900	1050	640
Charcoal	21	21				30
World total	8700	8700	5390	5840	6840	3940

Savanna fires consist mostly of flaming combustion characterized by rather low CO/CO_2 ratios in plumes (Table 4.3). For various types of savannas, emission factors have been measured and biomass burning patterns have been documented (Lacaux *et al.*, 1993) (FOS/DECAFE) and in South Africa (STARE/SAFARI). Emission factors are highly dependent on the chemical composition of plants and it is therefore important to correctly identify the time of burning in vegetation seasonal development. This is especially true for nitrogen compounds, since the N content of plant material may vary with the types of plants and with the season.

The atmospheric impact of boreal forest fires has been re-estimated through the case study of the big fire which occurred in China and Russia in 1987. On the basis of satellite measurements of fire scars (NOAA AVHRR imagery), more than 1.3 million hectares were found to have burned on Chinese territory, and altogether 14.4 million hectares in China and Siberian were burned in that event. The Chinese and Siberia fires in 1987 are estimated to have contributed about 20% of the total carbon dioxide, 36% of the total carbon monoxide, and 69% of the total methane emitted by savanna burning (Cahoon *et al.*, 1992). On the basis of ground survey before and after the 1987 big Chinese fire, Wang *et al.* (1995) reported a direct carbon release of 18.8 Tg.

Table 4.2 shows that rather large discrepancies exists between the various estimates. The estimates of global biomass burned annually still differ by more than 38%. The difference mainly comes from the uncertainty in agricultural residue burning, to which substantially less attention has been given. Approximately 350 million tonnes of crop residues are burned every year under small-scale

Table 4.3 *Characteristics of savanna fires*

Species	Molar emission Ratio to CO_2 (10^{-3}) FOS/DECAFE*	Savanna emission (10^6 tonnes/yr)		Contribution of Africa savannas (%) to all sources–world
		Africa	World	
Biomass burned		1300 – 2500	2500 – 3700	
Carbon burned		550 – 1050	1050 – 1550	6 – 11
Gases				
CO_2		480 – 920	920 – 1360	6 – 11
CO	61	29 – 56	56 – 83	3 – 5
CH_4	3.1	1.5 – 2.9	2.9 – 4.2	0.4 – 1
NMHC	7.0	3.4 – 6.4	6.4 – 9.5	3 – 6
NO_y	1.91	1.1 – 2.1	2.1 – 3.0	
NO_x	1.37	0.8 – 1.5	1.5 – 2.2	2 – 4
NO_2	0.14	0.1	0.1 – 0.2	
N_2O	0.145	0.2 – 0.3	0.3 – 0.5	15 – 23
NH_3	0.057	0.0 – 0.1	0.1	0
SO_2	4.7×10^{-2}	0.06 – 0.12	0.12 – 0.17	0
COS	1.7×10^{-3}	0.02 – 0.04	0.04 – 0.06	1 – 3
O_3	30	270	410	25
Aerosols				
C	9	5.0 – 9.5	9.5 – 14.0	7 – 13
CCN	2×10^{11}	2.5 – 5.0 (×10)	5.0 – 7.4 (×10)	

*FOS–DECAFE: 1991 experiment, Fire Of Savanna–Dynamique et Changement de l'Atmosphère en Forêt Equatoriale (Lacaux, *et al.*, 1995).

closed or semienclosed conditions for cooking and heating. Perhaps 90% of the firewood and a large fraction of the crop residues in the world are burned under such conditions but there is still much uncertainty about the associated emissions which are quite different from open large-scale combustion (Smith *et al.*, 1992).

Emissions resulting from biomass burning

Biomass burning is a major global source of many environmentally significant radiatively and chemically active species (Table 4.4). On an annual basis, biomass burning may lead to the chemical production of 40% of gross carbon dioxide and about 38% of tropospheric ozone, and produce about 32% of the world's carbon monoxide, about 39% of the particulate organic carbon, and more than 20% of the world's hydrogen, non-methane hydrocarbons, methyl chloride, and oxides of nitrogen. Biomass burning is also a major source of nitriles (HCN and CHCN). Recently, methyl bromide has been discovered in the emissions from boreal forest and savanna fires (FIRESCAN–93 and SAFARI–92 experiments, Mano & Andreae, 1994). Methyl bromide is the major source of stratospheric bromine. Mano and Andreae estimate its annual emission rate from biomass burning to be 10~50 Gg $year^{-1}$. More than 80 compounds of non-methane hydrocarbons (including isomers of C2~C10 alkanes, alkenes, aromatics and monoterpenes, together with some sesquiterpenes) have been identified in the volatile portions of emissions (Smith *et al.*, 1992). Simoneit *et al.* (1995) have examined the organic compounds in carbonaceous particles from forest fires. The dominant chemical species found are derived from plant lignin, cutin, suberin, wax, gum and resin by pyrolysis and volatilization. Several compounds are potential key indicators or biomarkers. Polycyclic aromatic hydrocarbons and fatty acids are also implicated as biomarkers for atmospheric transport (Ballentine *et al.*, 1996).

Atmospheric impacts of biomass burning

The first observational data convincingly supporting the postulated impact of biomass burning were produced by air-borne measurements of trace gases and aerosols. Satellite measurements of enhanced *carbon monoxide (CO) and ozone* over the Amazon Basin, South Atlantic Ocean and portions of Africa indicated that biomass burning associated with tropical land clearing and agricultural activities were a likely source of these gases. Intensive scientific campaigns (pre-TRACE, TRACE-A, STARE/SAFARI–92), using both ground-based and airborne measurements, confirmed the linkage between biomass burning and chemical change during the dry season in Africa and South America. The importance of biomass burning as a significant source of *tropospheric ozone* was first proposed by Crutzen *et al.* (1979).

Although continuous data records for *CO* from 1978 to 1990 did not show a secular trend, two recent studies have shown that CO concentrations have de-

Table 4.4 *Estimates of global missions from biomass burning*

Species	Biomass burning (Tg element yr^{-1})	All sources (Tg element yr^{-1})	Contribution of biomass burning (%)
Carbon dioxide (gross from biomass plus fossil fuel combustion)	3500	8700	40
Carbon dioxide (net from deforestation)	1800	7000	26
Carbon monoxide	350	1100	32
Methane			
Non-methane hydrocarbons ($C_2 - C_{10}$, excluding isoprene and terpenes)	38	380	10
Nitrous oxide	0.8	13	6
Nitric oxide	8.5	40	21
ammonia	5.3	144	12
SO_x (SO_2 + aerosol sulfate)	2.8	150	2
Carbonyl sulfide	0.09	1.4	6
Methyl chloride	0.51	2.3	22
Hydrogen	19	75	25
Tropospheric ozone	420	1100	38
Total particulates	104	1530	7
Particle organic carbon	69	180	39
Elemental carbon (soot)	19	< 22	>86

creased significantly between 1990 and 1993 (Khalil & Rasmussen, 1994). There are several possible causes of the downward trend. Prinn *et al.* (1992) deduced globally averaged lower atmospheric hydroxyl radical concentrations over the 1978–1994 period from the trichloroethane (methyl chloroform) measurements. He concludes that the weighted average concentration is determined to differ insignificantly from zero. This implies the oxidation capability of the lower atmosphere was not changed significantly over this period. This rules out increasing OH levels as an explanation for the recently observed decreases in CO and methane. The remaining possible causes are a decrease in the CO sources from tropical biomass burning, and from fossil fuel consumption in western and eastern Europe. Another example of the tropospheric oxidant cycle is the detection of *nitrate* in the remote southern Indian Ocean (Moody *et al.*, 1991). Biomass burning is probably the dominant source of nitrate in this region.

Biomass burning is also a source of *acid deposition*. In contrast to the fossil fuel-induced acid rain, organic acids (especially formic and acetic acids) and nitric acid account for a large part of the acidity (Lacaux *et al.*, 1993). The ubiquitous presence of black carbon in all rain water samples collected at a remote site in Congo also traces the influence of biomass burning on rain water throughout the year. The effect of *aerosols* produced by biomass burning is beyond current understanding, due to the complex nature of the interactions involved. Recently, pyrogenic aerosol particles have been highlighted owing to their potential climatic impact (Charlson *et al.*, 1992). Aerosol particles affect the Earth's radiative balance, both directly through the back-scattering of solar radiation and indirectly through their role as cloud condensation nuclei, changing precipitation efficiency. Aerosols have the potential to rapidly gain a hydrophilic coating through interaction with other gaseous species produced in smoke plumes (Liousse, 1993).

In an undisturbed ecosystem, *nitrogen* usually limits plant production, and mechanisms exist to reduce losses from the system. Disturbance to the ecosystem, whether natural (e.g. fire, grazing, disease) or anthropogenic (e.g. fire, grazing, harvesting, fertilization), tend to disrupt these natural controls, leading to increased volatilizations of soil nitrogen and injection into the troposphere (Levine, 1990). The enhanced nitric oxide biogenic emissions from soils are related to enhanced ammonium concentrations in the soil, which is a component of the burn ash. *Carbon monoxide* fluxes were always observed to be from the soil to the atmosphere, and they rose nearly an order of magnitude after burning.

Other impacts

Economic impacts

Economic losses due to fire are, as a rule, highly localized and, despite the spectacular and often dramatic nature of those highly publicized events, they have so far not been considered in a global framework. Australia, California, the Mediterranean Basin and parts of the southeastern Andes are areas of concentration of those highly damageable fires. Unexpected fires in tropical rainforests can severely damage the exploitable timber resources; Indonesia has reported losses of up to $US 150 million in its 1991 forest fires (Directory of Forestry, 1996). Regional economic impacts have also been brought to light by the fires in Sumatra and Borneo, which have created harsh living conditions and impeded air traffic in the Malaysian Peninsula and Singapore. The economic impact of burning in Siberian forests will surface as a prime issue as attempts are made to increase exploitation of the vast resources they contain.

Economic gain due to the use of fire must also be considered; it is directly associated with agricultural and land clearing practices. Practices which are made possible and even benefit from fire must be identified in a global stratification, because they are those influenced most directly by a range of external driving forces. A direct link between economic policy and biomass burning was recently illustrated by the high level of fire activity experienced in the 1995 dry season in Brazil; it is, indeed, believed that investments in plantations and ranching were favored by monetary and fiscal measures at the national level (the so-called Plano Réal). Fire was used as the cheapest way to bring abandoned land back to cultivation (John & de Miranda, 1996). Again, it is the change in those practices which will be of interest in the global picture, and trends in biomass burning such as those associated with agricultural intensification and change in rural energy consumption patterns will be important to assess.

Ecological impact

Fire has contributed to shaping and maintaining many land systems of economic importance to man (Goldammer, 1991). In temperate and boreal forests, fire is often the most important factor controlling age structure, species composition and regeneration. The world can thus be stratified according to the role played by fire in ecosystem maintenance or degradation. The use of this criterium would, in a first step, separate most savanna and temperate/boreal forests (fire "prone" ecosystems) from areas where the occurrence of fire is not "natural" and drastically changes the characteristics of the land cover. This latter category includes all areas where the regrowth of a tree cover is prevented by the repeated use of fire. Examples are the *alang-alang* areas in South East Asia, parts of the Brazilian Cerrado, or the "savannes incluses" of the Central African Basin. "Catastrophic" fire events, which are those creating important economic damage or those occurring in exceptional circumstances (long-term frequency of return or large area affected), are important to consider in a global change perspective. Large-scale fires in non- "fire ecosystems" are indeed important to detect as they often indicate changes in the nature of the relationships between population, vegetation and climate (Malingreau *et al.*, 1985). The capability to detect and interpret catastrophic events in a systematic way and at global scale requires, however, a powerful information system which can, at the same time, identify "normal" and "abnormal" events and provide a base for comparison.

The three-pronged approach presented above leads to a multiple entry classification of fire areas. Such classification can be used to support a systematic analysis of processes and impacts associated with biomass burning around the world (Fig. 4.2). The approach may also help in designing better research strategies focusing upon the most significant biomass burning areas and issues.

Figure 4.2 *A general stratification of fire areas according to the relevance of the process in various fields of investigation.*

Biomass burning in Asia

Local accounts of the role of fire in forest, agriculture and other ecosystems of South and Southeast Asia are found in the literature (for review see Goldammer, 1990) and there is a host of information, for example, on the role of burning in shifting cultivation practices, traditionally one of the most important forms of agriculture in large parts of the Asian subcontinent (Kunstadter, 1980). Yet, comprehensive regional assessments of biomass burning have yet to be produced for that part of the world. A few major characteristics and trends can, however, be discerned which are highly relevant to the study of global change in that region.

It is important to underline at the outset that Southeast and East Asia are regions experiencing rapid economic and societal transformations leading to major changes in the use of natural resources. Patterns of land cover and land use have evolved rapidly in the region and it is likely that, given the rapid development pace of many countries, further drastic changes in landscape conditions will be experienced in the near future. It is thus critical to examine biomass burning in connection with changing patterns of land use. Clarification of the role of fire in an ecosystem not only as a possible consequence of changes in climate or agricultural practices but also as a tool which contributes to those changes must now be sought in a more systematic manner. The predictive capabilities of a regional biomass burning model will be enhanced further by an improved capability to identify the various types of association which exist between fire and land use practices. The salient features of the biomass burning situation in South and Southeast Asia are summarized as follows.

- Given its *large population* in comparison to other tropical and subtropical regions, Asia has the highest consumption of fuel wood and charcoal (67% and 80% of all tropical areas, respectively; Andreae, 1991).
- In many countries of the region, *agricultural waste* represents an appreciable source of domestic energy; open agricultural fires, used to clear the fields of crop residues, are likely to increase with the changeover to other sources of domestic energy that is associated with rural development. Quick fires set to sugar cane to facilitate harvest do not appear to be used as commonly as in other parts of the world.
- Fires in non-intensive agricultural systems are associated mainly with *shifting cultivation practices* which rely on burning for land clearing and nutrient recycling; such practices are widespread throughout the highlands of most of the countries of the region. A clear concentration of fires in the upland areas of Myanmar, Northern Laos, Northern Thailand, Vietnam and southern China is readily noticed on satellite data (Jones, 1996). A large concentration of fires in the uplands of the Kwangsi (Guangsi) and Kweichow (Guizhou) Provinces of China have recently been noted on satellite data and smoke

generated by those burns can be observed traveling all the way to the Chengdu Basin in Sichuan. Such fires, associated with the traditional cultivation practices of minorities, are little reported by the Forestry Services, yet they are likely to represent regular emission sources which are larger and regionally more significant than the emissions from the "official" forest fires (Wang *et al.*, 1995).

- Fire is very frequent in *deciduous (seasonal) forests*. On a 1992–1993 satellite data set, Jones (1996) reported that more than one-third of the visible fire over the Southeast Asian subcontinent occurred in seasonal forests. It is also reported that, during the colonial period, all Indian deciduous forests burned every year (Goldammer, 1991). Current fire accounts over India suffer, however, from very poor quality information with respect to both the total extent and the patterns of fire (Joshi, 1991). Fires in deciduous forests are closely associated with shifting cultivation since dry forest areas usually offer the best sites for such practices. The analysis of the satellite data (see further section) shows a consistent relationship between the fire patterns and the distribution of deciduous forests over South and Southeast Asia. Bamboo fires are commonly reported in the central and northern part of Southeast Asia.
- Tropical *pine forests* (*Pinus mercuzii* and other species) have a fire-related ecology with a direct influence of fire on their distribution; such fires are prevalent in Northern Sumatra and, on satellite data, large-scale fires have been observed over the highlands of Central Vietnam. The *Melaneuca* spp. formations of the Cai Peninsula in southern Vietnam are also affected by large scale fires.
- Large tracts of what was once rainforest are now being covered by *savanna*-like formations (mainly alang-alang, *Imperata cylindrica*) which are maintained by annually recurring fires. These fires are being lit in order to favor grass regrowth and grazing, and for hunting. Extensive alang-alang areas are found in Sumatra and Borneo; they represent major sites of recurrent biomass burning on or near the Equator itself. Other non-woody areas covered with tall grass and shrub (such as the ubiquitous *Eupatorium odoratum* in Southeast Asia) also experience annual fires.
- Fires in the *humid rainforests* of Asia are unusual because of the high humidity of the fuel and the frequent and regular rainfall. When occurring in this biome, fire indicates the occurrence of shifting cultivation or clearing activities for agricultural expansion. Trees and vegetation are felled and burned after a short drying period. The larger trunks can smoulder for several days after the main fire. Swiddens can be burned more than once since the fire efficiency can be rather low in such a wet climate.

- The Southeast Asia fire pattern is strongly influenced by the *El Niño* as shown by the occurrence of large-scale fires in the rainforests of Kalimantan and Borneo during El Niño droughts (Malingreau *et al.*, 1985). Severe El Niño years (such as 1982–1983 and 1991–1992) present strong evidence for the close connection between fire and climatic events, vegetation conditions and human activities. Indeed, it has been shown that local populations or new migrants take the opportunity provided to them by exceptionally dry years to clear the land, which would otherwise be difficult to cultivate. When humid rainforest formations become abnormally dry, they offer little resistance to the progression of fire since species and community have not evolved characteristics that enable them to withstand fire (Uhl & Buschbacker, 1985). This unique relationship between a well-identified and recurring climate event and fire outbreaks provides opportunities for the study of ecosystem interactions in exceptional (catastrophic) circumstances.
- The *smoke* generated by such major biomass burning events has been shown to travel several thousands kilometers and even to impede air traffic in Singapore and Malaysia more than 1500 km east (Malingreau *et al.*, 1985). The economic consequences of fire-related atmospheric pollution are considerable in a region where air and sea communications are vital.
- *Transnational boundary differences* can be noted in the calendar and spatial patterns of biomass burning in regions like central Southeast Asia; the frontier between Laos and Vietnam or between Burma and Yunnan is, for example, well marked on satellite images by the difference in fire density at specific dates (Jones, 1996). Such spatial characteristics reflect the presence of natural boundaries, but also the influence of local practices and land management policies on the use of fire.
- Several regions of Kalimantan and Borneo experience *underground peat and coal* fires which can burn unabated for years as in Kalimantan. These underground fires provide ready sources of ignition in vegetation ecosystems during drought. Underground coal seam burning is also reported in some regions of northern India (Saraf *et al.*, 1995).

The following section summarizes a set of observations made using a global fire data set developed in the framework of the IGBP-DIS Global Fire Product Activity (Justice and Dowty, 1993; Malingreau *et al.*, 1997). The pattern of active fires detected by the NOAA AVHRR instrument has been followed for 12 months between April 1992 and March 1993. In order to properly assess the results of such analysis, the principles of fire monitoring by satellite are first reviewed.

A global fire monitoring system; the role of satellite observations

With respect to the acquisition of an improved database on global biomass burning, three alternatives can then be envisaged.

- The monitoring system is designed to account for all biomass burning events at the surface of the earth at all time.
- The system addresses only events which are "significant" in the perspective of global change (i.e. large fires, large biogenic emission sources, exceptional fire situations, critical ecosystems, trends, etc.).
- The system collects information on any biomass burning event detected by a given sensor or survey (this is essentially a technologically driven approach).

The first alternative can only be achieved at great cost and through the systematic compilation of a wide variety of data; it is essentially unpractical and will not be considered here. The question of addressing only significant events requires that a series of criterias be identified for their detection and their evaluation. Only when the question "significant for what and for whom? " has been answered can a data collection strategy be designed. This approach can be supported by a classification of biomass burning situation such as the one proposed in the following section.

Finally, remote sensing techniques, which have been very instrumental in bringing about a realization of the global importance of biomass burning in the world biomes, can be used (Robinson, 1992; Malingreau & Grégoire, 1996; Cahoon *et al.*, 1992; Belward *et al.*, 1994; Chuvieco & Pilar Martin, 1994). Satellite observations have brought to Earth very dramatic views of fire in natural and artificial ecosystems. They have demonstrated that the process is widespread and that it can affect very large areas. Satellites have a global coverage and maintain their systematic data gathering activities for a long period of time. While, so far, no biomass burning dedicated sensor is available, a range of systems which all present their advantages and shortcomings has been tapped for fire-related information; their use is reviewed later in the chapter.

The AVHRR instrument of the NOAA satellite series and its capability to detect elevated heat sources at the surface of the Earth has been exploited intensively in many parts of the world during the last decade. Despite technical and operational limitations, the AVHRR has played a major role in alerting the scientific community to the global scale of biomass burning (Matson & Dozier, 1981; Malingreau *et al.*, 1985; Robinson 1992, Setzer & Pereira, 1991; Grégoire *et al.*, 1993). Before describing the current state of the art in the use of such technology for global fire assessment, a few caveats are in order. The limitations of satellite measurement pertain to the spatial sampling associated with the geometry and resolution of the instrument and, more importantly maybe, to the effect of the temporal sampling. The fire detection threshold is very much dependent upon the radiative power and the size of the individual fire but also to the background temperature. As a rule, fires as small as a few tens of meters, but of high temperature, will lead to the identification of a fire "pixel". The representation of such a pixel is, however, misleading since generally an active fire is smaller by an order of magnitude or two than the pixel in which it is contained (1.1 km^2). Misclassification of satellite data can also occur in especially high reflective environments (ex: sun glint, bright very warm soils, etc.).

A large error term in current satellite-based assessments may also be introduced by the current inability to cope with diurnal cycles. Orbiting satellites acquire data at specific and relatively constant times of the day for any suborbital point. The overall ecosystem dynamics will determine whether such frequency and timing appropriately represent a temporary event such as biomass burning. This question of sampling the diurnal cycle of fire activity has so far been relatively little examined with respect to satellite assessments. Atmospheric conditions (wind and air humidity), fuel moisture and local practices all determine the scheduling of field activities. In satellite analyses using the AVHRR approach, it is the afternoon pass (nominally 14:30 local time but drifting to later in the afternoon during the satellite's lifetime) which is often retained. Morning or night passes are rarely considered except for *ad hoc* campaigns. To accept AVHRR fire assessment at face value one must therefore assume (1) that few subresolution fires are present in the scene, (2) that the 14:30 overpass has picked up most of the active fires for that day and (3) that extrapolation to cloudy areas and missing days has been made. Anyone familiar with biomass burning situations in the field will quickly realize that the first two assumptions are not always valid. Extrapolation to cloudy areas is, furthermore, very rarely performed (Arino & Melinotte, 1995). The following equation summarizes the detection capability of the system:

$$(\text{n fires})_t = (\text{n fires})_{\text{detected}} + (\text{n fires})_{\text{undetected}}$$

where

$$(\text{n fires})_{\text{undetected}} = (\text{n fires})_{<\text{resolution}} + (\text{n fires})_{\text{outside temporal window}} + (\text{n fires})_{\text{in cloudy areas or gaps in coverage}}$$

The AVHRR being one of the main current source of information on biomass burning at the scale of interest in global change analysis and, in the absence of dedicated instruments, work has continued to refine its utilisation. Despite shortcomings, the AVHRR approach has been used to produce continental and, more recently global assessment of vegetation fires around the world (Dwyer *et al.*, 1996). Africa has been well covered in such respect with continental or regional fire maps produced by Cahoon *et al.* (1992), Koffi *et al.* (1996), Arino & Melinotte (1995), and Belward *et al.* (1994). The following analysis is derived from that data source.

South and Southeast Asian active fire distribution 1992–1993

Fig. 4.3 is extracted from the IGBP Global Fire Product for the February 20–28, 1992 and the August 23–29, 1993 periods over Asia. The two dates illustrate the distribution of active vegetation fires seen by the NOAA satellite during its afternoon overpass over the region. They clearly show the intense burning activity in Continental Southeast Asia during the northern winter, and the occurrence of numerous fires just south of the Equator in Sumatra and Kalimantan in the Southern Hemisphere. The following summarizes a recent analysis of a one-year time series of such satellite images over the Asian continent (Malingreau *et al.*, 1986).

- October 1992. Little fire activity is noted at the beginning of the month over the region with the notable exception of an intense and well-circumscribed area of fire activity in the eastern part of Punjab (India); the fire pattern stops at the border with Pakistan. The island of Timor as well as the northern tip of Australia (where it is the end of the dry season) are also marked by a dense fire pattern. The main feature of the regional image in late October is the distribution of fires over southern and eastern China; fires are spread along the Kwangsi–Hunan axis and in the Jiangsi–Fujian Provinces. It is believed that most of the fires in this region are related to crop residue burning at the end of the second summer crop. October fires in East and Central Kalimantan are somewhat surprizing, considering that the burning season is more likely to be August–September but the pattern still loosely corresponds to transmigration and shifting cultivation areas. The Nusatenggara Islands of Indonesia are covered with fire pixels, but a hot, dry and barren soil could lead to erroneous detection. The high fire activity on Java is located in a sugar cane growing area of East Java. Fires near Jambi in Sumatra are probably linked to plantation management (clearing).
- December 1992. Fire appears in the Southeast Asian subcontinent with a very clear concentration in central andastern Thailand announcing the start of the December to April burning season. Individual AVHRR images have shown that the fires straddling the Kweichow and Kwangsi provinces of southern China are related to hill-shifting cultivation in ethnic minority areas. In southern Yunnan, fires are related to shifting cultivation and, possibly bamboo fires. Fire activity is also observed in the northern part of the western Ghats of India; these fires are the precursors of extensive fire activity in the State of Maharastra a few weeks later in early January. Isolated fires in Haryana and Uttar Pradesh are likely to be related to crop residue burning during the dry winter, although burning in underground coal seams is also reported in that region (Saraf *et al.*, 1995).

- January 1993. A high concentration of open vegetation fires is clearly visible on satellite data over central Thailand and Cambodia. The dry, deciduous forests of eastern Cambodia appear particularly affected by burning. Note that the fire pattern stops at the border between Thailand and northern Laos. In Myanmar, the Irrawady Valley (or its flanks mainly occupied by deciduous forests) is also delineated by fire points. The almost total absence of fires in China is probably due to persistent cloud cover during the observation period. Later in January, intense biomass burning activity continues over the agricultural plains and hills of central Thailand. Fires now appear in northern Vietnam and are still seen, but to a lesser extent, in the previously reported area of southern China.
- February 1993. The images are taken at the peak of the burning season in the subcontinent: Thailand, Laos, Cambodia and Vietnam experience widespread biomass burning (Fig. 4.3). Again, spatial patterns are recognized: fires are concentrated in the dry forests of Cambodia and on the footslopes of the western rainforest range (Cardamon Mountains). The two most fire-prone areas of Vietnam are clearly visible: the Central Plateau-Dac Lac Province and the Northwestern province of Lai Chau. In Thailand, high intensity fire areas are found in the central–north region with a decrease of fires in the Korat plateau. In Myanmar, fire is on the increase in the southern Irrawadi Valley and over the dry forests of the Pegu Range. A few fires are still noted in the part of Sumatra north of the Equator. Late in February, a higher concentration of fires appear in the dry, seasonal forests of western India (mainly Orissa and Andra Pradesh at this time); the crescent-shaped area of high density of burning will be visible until the end of March.
- March 1993. While biomass burning is still very active in the Thailand, Cambodia, Laos and Vietnam region, fires are now extending more to the north, and northern Thailand and Myanmar show a maximum occurrence. The hilly area between the Kweichow and Kwangsi Provinces of southern China again stands out with a high fire activity which we interpret as associated with shifting cultivation. Starting in late February and by March, fires are seen in many parts of the Philippines and especially in Luzon and Mindanao. The clear transition in the fire distribution pattern between Thailand and Laos is notable as it illustrates again the possible differences in burning practices or calendar from one country to another. Intense shifting cultivation activities in the forest of northeastern Myanmar explain the high concentration of burning in that area in March. Higher still is the biomass burning activity in Nagaland, East India. A well-delineated fire pattern is seen at the border between India and Bangladesh and follows the distribution of deciduous forests on the Indian side (Blasco *et al.*, 1996), while the

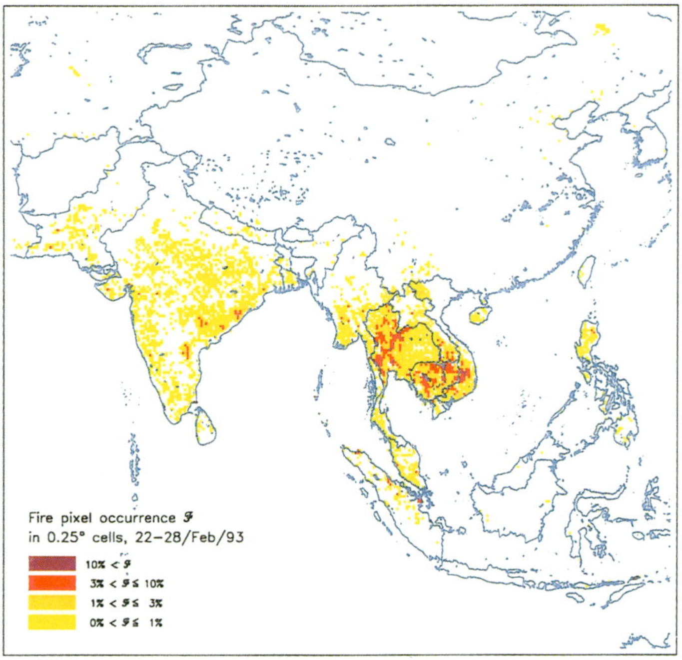

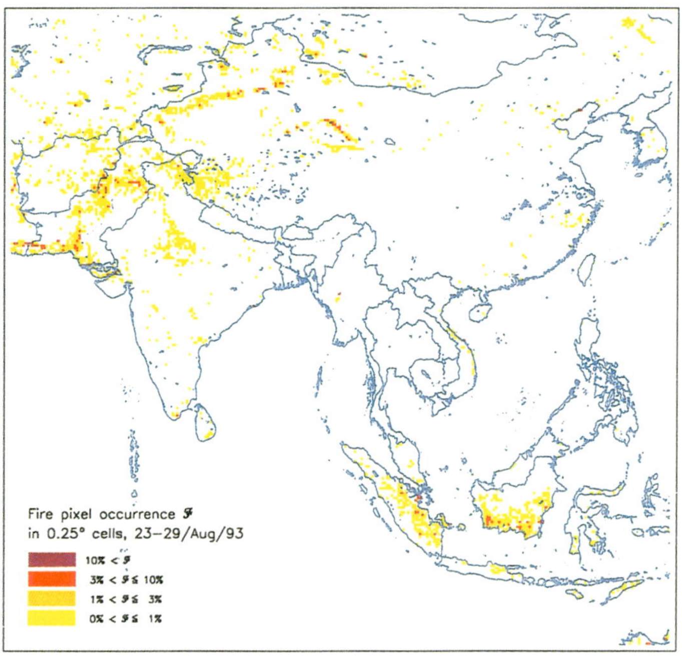

Figure 4.3 *Fire activity in Asia as from the NOAA-AVHRR satellite data during the February 20–28 and August 23–29, 1993 periods. This image is extracted from a time series of Global Fire Products covering the April 1992–September 1993 period. The information on regional active fire patterns derived from such documents is described in the text.*

Bangladesh side is occupied by agriculture. As already indicated, the dry forests of India experience extensive fires at this time of the year; later in March, fires extend to the southern part of the subcontinent.

- August 1993 (see also Fig. 4.3). A gap in the 1993 data set forces a hiatus in the time series used in the present analysis. However, June and July data available for the year 1992 indicate that, in this intervening period, burning is still active in Central Myanmar, in the Terai Region of Southern Nepal and, in early June, in the dry forests of the Deccan Plateau of India. High temperature readings are obtained consistently throughout these northern summer months over Pakistan; these thermally distinct areas somewhat follow the vegetation distribution (shrub and steppe) and consistently delineate the Indus valley for the months of July and August. Interpretation is difficult because false fire detection could be related to the arid nature of the area and the presence of warm soil. Agricultural fires, on the other hand, cannot be excluded. Two major features are to be noted over continental Asia in August. The first one is the appearance of fire points in the Tian Shan Mountain range of China and further to the northeast in southern Mongolia; the second is the ring of warm cells found around the fringe of the Taklamakan Desert. In the first case, fires in larch and pine forests, such as those which have occurred in Mongolia in 1996, are to be considered. In the second case, spurious effects induced by the radiometry of highly reflective and warm soils are probable. Agricultural fires should, however, not be discounted since they are known to occur in oases and irrigated valleys. Kashmir retains a "hot spot" status throughout the month of August. Of more importance, however, in the global picture are the fires appearing south of the Equator in what is now the middle of the local dry season. Patterns of fire in Sumatra show a high concentration in the settlement areas of South Sumatra (Lampung and south of Palembang); here, clearing is taking place in secondary vegetation, and fire is put to grass (alang-alang) areas for grazing. A clear pattern also emerges over Borneo where fires are concentrated in the settlement (transmigration) and shifting cultivation areas of East, South and West Kalimantan.

Conclusions

Current research has established the importance of the biomass burning process in atmospheric chemistry; such work must clearly be supported by an extension of investigations to a broader range of ecosystems and over longer periods of time. Of particular relevance for emission assessments are the improvements to be brought to the identification and monitoring of the occurrence and impact of fires in a large number of vegetation and land use situations. Given the large diversity associated with biomass burning around the globe and the wide variety of interests requiring a global analysis, it has been suggested that the use of a framework can enhance the relevance and effectiveness of the data collection and interpretation process. A series of criteria for asserting whether a particular fire situation is of interest in a defined context have been presented. A stratification of vegetation fires is proposed, which can guide the elaboration of an analytical framework.

The present review has identified a lack of systematic studies on the driving forces leading to a changing role of biomass burning in land management. Such understanding is vital if the results of global change research are to be translated into practical policy decisions. The stratification of ecosystems, according to the role which fire plays in their management, appears especially relevant in this respect.

A global product derived from satellite instruments and their daily passes over the entire land masses of the Earth is proposed as a starting point for a global analysis. Limitations of existing instruments have been noted. The analysis of satellite-derived vegetation fire patterns over Asia shows a picture consistent with what is known about the distribution of fire prone ecosystems in the region. Above all, the satellite approach allows the identification of regional patterns not only linked to seasonality and vegetation conditions but also to cross-border differences. As such, it draws the attention upon the widespread occurrence of the phenomenon and upon its relationship with changing land use practices.

References

Albini, F. A. (1984). Wildland fires. *American Scientist*, **72**, 590–97.

Andreae, M. O. (1991). Biomass burning: its history, use, and distribution and its impact on environmental quality and global climate. In *Global Biomass Burning: Atmospheric, Climatic and Biospheric Implications*, ed. J. S. Levine, pp. 3–21. Cambridge, MA: MIT Press.

Arino, O. & Melinotte, M. (1995). Fire index atlas. *Earth Observation Quarterly*, No. 50, pp. 11–16. Paris: ESA.

Ballantine, D. C., Macko, S. A., Turekian, V. C., Gilhooly, W. P., & Martincigh, B. (1996). Chemical and isotopic characterisation of aerosols collected during sugar cane burning in South Africa. In *Biomass Burning and Global Change*, ed J. S. Levine, pp. 460–5. Cambridge, MA: MIT Press.

Belward, A. S., Kennedy, P. J. & Grégoire, J-M. (1994). The limitations and potential of AVHRR GAC data for continental scale fire studies. *International Journal of Remote Sensing*, **15(11)**, 2215–34.

Blasco, F., Bellan, M. F. & Aizpuru, M. (1996). A vegetation map of tropical continental Asia at scale 1 : 5 million. *Journal of Vegetation Science*, **7**, 623–34.

Cahoon, D. R., Stocks, B. J., Levine, J. S., Cofer, W. R. & O'Neill, K. P. (1992). Seasonal distribution of African savanna fires. *Nature*, **359**, 812–15.

Charlson, R. J., Schwartz, S. E., Hales, J. M., Cess, R. D., Coakely, J. A., Hansen, J. E. & Hofmann, D. J. (1992). Climate forcing by anthropogenic aerosols. *Science*, **255**, 423–30.

Chuvieco, E. & Pilar Martin, M. (1994). Global fire mapping and fire danger estimation using

AVHRR images. *Photogrammetric Engineering and Remote Sensing*, **60(5)**, 563–70.

Cooke, W. F., Koffi, B. & Grégoire, J-M. (1996). Seasonality of vegetation fires in Africa from remote sensing data and application to a global chemistry model. *Journal of Geophysical Research*, **101(D15)**, 21 051–65.

Crutzen, P. J. & Andreae, M. O. (1990). Biomass burning in the tropics: impact on atmospheric chemistry and biogeochemical cycles. *Science*, **250**, 1669–73.

Crutzen, P. J., Heidt, L. E., Krasnec, J. P., Pollock, W. H. & Seiler, W. (1979). Biomass burning as a source of atmospheric gases CO, H_2, N_2O, NO, CH_3Cl and COS. *Nature*, **282**, 253–56.

Delmas, R. A., Marenco, A., Tathy, J. P., Cros, B. & Baudet, J. G. R. (1991). Sources and sinks of methane in African savanna: methane emission from biomass burning. *Journal of Geophysical Research*, **96**, 7287–97.

Directory of Forestry, Indonesia (1996). National Coordination Team on Land and Forest Fire Management. *International Forest Fire News*, **14**, 29–30. Geneva: UN-ECE.

Dwyer, E., Grégoire, J-M. & Malingreau, J-P. (1996). A global analysis of vegetation fires: spatial and temporal dynamics. Submitted to *Ambio*).

Fearnside, P. M. (1990). Fire in the tropical rain forest of the Amazon basin. In *Fire in Tropical Biotas, Ecosystems Processes and Global Challenges: Ecological Studies*, ed. J. G. Goldammer. Vol. 84. Berlin: Springer-Verlag.

Fishman, J., Watson, C. E., Larsen, J. C. & Logan, J. A. (1990). Distribution of troposheric ozone determined from satellite data. *Journal of Geophysical Research*, **95**, 3599–617.

Goldammer, G. J. (ed.) (1990). *Fire in the Tropical Biota, Ecosystem Processes and Global Challenges*. Berlin: Springer Verlag.

Goldammer, G. J. (1991). Tropical wild-land fires and global changes: prehistoric evidence, present fire regimes, and future trends. In *Global Biomass Burning: Atmospheric, Climatic and Biospheric Implications*, ed. J. S. Levine, pp. 83–91. Cambridge, MA: MIT Press.

Grégoire, J-M., Belward, A. S. & Kennedy, P. J. (1993). Dynamique du saturation du signal dans la bande 3 du senseur AVHRR. *International Journal of Remote Sensing*, **14**, 2079–95.

Hao, W. M. & Liu, M. H. (1994). Spatial and temporal distribution of tropical biomass burning. *Global Geochemical Cycles*, **8(4)**, 495–503.

Hao, W. M. & Ward, D. E. (1993). Methane production from global biomass burning. *Journal of Geophysical Research*, **98**, 20 657-61.

Houghton, R. A., Boone, R. D., Melillo, J. M., Palm, C. A., Woodward, G. M., Myers. N., Moore III, B. & Skole, D. L. (1995). Net fluxes of carbon dioxide from tropical forests in 1980. *Nature*, **310**, 617–20.

IGAC (1994). The International Global Atmospheric Chemistry Project . An Overview. Princeton: IGAC Project Office.

John, L. & de Miranda, E. E. (1996). Brazil: Fires heat up after a four-year decrease. *International Forest Fire News*, **14**, 13–5. Geneva: UN-ECE.

Jones, S. H. (1996). Vegetation fire in mainland southeast Asia: spatio-temporal analysis of AVHRR 1 km data for the 1992–93 dry season. Joint Research Centre Technical Publication, EUR 17282.EN, 48 pp. Italy: Ispra.

Joshi, V. (1991). Biomass burning in India. In *Global Biomass Burning: Atmospheric, Climatic and Biospheric Implications*, ed. J. S. Levine, pp. 185–93. Cambridge, MA: MIT Press.

Justice, C. O. J. & Dowty, P. (eds.). (1993). IGBP-DIS Satellite Fire Detection Algorithm Workshop. Technical Report. IGBP-DIS Working Paper No. 9, 88 pp., Paris.

Kasischke, E. S., Bourgeau-Chavez, L. L. & French, N. H. F. (1994). Observations of variations in ERS-1 SAR image intensity associated with forest fires in Alaska. *IEEE Transactions Geoscience and Remote Sensing*, **32**, 206–10.

Kaufman, Y., Tucker, C. J. & Fung, I. (1988). Remote sensing of biomass burning in the tropics. *Journal of Geophysical Research*, **95**, 9927–34.

Khalil, M. A. K. & Rasmussen, R. A. (1994). Global emissions of methane during the last several centuries. *Chemosphere*, **29(5)**, 833–42.

Kirchoff, V. W. J. H. & Rasmussen, R. A. (1990). Time variations of CO and O_3 concentrations in a region subjected to biomass burning. *Journal of Geophysical Research*, **95**, 7521–32

Koffi, B., Grégoire, J-M., Mahé, G. & Lacaux, J-P. (1995). Remote sensing of bush fire dynamics in central Africa from 1984 to 1988: analysis in relation to vegetation and pluviometric regional patterns. *Atmospheric Research*, **39**, 179–200.

Koffi, B., Koffi, E. & Grégoire, J-M. (1996). Atlas of fire seasonality and its interannual variability for the African Continent. Joint Research Centre EUR 16407. Luxembourg: ECSC-EC.

Kunstadter, P. (1980). The impact of economic development on southeast Asian tropical forests. In *Tropical Ecology and Development*, ed. J. J. Furtado, pp. 65–72. Kuala Lumpur: The International Society of Tropical Ecology.

Lacaux, J-P., Cachier, H. & Delmas, R. (1993). Biomass burning in Africa: an overview of its impact on atmospheric chemistry. In *Fire in the Environment: The Ecological and Climatic Influence of Vegetation Fires*, ed. P. J. Crutzen & J. G. Goldammer, pp. 159–91. New York: John Wiley.

Lacaux, J-P., Brustet, J. M., Delmas, R., Menaut, J. C., Abbadie, L., Bonsang, B., Cacheir, H., Baudet, J., Andreae, M. O. & Helas, G. (1995). Biomass burning in tropical savannas of the Ivory Coast: an overview of the field experiment Fire Of Savannas (FOS/DECAFE). *Journal of Atmospheric Chemistry*, **22**, 195-216.

Levine, J. S., Cofer, W. R., Sebacher, D. I., Rhinehart, R. P., Winstead, E. L., Sebacher, S., Hinkle, C. R., Schmalzer, P. A. & Koller, A. M. (1990). The effects of fire on biogenic emissions of methane and nitric oxide from wetlands. *Journal of Geophysical Research*, **95**, 1853–64.

Levine, J. S., Cofer, W. R., Winstead, E. L., Rhinehart, R. P., Cahoon, D. R., Sebacher, D. I., Sebacher, S. & Stocks, B. J. (1991). Biomass burning: combustion emissions, satellite imagery and biogenic emissions. In *Global Biomass Burning: Atmospheric, Climatic and Biospheric Implications*, ed. J. S. Levine, pp. 264–71. Cambridge, MA: MIT Press.

Liousse, C. (1993). Emissions carbonees particulaires des feux de savane d'Afrique: mesures au sol et teledetection spatiale des panaches, PhD Thesis, Paris VII University.

Malingreau, J-P. & Grégoire, J-M. (1996). Fire monitoring system for global change studies: a framework. In *Biomass Burning and Global Change*, ed. J. S. Levine, pp. 14–24. Boston: MIT Press.

Malingreau, J-P., Stephens, G. & Fellows, L. (1985). Remote sensing of forest fires: Kalimantan and North Borneo in 1982–83. *Ambio*, **14(6)**, 314–21.

Malingreau, J-P., Jones, S. H., Dwyer, E. & Pinnock, S. (1986). Regional vegetation fire patterns in south and southeast Asia: a satellite-based assessment. Conference on Transboundary Pollution and the Sustainability of Tropical Forests, December 2–4. Kuala Lumpur, Malaysia: Asian Institute of Forest Management.

Malingreau, J-P., Laporte, N. & Grégoire, J-M. (1990). Exceptional fire events in the tropics. Southern Guinée January 1987. *International Journal of Remote Sensing*, **11**, 2121–3.

Malingreau, J-P., Justice, C.O., & Pereira, J. (1997). *The IGBP-DIS Global Fire Product derived from the AVHRR 1 km data*. Working Group Report. IGBP-DIS Working Paper #17. Toulouse.

Mano, S. & Andreae, M. O. (1994). Emission of methyl bromide from biomass burning. *Science*, **263**, 1255–7.

Matson, M. & Dozier, J. (1981). Identification of sub-resolution high temperature sources using a thermal infrared sensor. *Photogrammetric Engineering and Remote Sensing*, **47(9)**, 1311–18.

Menaut J-C., Abbadie, L., Lavenu, F., Loudjani, P. & Podaire, A. (1991). Biomass burning in West African Savannas. In *Global Biomass Burning: Atmospheric, Climatic and Biospheric Implications*, ed. J. S. Levine, pp. 133–42. Cambridge, MA: MIT Press.

Moody, J. L., Pszenny, A. A. P., Gaudry, A., Keene, W. C., Galloway, J. N. & Polian, G. (1991). Precipitation composition and its variability in the southern Indian Ocean: Amsterdam Island, 1980–1987. *Journal of Geophysical Research*, **96(20)**, 769–86.

Moula, B. (1996). Modélisation des feux de brousse en savane africaine et évaluation des émissions dans l'atmosphère des constituants en trace. Thèse. Univ. Paul Sabatier. Toulouse.

Mueller-Dombois, D. (1981). Fire in tropical ecosystems. In *Fires, Regimes and Ecosystem Properties*, General Technical Report WO-26, pp. 137–76. Washington, DC: U.S. Department of Agriculture and Forest Service.

Prinn, R.G., Cunnold, D., Simmonds, P., Alyea, F., Bokli, R., Crawford, A., Fraser, P., Gutzler, D., Harley, D., Rosen, R. & Ramussen, R. (1992). Global average concentration and trend for hydroxyl radicals deduced from ALE/GAGE trichloroethane (methyl chloroform) data for 1978–90. *Journal of Geophysical Research*, **97(D2)**, 2445–61.

Robinson, J. M. (1992). Fire from space: global fire evaluation using infrared remote sensing. *International Journal of Remote Sensing*, **12**, 3–24.

Saraf, A. K., Prakash, A., Sengupta, S. & Gupta, R. P. (1995). Landsat TM data for estimating ground temperature and depth of subsurface coal fire in the Jharia coalfield, India. *International Journal of Remote Sensing*, **16(12)**, 2111–24.

Setzer, A. W. & Pereira, M. C. (1991). Amazon biomass burning in 1987 and their tropospheric emissions. *Ambio*, **20**, 19–22.

Sieler, W. & Crutzen, P. J. (1980). Estimates of gross and net fluxes of carbon between the biosphere and the atmosphere from biomass burning. *Climatic Change*, **2**, 207–47.

Simoneit, B. R. T., Radzi bin Abas, M., Cass, G. R., Rogge, W. F., Mazurek, M. A. & Hildemann, L. M. (1995). Natural organic compounds as tracers for biomass combustion in aerosols. In *Abstracts of Chapman Conference "Biomass Burning and Global Change"*, 33 pp. Williamsburg, Virginia.

Smith, K. R., Rasmussen, R. A., Manegdeg, F. & Apte, M. (1992). Greenhouse Gases from Small-scale Combustion in Developing Countries, US EPA Report EPA-600-R-92-005. Washington, DC: US EPA.

Steffen W. L. & Shvidenko, A. Z. (eds.) (1996). The IGBP Northern Eurasia Study. Prospectus for Integrated Global Change Research. The International Geosphere Biosphere Programme. IGBP Stockholm (in press).

Turner II, B. L., Meyer, W. B. & Skole, D. L. (1991). Global land use/land cover change: towards an integrated study. *Ambio*, **23**, 91–5.

Vickos, J. B. (1986). Etude des feux de savanes. Thèse DEA. Universite Paul Sabatier, Toulouse, France.

Uhl, C. & Buschbacher, R. (1985). A disturbing synergism between cattle ranch burning practices and selective tree harvesting in the eastern Amazon. *Biotropica*, **17(4)**, 265–8.

Wang, X. K., Zhuang, Y. H. & Feng, Z. W. (1995). Forest fires in China: carbon dioxide emission to the atmosphere. In *Abstracts of Chapman Conference "Biomass Burning and Global Change"*, 45 pp., March 13–17, 1995. Williamsburg, Virginia.

Ward, D. E. & Hardy, C. C. (1991). Smoke emissions from wildland fires. *Environment International*, **17**, 117–34.

5 Asian grassland biogeochemistry: factors affecting past and future dynamics of Asian grasslands

D. S. Ojima, X. Xiao, T. Chuluun and X. S. Zhang

Introduction

Natural grasslands play a major role in the global carbon cycle. Over the centuries, the grassland ecosystems have been modified extensively through agricultural conversion, resulting in large release of CO_2 to the atmosphere (Houghton, 1994). These ecosystems have also undergone dramatic declines of soil fertility, due to over-utilization of these ecosystem under increased grazing pressures. Grasslands are one of the most widespread vegetation types world-wide, covering nearly one-fifth of the Earth's land surface (24×10^6 km^2), and containing >10% of global soil C stocks (Anderson, 1991). They range from the savannas of Africa to the North American prairies, and include the converted grasslands of Latin America and Southeast Asia. Fluxes of carbon (C) through plant and soil organic matter (SOM) in grasslands, especially in the tropics, may have been underestimated in the past (Long *et al.*, 1989).

In Asia, grasslands extend from continental regions west of the Urals to the tropical regions of China and India. One of the most extensive grasslands in the world exists in the Mongolian Plateau region and is a critical resource for meat, milk, leather, wool, and cereal grain production. Historically, nomadic pastoralists in the region have utilized the rich grassland region to graze their mixed herds of cattle, sheep, goats, horses and camels. Grazing patterns were dictated more by intra- and interannual climate variability than by political or economic factors. However, there have been significant changes in the traditional pastoral systems and in livestock population patterns in the Mongolian Plateau region. During the last 40 years, many aspects of the traditional nomadic culture were replaced by socialistic practices (Mearns, 1993). Herdsmen were commonly organized into collectives and were allowed only a small number of animals for private ownership. In addition, human population in the region has risen dramatically during the past several decades (NRC, 1992). This has led to increased grazing intensity and rates of cropland conversion. The most fertile land areas have been converted to croplands, and grazing has been shifted to less productive lands. This practice has intensified grazing on the remaining areas and has, in many cases, reduced ecosystem productivity. In the more humid regions of the Asian grasslands, agriculture has become a dominant factor. These land use changes have resulted in soil losses and reduced fertility of the soils.

The land use systems of grassland areas are tightly coupled to the climatic system. The seasonal distribution of rainfall is a major determinant of plant production in most semiarid and arid regions. In semiarid regions, such as the Asian steppe, most of the rainfall occurs during the growing season, May through August (Xiao *et al.*, 1995; Ojima *et al.*, 1993; Zhang, 1992). This rainfall pattern permits greater biological productivity, less evaporation and less runoff than a rainfall pattern more evenly distributed throughout the year. This synchrony of rainfall during the growing season accounts for a relatively productive system despite the modest annual input of rain (120–560 mm $year^{-1}$). Simulation of ecosystem responses to climate change in the Mongolian steppe and the central grasslands of North America has demonstrated the sensitivity of soil carbon storage and grassland biogeochemistry processes to changes in seasonal distribution of precipitation (Schimel *et al.*, 1990; Ojima *et al.*, 1993).

Arid and semiarid lands (ASALs) may be among the earliest ecosystems to exhibit the effects of climatic change (OIES, 1991). Sensitivity to climatic change may be a reflection of inadequate reserves of water and soil nutrients. In addition, vulnerability to climate variability and directional climate change in the ASALs is elevated by human-induced land use changes (Ellis & Galvin, 1994). Changes in land use arise from the changes in demographic patterns, economic incentives, and pastoral management issues related to land ownership and land use. These factors vary across the Asian region owing to political and environmental constraints on land use and ecosystem dynamics. The drought experienced during the late 1950s, coupled with increased human population levels in Inner Mongolia, led to land degradation and human suffering.

Modifications of land cover have potential impacts on the climate system of ASALs (Pielke & Avissar, 1990; Fu & Wei, 1993; Ellis & Galvin, 1994). Changes in land surface characteristics due to cropland conversion or due to desertification, will modify surface albedo and water exchange that can potentially alter regional rainfall patterns (Schlesinger *et al.*, 1990; Pielke & Avissar, 1990; Fu & Wei, 1993). These regional climate changes may also increase the amount of particulates that are put aloft.

The conversion of the grassland ecosystems to croplands has modified biological processes, such as plant production, nutrient cycling, and grazing (Schlesinger *et al.*, 1990; Ojima *et al.*, 1991) and has affected the grassland utilization of Asian grasslands (NRC, 1992; Sheey, 1988; Wu & Loucks, 1992; Zhang, 1992; Zhu & Wang, 1993). The interaction between climate and land use changes will determine how the land cover and thereby carrying capacity of various regions will respond to these physical and social–economic-caused modifications (Riebsame *et al.*, 1994). The objective of this chapter is to characterize the factors controlling ecosystem integrity of the Asian grasslands, especially in the large grassland region of the Mongolian steppe. In addition, a discussion of the important socio-economic considerations will be discussed relative to land use management of the Asian grasslands. In this context, we define ecosystem integrity as a measure of the percentage departure of the state of the system from a pristine state. Thus, for soil organic matter (SOM), it could be percentage loss of SOM relative to an unexploited state. We are also interested in determining how socio-political and economic factors can be managed to maintain ecosystem integrity and sustain the economy.

Key issues of carbon storage in grassland ecosystems

The seasonal distribution of rainfall is a major determinant of plant production in many semiarid and arid regions. Annual rainfall for many of these sites tends to be highly seasonal, with most of the rainfall occurring in a three- to four-month period. The Intergovernmental Panel on Climate Change (IPCC) estimates (based on UKMO general circulation model climate projections, Houghton *et al.*, 1990) indicate that potential changes in seasonal rainfall and temperature patterns in Asia will have a greater impact on biological response and feedback to climate than changes in the overall amount of annual rainfall. Changes in plant species composition observed during extreme drought events are likely to alter ecosystem processes (e.g. through modified resource use efficiencies).

Modeling framework

The grassland analysis presented in this chapter has divided the grassland areas of the world into ecoregions, based on the global vegetation description of Bailey (1989). The Mongolian steppe falls into the category of dry cold steppe region; the climate is characterized as being extreme continental, with annual rainfall ranging from 120 to 500 mm, and a mean annual temperature ranging from <0 to 10 °C. This ecoregion is noted for its highly variable weather, with severe droughts and winter storms being common.

The analysis of the factors controlling soil organic matter and grassland ecosystem dynamics was conducted using the CENTURY model (Parton *et al.*, 1987). This model is capable of simulating the plant–soil ecosystem, and is able to simulate carbon and nutrient dynamics for a variety of ecosystems (e.g. grasslands, forest, crops, and savannas). A brief description of the model structure and the scientific basis for the model is given here, and a more detailed description of the model is contained in different papers (Parton *et al.*, 1987; 1993).

The CENTURY model is set up to simulate the dynamics of carbon (C), nitrogen (N), phosphorus (P), and sulfur (S) for different plant–soil systems. The model can simulate the dynamics of grassland systems, agricultural crop systems, forest systems, and savanna systems. The different plant production submodels are linked to a common soil organic matter submodel. The soil organic matter submodel simulates the flow of C, N, P, and S through plant litter and the different inorganic and organic pools in the soil. The model runs using a monthly time step and the major input variables for the model include monthly average maximum and minimum air temperature, monthly precipitation, plant type (e.g. C_3 or C_4 plant physiology), lignin content of plant material, plant N, P, and S content, soil texture, atmospheric and soil N inputs, and initial soil C, N, P, and S levels.

The soil organic matter (SOM) submodel simulates the dynamics of C, N, P, and S in the organic and inorganic parts of the soil system. The soil organic C in the model is divided up into three components which include active (microbe), slow, and passive soil C. Flows of carbon between these pools are controlled by decomposition rate and microbial respiration loss parameters, both of which may be a function of soil texture. Typical turnover times for a grassland site are 2, 40, and 2000 years, respectively, for the active, slow, and passive pools.

The CENTURY model includes a simplified water budget model which calculates monthly evaporation and transpiration water loss, water content of the soil layers, snow water content, snow melt and saturated flow of water between soil layers. Sublimation and evaporation of water from the snow pack occur at a rate equal to the potential evapotranspiration rate (PET). Average monthly soil temperature near the soil surface is calculated from air temperature and corrected for above-ground biomass levels. The actual soil temperature used for decomposition and plant growth rate functions is the average of the minimum and maximum soil temperatures.

For each grassland site, the model was run to equilibrium for 5000 years, using a repeated 25-year pattern of observed current weather data. Site-specific patterns of grassland management (i.e. burning and grazing) was incorporated in these long-term runs as described in Parton *et al.* (1993).

Modeling the effect of climate change and increased CO_2

Ecosystem sensitivity to climatic change was evaluated by driving an ecosystem model with climate scenarios generated by general circulation models (GCM). We evaluated changes in above-ground net primary productivity (NPP) and soil organic carbon (SOC) for about six sites across the grassland regions of Asia.

The effect of increased atmospheric concentrations of CO_2 on the photosynthetic pathway has been well documented in C_3 plants. In addition to the direct effects of CO_2 enhancement on C uptake, increases in water- and nitrogen-use efficiency (WUE and NUE) in both C_3 and C_4 plants (Owensby *et al.*, 1993*a*,*b*), have been observed under elevated CO_2 conditions. In CENTURY, we modified plant production parameters under a "double-CO_2" climate so that increased WUE (i.e. decreased PET rates) and NUE (i.e. greater C : N ratio of plant material produced), for both C_3 and C_4 grasslands, were simulated. We allowed a 20% decrease in total PET and a 20% decrease in N content, with the change in atmospheric CO_2 concentration from 350 ppm to 700 ppm, via a simple linear function based on results from a tallgrass prairie (Owensby *et al.*, 1993*a*,*b*).

Model runs under climate change conditions

The doubled CO_2 climatologies used to drive the CENTURY model in the present study were derived from the Canadian Climate Centre (CCC) and Geophysical Fluid Dynamics Laboratory, High Resolution (R30) Scenario (GFHI) models of global climate change, which provide monthly projections of temperature and precipitation for 2.5° grid cells world-wide. These monthly projections were incorporated into the observed climate from each site in order to represent the seasonal change in climatic factors critical to assessing ecosystem responses (Ojima *et al.*, 1991). We took the values projected for a doubling of atmospheric CO_2, and assumed that these would be reached within 50 years from the present, corresponding roughly to IPCC Scenario A (Houghton *et al.*, 1990; Fig. 5.1).

The two GCMs simulate similar changes to mean annual air temperature (5–7 °C); changes in precipitation were different between the models. For most regions in Asia, an increase in rainfall is projected. The tropical regions of the CCC model is projected to be drier than it is today. Overall, the CCC model appears to be wetter than the GFHI projections for the Asia region.

We assumed that changes in precipitation and temperature at each site were linear over a 50-year period, at which point they stabilized at the modified "double CO_2" climate, as did atmospheric CO_2. The results from the 25-year period immediately following this stabilization of the double CO_2 climate represent the transient response to climate change. To determine a "near-equilibrium" long-term response to change, we continued simulation for an additional 150 years of the stabilized modified double-CO_2 weather pattern. Model results were analyzed for changes in the level of soil organic matter, plant productivity, N mineralization and the model-calculated effect of climate on decomposition (abiotic decomposition factor).

Results

Regional analysis: climate change sensitivity

Climate change impacts of the two GCM scenarios on above-ground plant production and soil carbon are shown in Table 5.1. These values represent the means for the 25 years of "double-CO_2" climate following the 150-year climate change period; the average percentage changes for each site are also given in Table 5.1. Results show that the cooler, drier sites had a substantial reduction in plant production (about 25%), while the more humid sites in India varied. The tropical grassland regions of Thailand showed negative changes in above-ground NPP (Table 5.1*b*). The CO_2-only response for all sites resulted in a positive increase in NPP.

Soil carbon results show substantial losses of soil C for most of the sites, with the largest losses occurring at the Tuva, Russia and the Xilinhot, China sites (Table 5.1*c*). Soil carbon losses are low in the more humid regions of India and Thailand, with greater losses being in the humid tropical region of Thailand. Except in the cold dry steppe and the temperate steppe, N mineralization rates tended to increase for both GCM scenarios, with the largest increases observed for the humid temperate ecoregions. The changes in soil carbon are negatively related to changes in decomposition and positively related to changes in productivity, with the greatest changes simulated in the drier ecosystems (Fig. 5.2*f*). This is reflected in the decomposition rates of these drier sites under climate change simulations compared with the relatively small changes in soil C simulated in the tropical regions of the study.

Regional effects of increased CO_2 and CO_2/climate interaction

We simulated the impact of CO_2 on grassland ecosystems with, and without, the inclusion of climate change. Overall, the effect of increasing atmospheric CO_2 concentrations under current climatic conditions was to increase total plant production and soil C storage (Table 5.1). This observed CO_2 fertilization effect increased plant production and was most noticeable in the cold, dry steppe region. The soil carbon storage increased most for the humid tropical region of Thailand. Net N mineralization rates decreased in most regions, resulting from the increased input of lower quality litter with a higher C/N ratio, but increased for the cold desert steppe region. Simulated increases in decomposition under elevated CO_2 resulted from improved soil water relationships at the cold, dry sites (cold, desert steppe), with smaller increases for the warmer and more humid sites.

The combined effects of CO_2 and climate change were additive for each ecosystem property studied. Relative changes in total plant production, soil C, N mineralization, and decomposition were compared statistically between (1) the

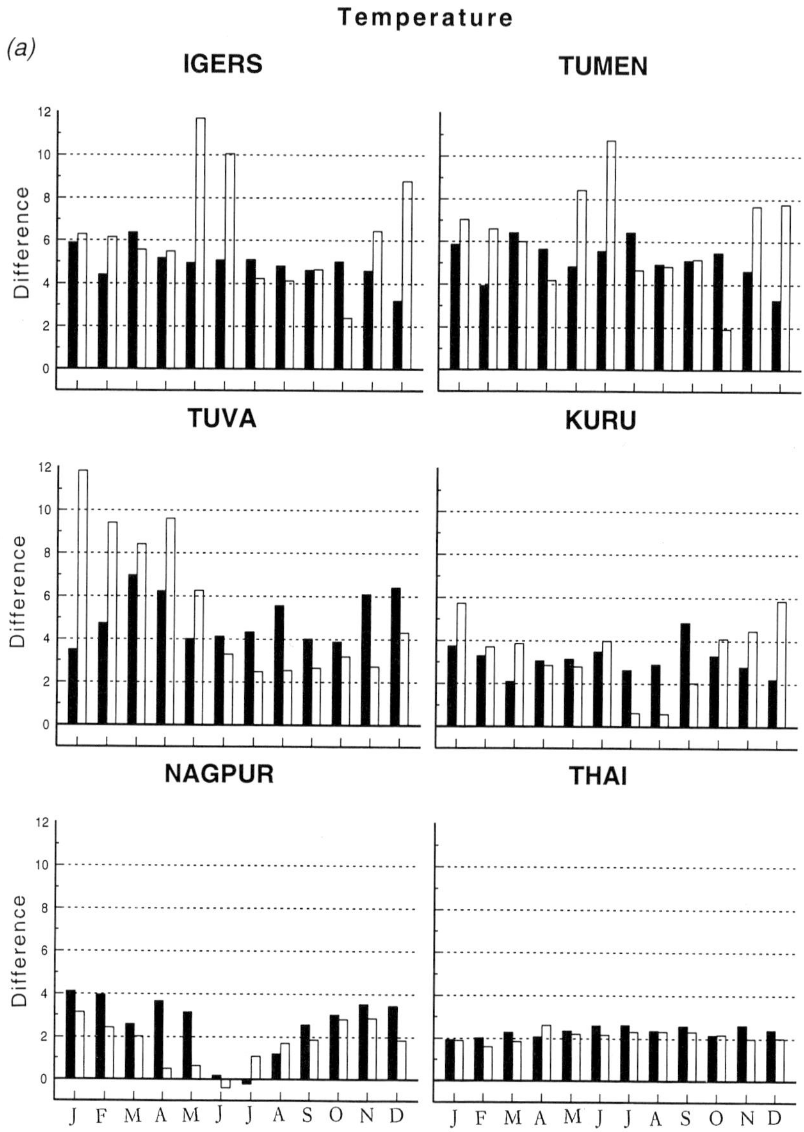

Figure 5.1 *(a) Seasonal changes in projected monthly temperatures for CCC (empty columns), GFDL (solid columns) and GCMs for the six sites simulated in the Asia region.*

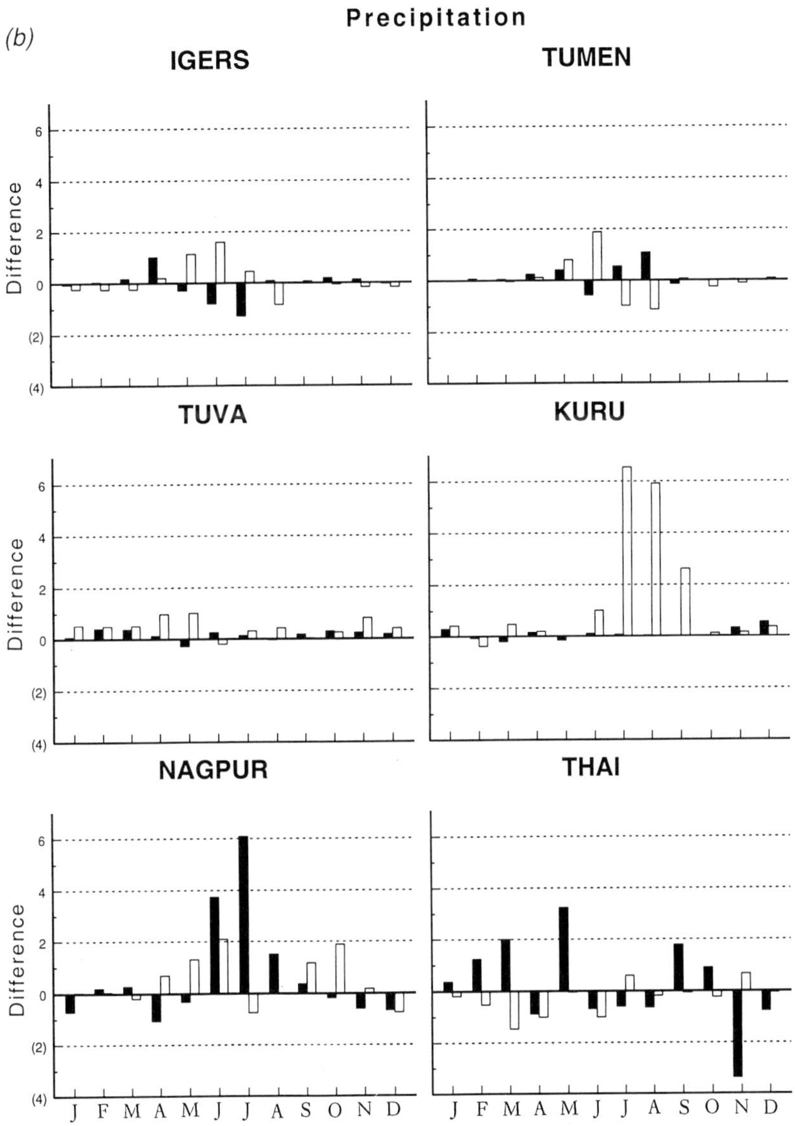

Figure 5.1 (*b*) *Seasonal changes in projected monthly precipitation for CCC (empty columns), GFDL (solid columns) and GCMs for the six sites simulated in the Asia region.*

Table 5.1 *Site characteristics, climate change and CO_2 effects of six Asian grassland sites*

(a) Site characteristics

	Latitude/ longitude (approx.)	Land area represented (10^6 km^2)	Annual precipitation (mm)	Mean annual temperature (°C)
Tuva, RUSSIA	52° N 94° E	0.109	214	−3.4
Tumentsogt, MONGOLIA	46° N 113° E	0.740	269	1.5
Xilinhot, CHINA	44° N 117° E	0.441	360	−0.1
Kurukshetra Ludhiana, INDIA	31° N 76° E	0.353	715	24.4
Nagpur, INDIA	21° N 79° E	1.184	1203	26.9
Hat Yai, THAILAND	6° N 101° E	0.797	1540	27.6

(b) Climate change and CO_2 effects of above-ground NPP

		Above-ground NPP (% change)				
	Current conditions (g C m^{-2} y^{-1})	2 × CO_2	ΔC		ΔC + 2CO_2	
			CCC	GFDL	CCC	GFDL
Tuva, RUSSIA	46.2	+33	−6	−44	+27	−21
Tumentsogt, MONGOLIA	35.3	+20	0	+5.7	+26	+31
Xilinhot, CHINA	84.3	+23	−36	−29	−17	−6.0
Kurukshetra Ludhiana, INDIA	143.4	+20	+13	−16	+32	+3.5
Nagpur, INDIA	228.6	+14	+4.4	0	+18	+12
Hat Yai, THAILAND	2.22	+13	−10	−1.5	+7	+17.5

(c) Climate change and CO_2 effects of soil carbon

	Current conditions (kg C/m^2)	Soil C % Change				
			ΔC		ΔC + CO_2	
		CO_2	CCC	GFDL	CCC	GFDL
Tuva, RUSSIA	4.02	4	−22	−29	−21	−29
Tumentsogt, MONGOLIA	4.15	2	−3	−3	−4	−4
Xilinhot, CHINA	6.56	1.6	−35	−22	−34	−19
Kurukshetra Ludhiana, INDIA	3.33	6.3	0	−9	+7.3	−2.9
Nagpur, INDIA	5.12	9.9	−3	+1.3	+6.6	+13.3
Hat Yai, THAILAND	2.22	22.9	−8.7	−5.0	+8.7	−13.6

combined impact of CO_2 and climate and (2) the independent impacts of CO_2 and climate change; a linear additive effect was demonstrated.

Increased levels of atmospheric CO_2 enhanced total plant production, regardless of the climate change impact. This enhancement in plant production reduces soil C losses resulting from climatically driven changes in SOM decomposition. The direct CO_2 effect reduces carbon losses throughout the grassland ecosystems and, in fact, results in a soil C sink region in the tropical savanna and humid savanna regions.

Site level analyses of the Mongolian Steppe

The Mongolian steppe region of Eurasia was characterized as a dry, cold steppe ecosystem type. These ecosystems are characterized climatically by a low mean annual temperature and rainfall, with a highly seasonal pattern in both. The beginning of the spring rainfall and spring warming are strongly correlated, and the onset of the growing season rainfall triggers the green-up in the region. The synchrony of optimal growing temperatures and adequate water results in greater rain-use efficiency in this region of Asia, despite the relatively small levels of annual rainfall (Xiao *et al.*, 1995). Comparisons of CENTURY simulated peak biomass, NPP, and soil C levels for these sites are very good (Fig. 5.2).

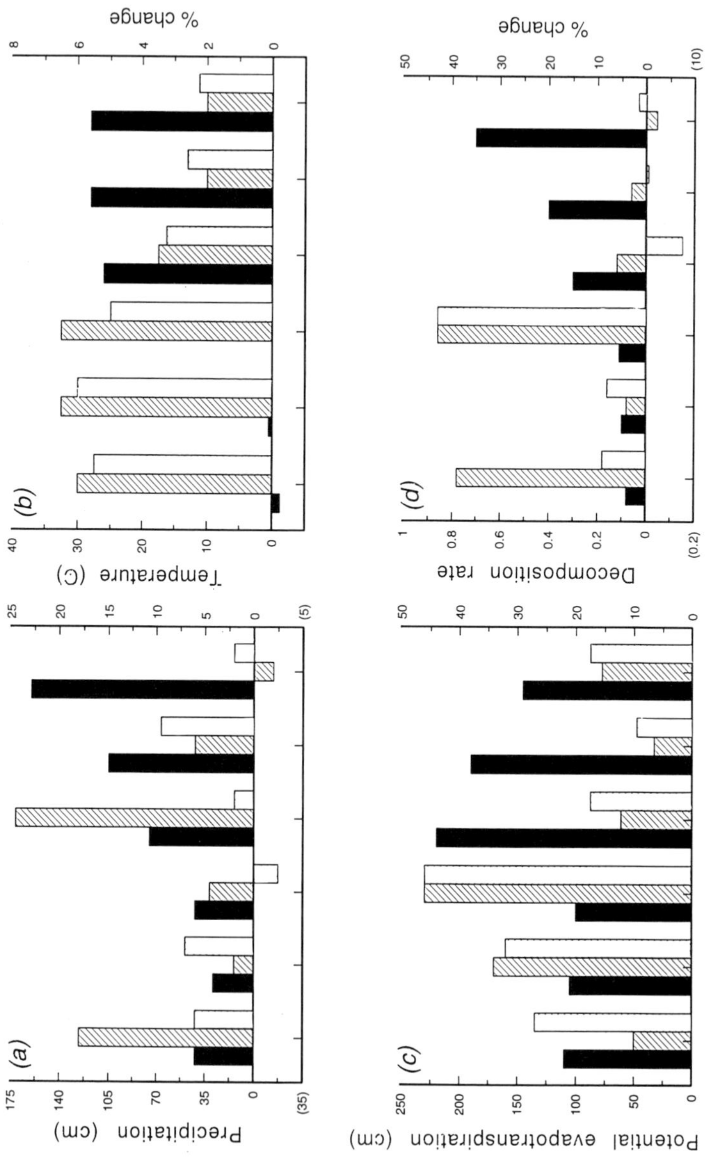
(a)
Precipitation (cm)
% change
(b)
Temperature (C)
% change
(c)
Potential evapotranspiration (cm)
% change
(d)
Decomposition rate
% change

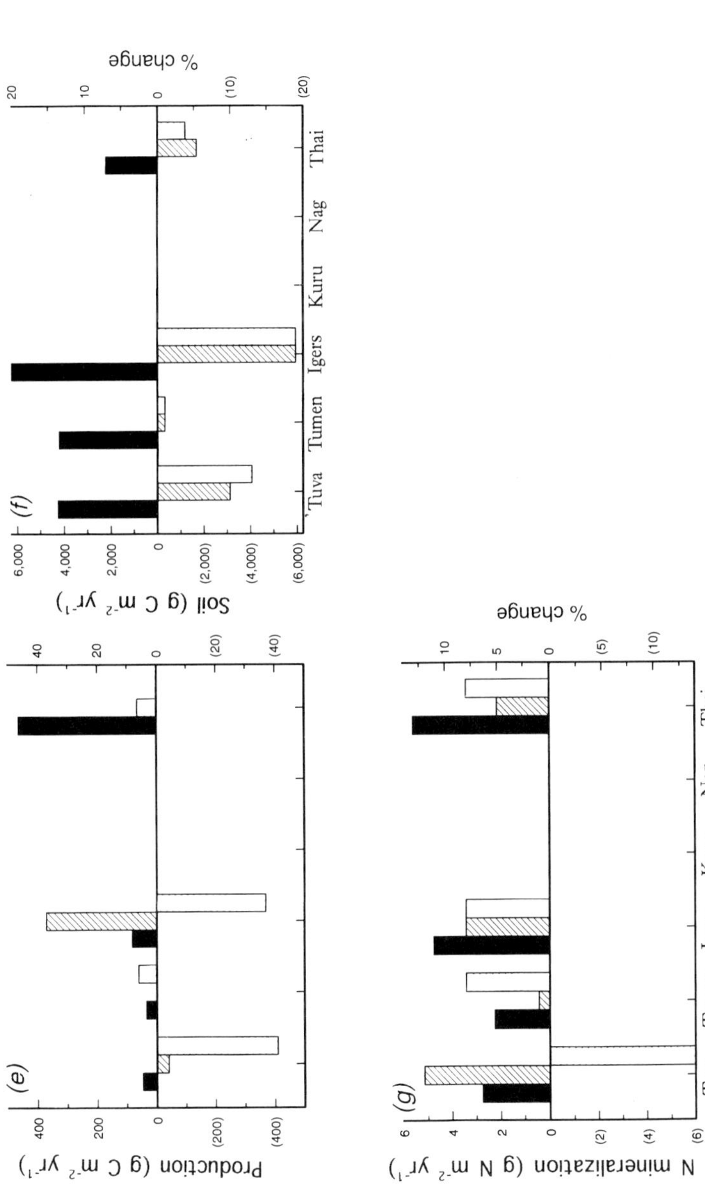

Figure 5.2 *Projected climate change effects on (a) annual precipitation, (b) mean annual potential evapotranspiration, (c) mean annual temperature, (d) annual relative decomposition rate, (e) annual above-ground plant C production, (f) soil carbon, and (g) annual net N mineralization. (Solid columns are current; hatched columns are percentage C change; empty columns are percentage g change.)*

The general response to the two climate perturbations of this ecoregion was a general decrease in NPP and soil organic C. These results suggest that these ecosystems are quite sensitive to the additional drought stress brought about by the combined effect of increased monthly temperatures and modest changes in rainfall.

The CO_2-only perturbation resulted in the largest relative increase in plant production or NPP. This response is due to the inherent drought conditions which normally prevail, and the effect of release from the drought conditions due to water conservation of the plants with elevated atmospheric CO_2 levels. An additional benefit is gained by the added N released from the decomposition of soil organic matter. This net increase in N mineralization and decomposition under elevated CO_2 levels was only observed in this ecoregion.

The Mongolian steppe region was one of the two regions which displayed a decline in NPP under the combined climate-elevated CO_2 perturbations. The region experiences the greatest warming in this study, while the rainfall increases are modest. This combination of climate change induced greater water stress resulting from decreases in soil C and NPP. The elevated CO_2 concentrations were able to ameliorate these climate change impacts, but only partially in most of the sites. The notable exception is the Tumentsogt site, in Mongolia. This site displayed small increases in plant production under the two climate change scenarios and, with the elevated atmospheric CO_2 levels, plant production increased by over 20%. Tumentsogt appears to be both temperature limited and water stress sensitive, and responds to the increased growing season temperatures and a slight increase in rainfall during the growing season (i.e. April through July). The plant production responds positively to the two climate perturbations and the CO_2 increases.

The results of the altered climate conditions and increased atmospheric CO_2 of grassland C fluxes indicate that the Mongolian Plateau is extremely sensitive to changes in the soil water availability. The arid conditions which prevail and the marked sensitivity to alterations in temperature and rainfall amounts are indicated in the sensitivity of plant production and decomposition of these grassland ecosystems due to slight changes in the weather patterns during the short growing season. Increased water stress in these ecosystems, as illustrated by the simulations of Xilinhot, Inner Mongolia, resulted in lower plant production and lower soil C levels in this region of Eurasia, except in Tumentsogt, Mongolia. The GCM projections used in this analysis have subtle differences in rainfall perturbations. At Tumentsogt, Mongolia, the May and June rainfall levels are enhanced over the base climate, and plant production is sensitive to rainfall levels during this time of year.

Projected climate changes appear to have greater negative impact in Xilinhot compared with Tumentsogt. The effect of increased atmospheric CO_2 on system response does not reverse the negative impact of climate changes at Xilinhot, Inner Mongolia. There appears to be little difference between the two GCM scenarios of the simulated soil organic matter levels, compared with the larger difference between plant production for the two GCM scenarios simulated.

Discussion

Sensitivity to climate and CO_2 changes

Climate change alone resulted in an increase in total above- and below-ground production for the mesic regions (humid temperate and Mediterranean), mainly attributable to increased N mineralization, and a decrease in the cold, desert steppe regions. Soil organic matter decreased in all the mesic and colder regions, owing to increased decomposition. In line with most GCM predictions, the tropical savanna regions were affected the least. Climate-driven redistribution of grassland regions was not modeled, but the world area of grasslands is expected to increase in the future or at least remain constant.

From our simulations, the CO_2 increase alone results indicated a significant increase in production in all regions, with the greatest proportional increase in soil organic matter in tropical savanna regions. When combined with predicted climate change, CO_2 had an additive effect, tending to ameliorate climate change effects. The net effect of climate change and CO_2 was a significant increase in NPP in mesic regions (attributable to N mineralization) as well as in dry savannas, with little or no net change in cold, desert steppe or humid, tropical regions. Overall, soil organic matter showed a decrease, especially in temperate steppes and cold, desert steppes owing to stimulation of decomposition by both climate change and CO_2, but tropical savanna and humid savanna regions were actually soil C sinks, regardless of the GCM scenario.

The Mongolian Plateau appears to be very sensitive to climate changes, especially those which would increase the level of water stress during the short growing season of these steppe ecosystems. The increased water–use efficiency under increased CO_2 conditions greatly enhances plant productivity and nutrient cycling rates. The slight changes in rainfall during the critical months of May through July in this region illustrate the sensitivity of these ecosystems to environmental changes. Growing human pressures in this region of the world may increase the sensitivity of these ecosystems to increased drought conditions induced by climate warming or reductions in rainfall during the plant growing season.

Impact of changed land use and tenure system

The land use and tenure system of the Eurasian region have undergone rapid changes. Examples of human dimensions affecting land use in this region include the October Revolution of 1917 in Russia and the Great Revolution of 1949 in China. These conflicts led to radical changes in land management in Inner Mongolia and the massive movement of ethnic Chinese into an area previously dominated by nomadic pastoralists. With the recent collapse of Soviet political and economic institutions, the political and economic situation has been modified greatly in Mongolia and in the former USSR. This has resulted in the privatization of the livestock and changing land use patterns in several of the new independent countries. In the early part of this century, land in Russia was declared as belonging to the people, but in reality it was owned by the state. Recently, the land privatization process was initiated in Russia. These political events are examples of the dynamic shifts in the political arena that have greatly altered human land use and thereby human–ecosystem interactions during the past century.

The political and economic liberalization in Mongolia is affecting the livestock sector in a number of ways. In 1991, the privatization of collectives began, with over 93% of all livestock now being privately owned (Mongolian Statistical Yearbook, 1996). The dissolution of the collective system and its associated top-down livestock and pasture management have led to the re-emergence of traditional institutions through which the pastoral economy is organized. However, their success is contingent upon other factors associated with liberalization such as the urban to rural migration, the terms of trade facing herders, the decreased availability of supplementary livestock feed, and so forth. Such factors are placing pressures upon the efficiency of traditional institutions to maintain sustainable pastoral livelihood systems. Mearns (1993) describes how traditional land tenure arrangements and pasture management are being undermined by the urban newcomers to herding, and how the restricted access to basic services is leading to the concentration of herds around district centres. They also found that many of the grazing territories now used by herding communities are no longer suitable as the availability of fodder and transport, which allowed the utilization of more marginal areas under the collective system, have seriously diminished. These problems are, to some extent, being addressed with the development of new land laws, which are being discussed by the new Mongolian Parliament. The final outcome remains to be seen; however, it is clear that, whatever the outcome, it will be an important factor in determining the sustainable utilization of these grasslands.

Acknowledgments

Model development and programming was assisted by Ms Rebecca McKeown. Data analyses were assisted by Mr Brian Newkirk and Ms Song Bo (visiting scholar sponsored by the China Committee of Scholarly Exchange). Graphics prepared by Becky Techau and Michele Nelson. Manuscript prepared by Kay McElwain. The model development and data analysis were supported by NASAs EOS Interdisciplinary Science Program by the US-Agricultural Research Service, US National Science Foundation (NSF) projects BSR 9013888 and BSR 9011659, and the US Department of Energy. The Chinese plant ecosystem data was collected by the Chinese Academy of Sciences Institute of Botany. The Mongolian data was supplied by the Mongolian Academy of Sciences. Measurement of monthly biomass dynamics, NPP and soil organic matter at the tropical grassland sites in Thailand was carried out under United Nations Environment Programme (UNEP) Project "Environment Changes and Productivity of Tropical Grasslands (1989–1992)". Analysis of data from the sites in the former USSR was carried out under the Russian National Scientific and Technical. Data synthesis and model validation was supported by the IUBS Scientific Committee on Problems of the Environment (SCOPE).

References

Anderson, J. M. (1991). The effects of climate change on decomposition processes in grassland and coniferous forests. *Ecological Applications*, **1**, 326–47.

Bailey, R. G. (1989). Explanatory supplement to ecoregions map of the continents. *Environmental Conservation*, **16**, 307–9.

Ellis, J. & Galvin, K. A. (1994). Climate patterns and land use practices in the dry zones of Africa. *Bioscience*, **44**, 340–9.

Fu, C. & Wei, H. (1993). Study on sensitivity of meso-scale model in response in land cover classification over China. *EOS Supplement* (October 26, 1993), 172 pp.

Houghton, J. T., Jenkins, G. J. & Ephraums, J. J. (eds.). (1990). *Climate Change. The IPCC Scientific Assessment.* World Meteorological Organization (WMO). Cambridge: Cambridge University Press.

Houghton, R. A. (1994). The worldwide extent of land-use change. *Bioscience*, **44**, 305–13.

Long, S. P., Garcia Moya, E., Imbamba, S. K., Kamnalrut, A., Piedade, M. T. F., Scurlock, J. M., Shen, Y. K. & Hall, D. O. (1989). Primary productivity of natural grass ecosystems of the tropics: a reappraisal. *Plant and Soil*, **115**, 155–66.

Mearns, R. (1993). *Pastoral Institutions, Land Tenure, and Land Policy Reform in Post Socialist Mongolia*, Policy Alternatives for Livestock Development, Research Report No.3, IDS, Brighton, Sussex.

Mongolian Economy and Society in 1995: Statistical Yearbook (1996). Ulanbaatar: State Statistical Office of Mongolia.

National Research Council. (1992). Grasslands and Grassland Sciences in Northern China. A report of the Committee on Scholarly Communication with the Peoples Republic of China. Washington, DC: National Academy Press.

Office for Interdisciplinary Earth Studies (OIES) (1991). *Arid Ecosystems Interactions: Recommendations for Drylands Research in the Global Change Research Program.* OIES–Report 6. 81 pp.

Ojima, D. S., Kittel, T. G. F., Rosswall, T. & Walker, B. H. (1991). Critical issues for understanding global change effects of terrestrial ecosystems. *Ecological Applications*, **1(3)**, 316–25.

Ojima, D. S., Parton, W. J., Schimel, D. S., Scurlock, J. M. O. & Kittel, T. G. F. (1993). Modeling the effects of climatic and CO_2 Changes on grassland storage of soil C. *Water, Air, and Soil Pollution*, **70**, 643–57.

Owensby, C. E., Coyne, P. I., Ham, J. M., Auen, L. M. & Knapp, A. (1993*a*). Biomass production in a tallgrass prairie ecosystem exposed to ambient and elevated levels of CO_2. *Ecological Applications*, **3**, 644–53.

Owensby, C. E., Coyne, P. I. & Auen, L. M. (1993*b*). Nitrogen and phosphorus dynamics of a tallgrass prairie ecosystem exposed to elevated carbon dioxide. *Plant, Cell and Environment*,**16**, 843–50.

Parton, W. J., Schimel, D. S., Cole, C. V. & Ojima, D. S. (1987). Analysis of factors controlling soil organic matter levels in Great Plains Grasslands. *Soil Science Society of America Journal*, **51**, 1173–9.

Parton, W. J., Scurlock, J. M. O., Ojima, D. S., Gilmanov, T. G. , Scholes, R. J., Schimel, D. S., Kirchner, T., Menaut, J-C., Seastedt, T., Garcia Moya, E., Kamnalrut, A. & Kinyamario, J. I. (1993). Observations and modelling of biomass and soil organic matter dynamics for the grassland biome worldwide. *Global Biogeochemical Cycles*, **7**, 785–809.

Pielke, R. A. & Avissar, R. (1990). Influence of landscape structure on local and regional climate. *Landscape Ecology*, **4**, 133–55.

Riebsame, W., Parton, W. J., Galvin, K. A., Burke, I. C., Bohren, L., Young, R. & Knop, E. (1994). Integrated modeling of land use and cover change. *BioScience*, **44**, 350–6.

Schimel, D. S., Parton, W. J., Kittel, T. G. F., Ojima, D. S. & Cole, C. V. (1990). Grassland biogeochemistry: links to atmospheric processes. *Climate Change*, **17**, 13–25.

Schlesinger, W. H., Reynolds, J. F., Cunningham, G. L., Huenneke, L. F., Jarrell, W. M., Virginia, R. A. & Whitford, W. G. (1990). Biological feedbacks in global desertification. *Science*, **247**, 1043–8.

Sheey, D. P. (1988). The desertification of Balin Right Banner, Inner Mongolia. In *Proceedings of International Development Symposium, Society of Range Management*, 10 pp. Denver, CO: Society of Range Management.

Wu, J. & Loucks, O. (1992). Xilingele. In *Grasslands and Grassland Sciences in Northern China*, pp. 67–84. Washington, DC: National Academy Press.

Xiao, X., Wang, Y., Jiang, S., Ojima, D. S. & Bonham, C. D. (1995). Interannual variation in the climate and aboveground biomass of *Leymus chinense* steppe and *Stipa grandis* steppe in the Xilin River basin, Inner Mongolia, China. *Journal of Arid Environments*, **31**, 283–99.

Zhang, X. (1992). Northern China. In *Grasslands and Grassland Sciences in Northern China*, pp. 39–54. Washington, DC: National Academy Press.

Zhu, Z. & Wang, T. (1993). Trends of desertification and its rehabilitation in China. *United Nation Environment Programme Desertification Control Bulletin*, **22**, 27–30.

Carbon and water exchange of terrestrial systems

E-D. Schulze and M. Heimann

Introduction

The climate system of the Earth is driven by the physical processes of mass and energy transfer, chemical transformations of gases, and biological processes such as transpiration, carbon dioxide assimilation and respiration. Although we have detailed knowledge of processes within the atmosphere and vegetation, it remains difficult to couple both in a predictive manner. At present, we cannot model with sufficient certainty the effect of climate and land use change on terrestrial ecosystems, or how a change in terrestrial ecosystems will feed back to the climate system (Melillo *et al.*, 1995). Another fundamental question which is still unanswered relates to the role of biodiversity in ecosystem function. We do not know if it makes a difference to the functioning of an ecological system if there are many species or a few, or if species which perform similar functions within an ecosystem are interchangeable with one another or if they differ in ways which matter to the working of the system and to the climate (Heywood & Watson, 1995).

It will be impossible to deal comprehensively with these issues in this short overview; however, we will try to point at certain processes for which a database exists, and we will point at other processes for which we need to know more for predictions of future global climate. We will concentrate on (1) the relation between plant nutrition, carbon assimilation and growth, (2) the transport of gases between plant cover and the atmosphere, and (3) processes which lead to carbon dioxide losses and long-term carbon immobilization. Since the emphasis of this book is aimed at an Asian context of global processes, we will relate to Asian examples whenever possible. The problem with respect to the global view is that most of the information about plant processes averages published information on the basis of plant functional types or vegetation types. It is impossible at this stage to make regional predictions, because the variation for each parameter and biome is large and there is not sufficient regional coverage of observations.

The relation between plant nutrition, carbon assimilation and water use

Rates of photosynthesis vary by a factor of 10 between different plant species with agricultural plants at the top of this scale and conifers at the bottom (Larcher, 1994). Although a classification of functional plant types may be convenient for global modelling of carbon assimilation, it remains unsatisfying (Körner, 1994). The limitation of any classification scheme is that it does not allow for a continuous change in biogeochemical parameters related to controls of plant production as was demonstrated by Field and Mooney (1983), who developed a linear relationship between net photosynthesis and leaf nitrogen concentration ($A_{max} = -76.1 + 149 \times N$). This simple relationship highlights the importance of plant nutrition for C assimilation, independent of plant life forms. Plant nutrition also affects water relations, since the uptake of CO_2 and losses of water vapor follow the same diffusion path through the stomata (Raschke, 1979). The plant internal regulation of both CO_2 and water vapor fluxes is the basis for the linear relationship between leaf conductance and nitrogen nutrition, leaf and canopy conductance, and canopy conductance and C assimilation which has been established for a broad range of plant species and vegetation types (Fig. 6.1; Schulze *et al.*, 1994).

When classifying and averaging the observed data according to vegetation type, and projecting this information on to a global vegetation cover, it emerges (Fig. 6.2*a*) that the centers of high nitrogen nutrition are located in the industrial and agricultural regions of the globe, especially in North America, Europe, India, and East Asia. Also, there are marked differences in nitrogen nutrition between the Northern and the Southern Hemisphere. Since photosynthesis is related to N-nutrition, a global map of photosynthetic capacity matches that of nitrogen (Fig. 6.2*b*), even though it is based on independent measurements (Schulze *et al.*, 1994).

It is not possible at this stage to calculate global transpiration from canopy conductance (see Fig. 6.1) because this would require information about daily courses of actual radiation and vapor pressure deficits, and knowledge of the water content of the soil surface for predicting soil evaporation. On the contrary, we know about the relative contribution of evapotranspiration during partitioning of rainwater (Fig. 6.3; Larcher, 1994). At high precipitation, evapotranspiration may only return less than 50% of the water to the atmosphere, with the remainder being lost as runoff or seepage to groundwater. However, below 1000 mm of rainfall, the contribution of evapotranspiration becomes increasingly significant. Land use will strongly affect evapotranspiration due to changes in leaf area index, nutrition and canopy conductance in the low rainfall range. Nevertheless, the overall effect may be very variable. Irrigation will exceed precipitation, drain surface and groundwater, and cause soil salinity. A change from forest into grain crop will alter the seasonality of water use, with higher transpiration in the

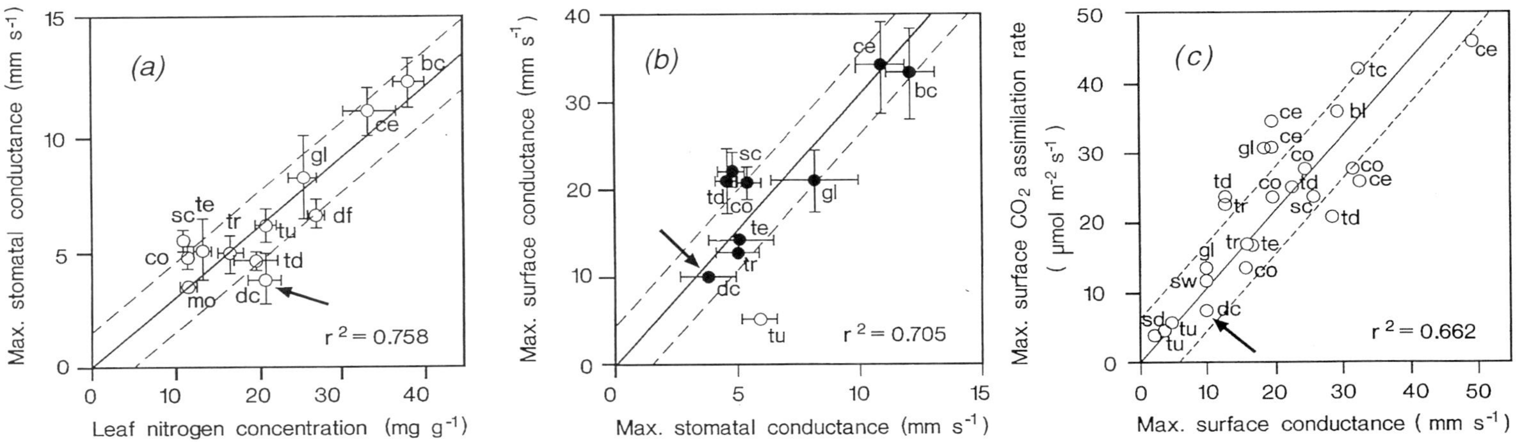

Figure 6.1 *Relationships between (a) maximum stomatal conductance and nitrogen concentration (solid regression line through the origin:* $y = 0.3012\,x$*; dashed line: s.e.of y:1.358,* r^2*:0.758; Table 6.1). (b) Maximum surface conductance and stomatal conductance (solid regression line through the origin:* $y = 2.996\,x$*; dashed line: s.e.of y:4.495,* r^2*:0.705, Table 6.1), and (c) maximum surface assimilation and maximum surface conductance (solid regression line through the origin:* $y = 1.048\,x$*; dashed line: s.e.of y: 6.445,* r^2*:0.684, Table 6.2). The arrow in Fig. 6.1c indicates the only data point that exists for whole Asia. Key: bc, broad-leaved crops; ce, cereal crops; co, evergreen conifer forest; dc, deciduous conifer forest; df, tropical deciduous forest; gl, grassland; mo, monsoonal forest; sc, sclerophyllous shrub; sd, savanna dry; sw, savanna wet; tc, tropical tree crop (Hevea); td, temperate deciduous broad-leaved forest; te, temperate evergreen broadleaved forest; tr, tropical rain forest; tu, herbaceous tundra. (After Schulze* et al., *1994.)*

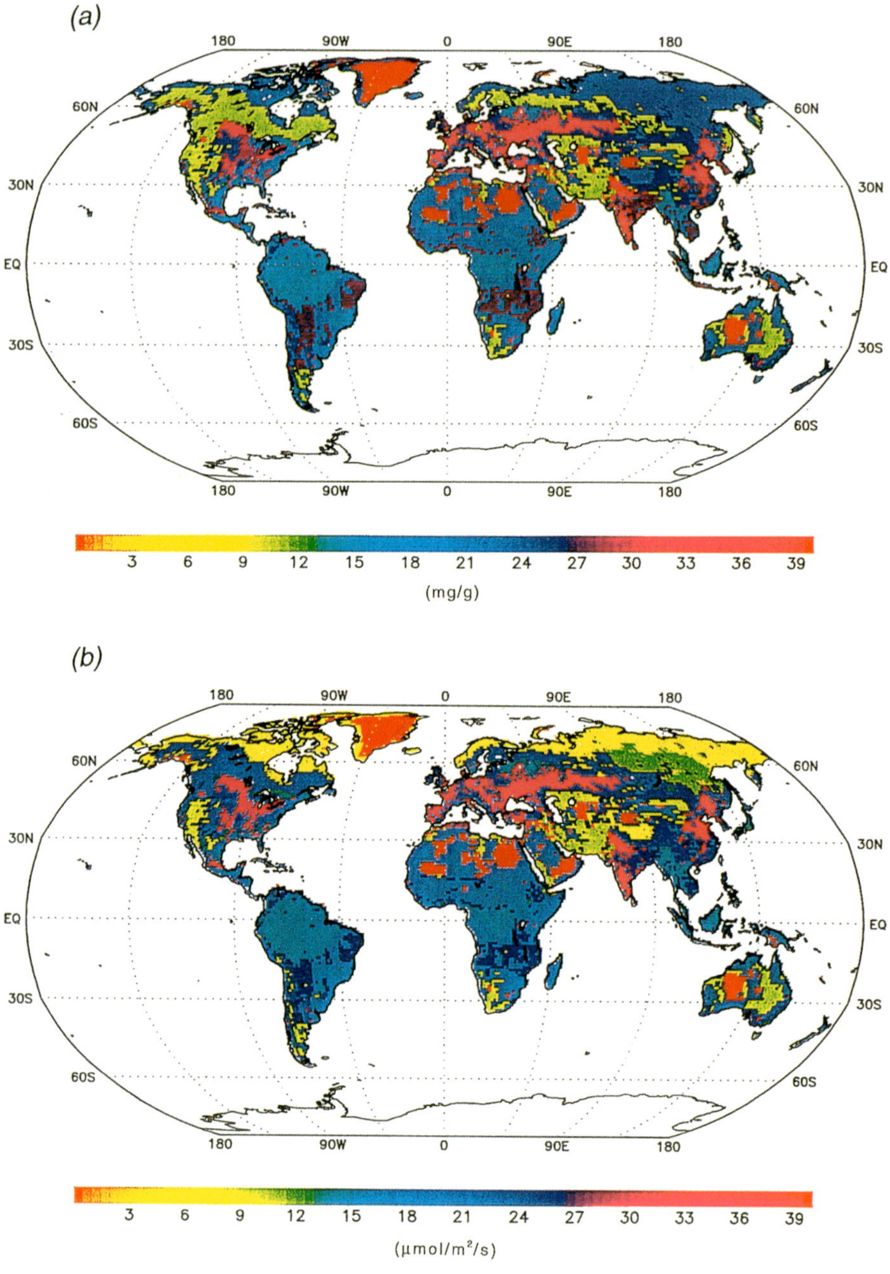

Figure 6.2 *Global map of (a) leaf nitrogen concentrations, and (b) maximum surface assimilation (A_{max}, µmol m^{-2} s^{-1}) of CO_2. The maps were prepared on the basis of values assigned to land-cover classes of primary and secondary vegetation types of Wilson and Henderson-Sellers (1985) and plotted on a 1 × 1 degree grid scale. White color indicates areas with insufficient information. (From Schulze* et al., *1994.)*

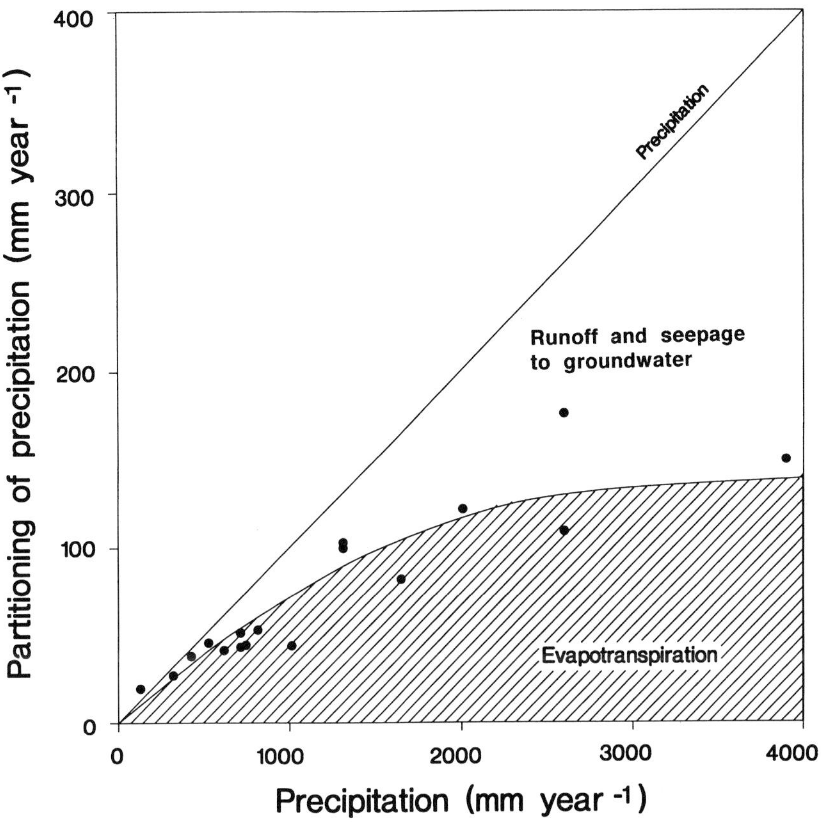

Figure 6.3 *Partitioning of rainwater into runoff and groundwater-recharge and evapotranspiration as related to precipitation. The data points represent regional averages of measurements in different global biomes (From Larcher, 1994.)*

growing season, but increased runoff from bare fields outside the growing season. In this case, the total amount of plant available water will decrease. Desertification will also decrease the returns to the atmosphere but contribute to runoff. Although it is tempting to speculate about the overall effects of land use change in a monsoon climate, such as East Asia, we are not ready to make such predictions because of the great number of alternative effects.

On the status of available data for Asia

Although predictions are made in Fig. 6.2 for the Asian continent, it should be made clear that these predictions are based on the vegetation map by Wilson and Henderson-Sellers (1985), which does not contain recent changes in land use, and that the databases are not restricted to Asian measurements. For example, the N-concentrations in tropical forest trees represents an average of all data which were measured in this vegetation type globally. It does not separate data according to geographic regions. This is not possible at this stage due to a lack of data coverage. If we inspect the actual databases for Asia, the lack of information on canopy processes becomes very obvious. For instance, with respect to flux measurements, to our knowledge the only existing Asian data point was taken by Kelliher *et al.* (1995) in a Siberian *Larix* forest, which entered into Fig. 6.1 representing deciduous conifers (see arrow in Fig. 6.1). We do not claim that other data do not exist, but they are not known to us despite extensive literature search. We are only aware of gradient studies (e.g. Cherkatuk *et al.*, 1991).

Xylem flux measurements may be a complementary method to eddy covariance measurements and useful for separation of canopy and soil water losses (Pearcy *et al.*, 1989). Again, the only measurements of this type were carried out in the *Larix* experiment (Arneth *et al.*, 1995).

Leaf level gas exchange data have been reviewed by Körner (1995) for stomatal conductance and by Woodward and Smith (1995) for assimilation. Again, both authors were not aware of gas exchange measurements in Asia. These data might exist (e.g. Cherkatuk *et al.*, 1991) but are not accessible in the West. Even major literature data banks (e.g. BIOSIS), which supply abstracts and translations also for Asian journals, did not contain any data on gas exchange in Asia (see Schulze *et al.*, 1994). In contrast, the most reliable data on net primary productivity (NPP) in tropical forest were generated in East Asia during the period of the International Biological Programme (see summary by Schulze, 1982), and this information entered into the global maps.

Relation between photosynthesis and growth at the plant and at the canopy level

The relationship between photosynthesis and growth is not linear because plants have numerous possibilities for compensation. These include changes in leaf size, variations in carbon partitioning between roots and shoot, and variation in longevity of leaves or plant individuals (Stitt & Schulze, 1994). By evolution, plant forms have emerged which cover the whole spectrum between short and long-lived plants, woody and herbaceous tissues, and deciduous and evergreen leaves (Schulze, 1982).

The specific effect of a change in photosynthesis on growth was studied using genetically manipulated plants of tobacco in which the activity of the carbon dioxide fixing enzyme Rubisco was manipulated genetically. In this case, a change in photosynthesis had very little effect on growth because of the capacity of plants for compensatory responses. A decreasing rate of photosynthesis changed plant internal CO_2 concentration and this led to a change in leaf conductance, water and ion uptake. The resultant osmotic effect on cell expansion produced bigger but thinner leaves which maintained the carbon gain per plant constant (Stitt & Schulze, 1994).

In addition, there is no simple relationship between soil fertility and plant nutrient status (Marschner, 1995). It is part of the evolutionary processes that allow plants to cope with adverse conditions with multiple options. Fast-growing species having high turnover and short life cycles coexist in deserts with slow growing species having low turnover and long life cycles. Also, the plant investment into roots is inversely related to the resource supply, but plants also have the option to increase the uptake capacity. Therefore, Schulze *et al.* (1994) correlated leaf assimilation to leaf nitrogen and not to soil nitrogen.

In order to investigate relations between fluxes (assimilation, nutrient uptake and growth) and pools (biomass and nutrient stores) and between pools and vegetation structure (plant height), we must consider an additional process operating at the whole-stand level, namely that of self-thinning (Westoby, 1984). Relating the above-ground biomass to the number of individuals per ground area in a log/log plot, a line with a negative slope of 0.5 emerges for very different species and vegetations, indicating competition for light. It can be interpreted by the fact that plant individuals of a certain biomass require a certain volume of space. Either many individuals of small size, or a few individuals of large size, may exist on the same ground area. An example of such a self-thinning response is given in Fig. 6.4 for a Siberian larch forest, where stand biomass increased until age 120 and remained constant from then on, while needle biomass decreased from age 50 (Fig. 6.4*a*). The nitrogen budget shows that an increasing amount of nitrogen is being immobilized in wood, and mineralization is not sufficient to maintain needle growth (Fig. 6.4*b*). Under these conditions, the self-thinning response does not follow the slope which was predicted for light competition, namely -0.5, but stem biomass and nitrogen content per ground area remain constant at variable stem number (Fig. 6.4*c*). This response resembles a situation of nutrient stress although nitrogen concentrations in needles remained constant (Schulze *et al.*, 1995). A major disturbance (such as wind or fire) is required to "thaw" the nitrogen cycle under these conditions. Thus, the analysis of self-thinning contains information about resource limitations and about changes in C pools with stand development, which may serve as an important tool in future global change research.

The productivity of plants, ecosystems and biomes

The large differences between plant species in photosynthetic capacity diminish if we compute productivity, thereby taking into account plant respiration and C allocation (Gifford, 1974). Net primary productivity depends not only on nutrient availability but also on length of growing seasons, leaf area indices and leaf

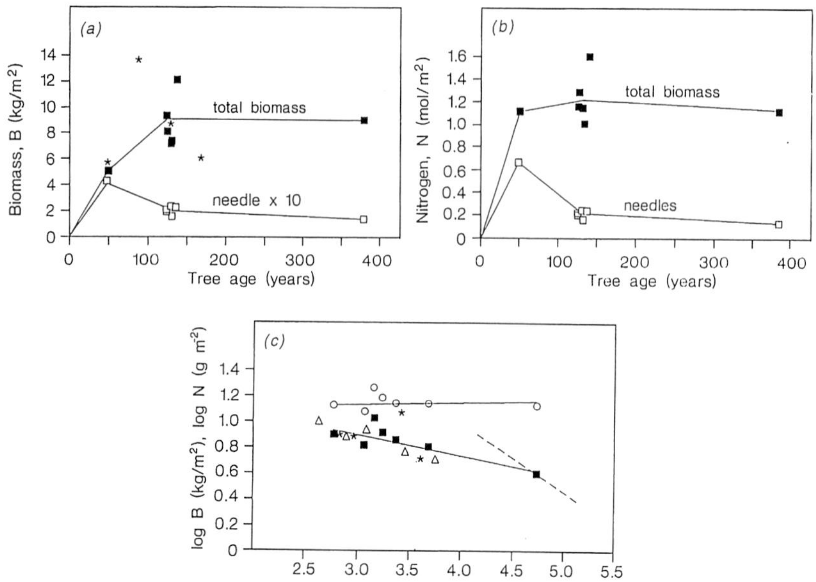

Figure 6.4 *(a) Relationships between above-ground biomass (solid squares) and needle biomass (multiplied by factor 10, open squares) and tree age for six stands of* Larix gmelinii *in Eastern Siberia. In (b), a second set of relationships, biomass is replaced by the quantity of nitrogen. Also shown in (c) are relationships between biomass (solid squares) and quantity of nitrogen (open circles) and stand density (n, trees m^{-2}) for* Larix gmelinii, *including data for* Pinus sylvestris *in West Siberia (open triangles). The two solid lines are linear regressions and the dashed line represents the theoretical self-thinning line (Westoby, 1981) originating at the pole stand data point. Data from Pozdnyakov (1983, 1987) are shown as asterisks in (a) and (c). (After Schulze et al., 1995.)*

structures and their turnover. Therefore, a global map of actual (not maximal) annual NPP (Fig. 6.5*a*) as calculated from an equilibrium simulation with the TEM model of Melillo *et al.* (1993) shows a distribution of NPP which is different from that of A_{max}. Despite lower rates of photosynthesis, but with a longer growing season and high leaf area indices, the tropical forest exhibits rates of NPP which are similar or even higher than those in temperate eastern North America, Europe or East Asia. Nevertheless, the high photosynthetic capacity of these latter regions results also in relatively high NPP. For most parts of the semitropical savannas, which are characterized by high photosynthesis of C_4 plants, the advantage of C_4 photosynthesis at the leaf level is lost in an annual budget of growth due to drought (Gifford, 1974). Also, the boreal forests of Asia and North America reach rates of NPP which are not much higher than arid and semiarid regions but as a result of nitrogen limitation (see Fig. 6.4). In contrast, the Asian coastal areas are among the most productive regions of the world, which coincide with high assimilation rates and with anthropogenic N deposition.

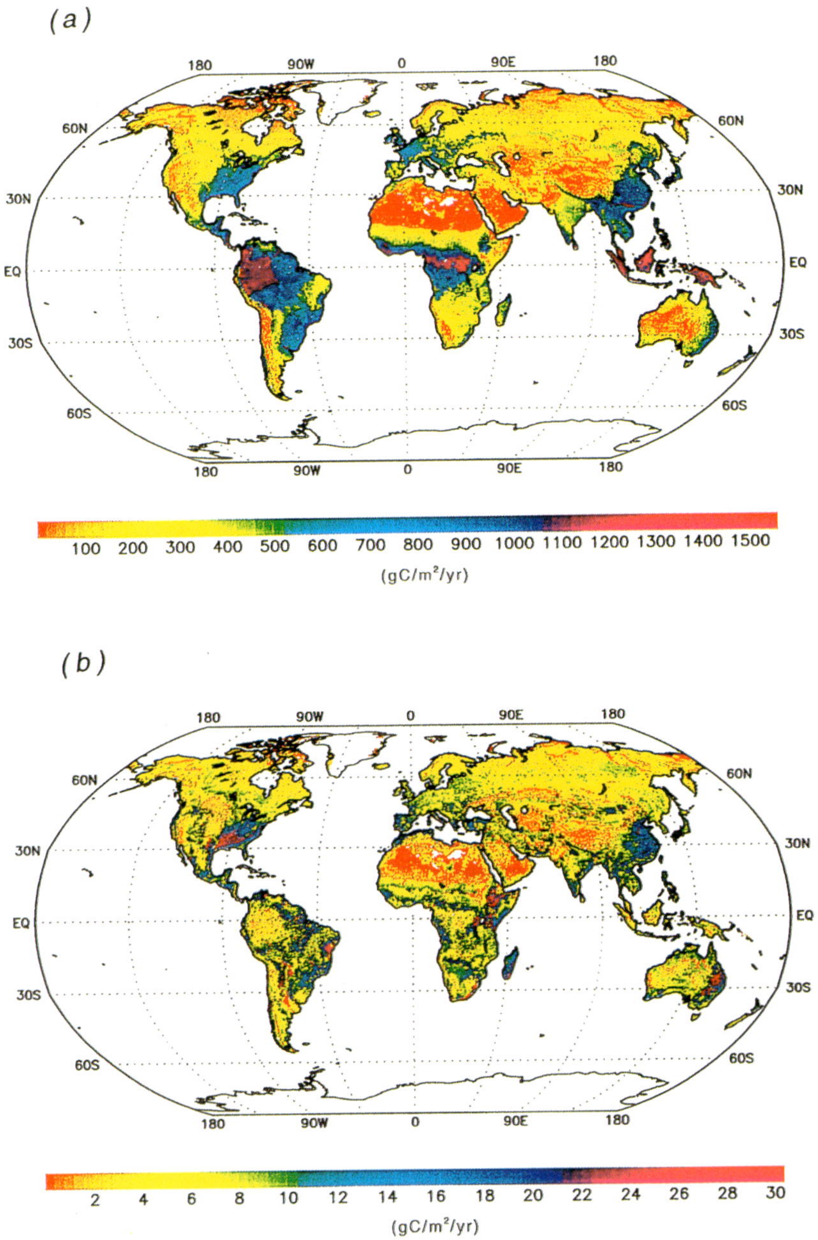

Figure 6.5 *(a) Global map of net primary production (NPP) (g m^{-2}) and (b) global map of net ecosystem production (NEP) as derived from the TEM model (Melillo* et al., *1993). NPP as shown in (a) is from an equilibrium model simulation with present-day climate and an atmospheric CO_2 concentration of 340 ppmv. NEP as shown in (b) is from a transient simulation with specified increasing atmospheric CO_2 concentration over the industrial period. The map shows the model predicted NEP for 1990. Both model simulations did not include any changes in land use, hence are representative for natural vegetation only.*

With respect to the global carbon balance, NPP is the primary input to various ecosystem C pools which are turned over by a large range of processes in variable times. Processes which consume organic carbon are the autotrophic and heterotrophic respiration, and disturbances.

Heterotrophic respiration (R_H) of microorganisms and animals in humus and soils responds mainly to temperature and soil moisture, and is directly dependent on the amount and quality (C/N ratio or lignin content) of material which is delivered to the ground as plant litter. In this process the total amount of circulating nitrogen and carbon is again a major regulator. An example illustrates this interaction (Schulze & Chapin, 1987). In a spruce forest only a small fraction (13%) of a large needle biomass (1.8 kg m^{-2}) is being replaced each year. This needle biomass has a low nitrogen concentration (5 mg g^{-1}) and this results in a low N turnover of 1.2 g N m^{-2} $year^{-1}$. This contrasts with a herbaceous vegetation, in which the total leaf biomass is low (0.5 kg m^{-2}) but total leaf turnover and litter fall are high (1.5 kg m^{-2} $year^{-1}$) with leaves of high N concentration (20 mg g^{-1}). The resultant N turnover is very high (30 g N m^{-2} $year^{-1}$). Despite this high supply, the total amount of humus accumulation may be lower in herbaceous vegetation than in forests, depending on secondary plant compounds, soil pH and water status. Thus, we may hypothesize that net ecosystem productivity (NEP), which includes autotrophic and heterotrophic respiration, may be low for an agricultural system of high NPP, high litter quality and rapid turnover, but high for a forest of low NPP, low litter quality and low turnover.

While respiratory processes are fairly well understood (Amthor, 1995), ecosystem disturbances contain a great variety of very contrasting events. They may act as a constant directional forcing for assimilatory and dissimilatory processes (e.g. changes in CO_2 and temperature). In contrast, disturbances may be periodic but only redistribute and activate existing C and N pools for respiration, and this could, in turn, supply resources for increased assimilation at a later time (e.g. windfall in forest, ploughing in agriculture). In addition, there are episodic disturbances which remove C and N and other resources from the system, and this affects the C pools but may or may not affect the permanent fluxes. Examples of such disturbances are fire and harvest.

Although the existing theory of net ecosystem productivity includes dissimilatory processes and disturbances (as defined by the variable turnover of pools) for the equilibrium situation, where NEP=0 by definition, this allows only limited insight into the effects of various forcings in the transient situation. Also, there is a problem of measurement. Scientists who carried out long-term measurements by the eddy covariance technique (Wofsy *et al.*, 1993) determined the net balance of C and N assimilation and they referred to this quantity as net ecosystem exchange (NEE), which does not include catastrophic and episodic disturbances, but does include climate forcing and disturbances which operate through changes in decomposition. We think that disturbances such as harvest and fire contain a different quality and time scale than directional forcings, such as climate change. Such episodic events have massive effects on all ecosystem pools, and modeling such events by modifying turnover time remains problematic. In the case of forest harvest, the turnover time will depend on whether timber was used for pulp or for furniture. Heart wood of trees does not respire and therefore does not contribute to NEE, but its turnover time and its effect on C storage will be quite different if it falls to the ground and decays over a longer period of time (Harmon *et al.*, 1990), or if it is burnt to charcoal, or if it goes onto the wood market.

NEP, in a model like TEM, goes to zero in an equilibrium setting. Conceptually, in this case, NEP includes disturbance. Hence the equilibrium established at any model grid point reflects only the long-term time average. NEP is expected to be different from zero in a transient situation, as induced by directional forcings from continually acting factors such as increasing CO_2 and N deposition. In this case the frequencies of episodic disturbance processes might very well change. In fact, in nature, an increase in NEP will most likely lead to an increase in episodic disturbances (e.g. due to biomass accumulation as fuel load for fire) which, in turn, may result in a reduced local carbon sequestration. It is even possible, that at a particular location, the sign of the response could be reversed, i.e. that an increase in NEP might be associated with a net carbon loss.

It is obviously difficult to calculate transient NEP and net carbon storage accurately, because each such model compartment may have different turnover times depending on the types of disturbances that are explicitly and implicitly modeled. One should be aware of this problem in ecosystem modeling in order to interpret the global NEP map correctly (Fig. 6.5*b*). This map shows the NEP for the year 1990 predicted by a simulation with the TEM model. In this simulation the TEM model was first run to an annually averaged equilibrium state under a prescribed preindustrial atmospheric CO_2 level of 278 ppmv. The model was then run through the industrial period up to 1990, whereby the CO_2 concentration increase was specified according to the observations from ice cores and direct atmospheric measurements. Fig. 6.5*b* shows the simulated difference NEP = NPP - R_H for the year 1990. The model predicts a globally integrated NEP of about 1 Pg C $year^{-1}$. This low NEP value results from a minimal response of the land surface to CO_2 enrichment. Notice also, that the model simulation did not include any specified changes in land use, hence it is representative of natural vegetation only. Furthermore, the model run was performed with a fixed, present-day climate, i.e. the simulation does not include any changes in carbon fluxes due to interannual or longer term trends in the driving climate or N deposition variables.

The highest values of the CO_2-dependent NEP are found in regions of high photosynthesis and/or of a high woody component in the vegetation. Surprisingly low is the predicted NEP in the Arctic, which appears at odds with observations of layers of peat which demonstrate major carbon storage and long turnover times under conditions of waterlogging. It must be realized, however, that the present model simulation started off from a pre-industrial equilibrium state, i.e. a state in which at every gridpoint, annually averaged heterotrophic respiration equals annually averaged NPP, hence with an NEP of zero. Therefore, the predicted NEP for 1990 only reflects any changes induced by the rising atmospheric CO_2 concentration, at least as far as represented in the model.

The fact that heterotrophic respiration is regulated by a different set of parameters than assimilation makes predictions difficult about the future role of ecosystems in the global carbon balance. There is consensus (Melillo *et al.*, 1995) that NPP will increase in the near term future due to elevated CO_2 (Fig. 6.6). This will reduce the capacity of ecosystems to sequester excess carbon dioxide. In contrast, heterotrophic respiration will not respond directly to elevated CO_2, but with some time lag it is expected that heterotrophic respiration will increase as well, due to the increased litter availability for decomposition and increased global temperatures. This will compensate the capacity of ecosystems to immobilize carbon dioxide. The increased capacity for storing carbon in ecosystems is most probably a temporary response which will dampen (or may even reverse) in the long term. A permanent non-zero NEP is not to be expected with the directional forcing of CO_2 and climate change. Considering land use changes as an additional disturbance, it is expected that NEP could be highly variable, even on a small scale.

Table 6.1 quantifies the role of Asia in the global carbon budget. Despite its large size, but due to its northern extension, the total NPP and NEP are smaller

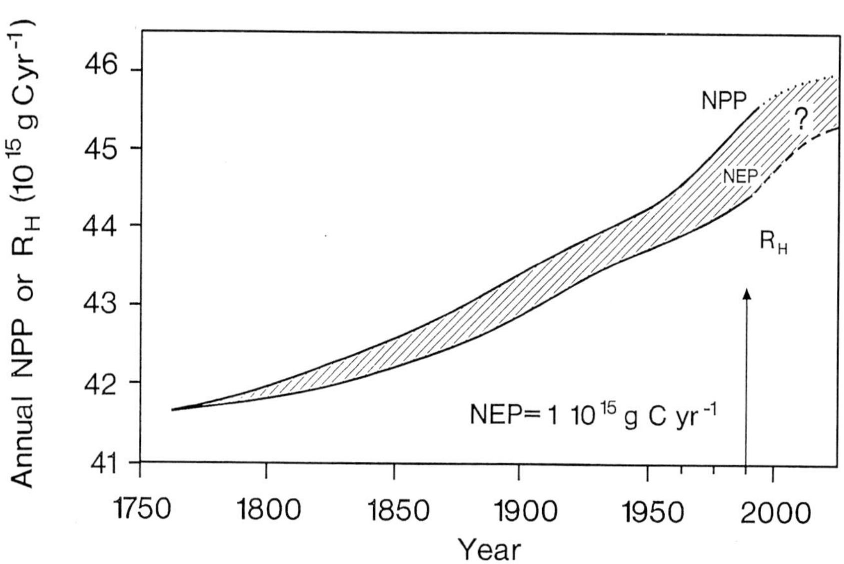

Figure 6.6 *Predicted changes in NPP and R_H over time. The area between the two curves represents NEP. (After Mellilo* et al.*, 1995.)*

Table 6.1 *Regional distribution of net primary productivity (NPP), heterotrophic respiration (R_H), and ecosystem productivity (NEP = NPP- R_H) as calculated for the preindustrial period (year 1765) to present day (year 1990) using the TEM model. The model run starts from an equilibrium state in 1765, and allows for increased CO_2 concentrations in the following period according to ice core and direct measurements. NEP does contain the increase in heartwood of trees, which does not respire, but which will decompose eventually.*

Year	1765			1960			1970			1980			1990		
process	NPP	R_H	NEP	NPP	R_H	NEP	NPP	R_H	NEP	NPP	R_H	NEP	NPP	R_H	NEP
N. America	6.33	6.33	0.00	6.58	6.50	0.08	6.62	6.52	0.10	6.68	6.54	0.14	6.74	6.57	0.17
S. America	9.53	9.53	0.00	10.36	10.26	0.10	10.41	10.29	0.12	10.47	10.32	0.15	10.52	10.35	0.17
Europe	2.15	2.16	0.01	2.21	2.19	0.02	2.22	2.23	0.01	2.24	2.20	0.04	2.26	2.21	0.05
Africa	7.48	7.49	0.01	8.09	7.99	0.10	8.17	8.02	0.15	8.26	8.07	0.19	8.35	8.13	0.22
Asia	11.54	11.54	0.00	12.03	11.89	0.14	12.10	11.92	0.18	12.20	11.96	0.24	12.31	12.01	0.30
Australia	1.71	1.71	0.00	1.85	1.80	0.05	1.88	1.84	0.04	1.92	1.86	0.06	1.95	1.88	0.07
Oceania	2.93	2.93	0.00	3.18	3.17	0.01	3.18	3.17	0.01	3.19	3.17	0.02	3.19	3.17	0.02
Globe	41.67	41.69	0.02	44.30	43.80	0.50	44.58	43.99	0.61	44.96	44.12	0.84	45.32	44.32	1.00

Units: 10^{15}g C year^{-1}.

than in America. Nevertheless, Asia's contribution is higher than that of Africa. The more worrisome is the lack of direct observations (see above) in this region.

Net biome productivity (NBP)

Besides the uncertainty in a differential regulation of NPP and R_H by climate and resources, predicting the effect of disturbance is very uncertain. Using fire as an example, it is obvious, that (1) fire is a quasi-periodic process with relatively low frequency, (2) it releases CO_2 in a burst and not continually as respiration, and it is therefore difficult to monitor, and (3) fire succession leads often to a change in the type of ecosystem. A lichen desert may follow a Siberian pine forest, a herbaceous vegetation alternates with a Siberian Larch forest and broadleaved forest follow after fire in a Siberian mixed conifer stands. We are dealing with step changes in vegetation structure and with successions, which include a complete change in species composition, i.e. we are dealing with different ecosystems, each of which will have a different NPP and NEP. The same problems occur with forest harvest, where logging may change deciduous forest into conifers, or tropical forest into pasture, or boreal forest into bogs.

We think that the episodic flush-type disturbances of low frequency, which result in products of different quality (charcoal), should be treated separately from NEP. We suggest combining such disturbances in the term "net biome productivity" (NBP) which reflects the mosaic of a landscape, where most patches may act as a carbon sink, but where variable plots are burnt or harvested each year releasing the accumulated carbon in a flush. In a boreal forest, due to periodic fire, the carbon sink at a landscape level may be close to zero, although eddy covariance would measure net ecosystem C-uptake at any plot for decades.

Seiler and Crutzen (1980) estimated that 2 to 4 Pg C are lost to the atmosphere annually by burning, which would bypass R_H. The estimate by Seiler and Crutzen is too low, because it only considered fires in the tropics. It becomes increasingly evident that fires are also the driving force in boreal forests (Kolchugina & Vinson, 1995), which in fact are ecologically important in the nitrogen cycle (Fig. 6.4). Considering the boreal forests of 1.7×10^{13} m^2 (Woodward & Smith, 1994) with 5×10^3 g C m^{-2} as standing crop and 3×10^3 g C m^{-2} in humus and litter (Schulze *et al.*, 1995; Shvidenko & Nielssen, 1994), a 50 to 100 year fire cycle (Schulze *et al.*, 1995; Kolchugina & Vinson, 1995), and a fire efficiency of charcoal production of about 5% (Kurz *et al.*, 1992) 0.6 Pg C $year^{-1}$ are expected to be immobilized in the Siberian forest (Kochugina & Vinson, 1995). Thus total boreal and arctic burning may reach the order of magnitude of biomass burning in the tropics. Therefore, biomass burning has received increasing attention in the global carbon cycle (Kahlbusch & Crutzen, 1995). It is a process which releases CO_2 to the atmosphere besides respiration. Its frequency may increase with human impacts and not only with global climate change. The model-predicted increase of NEP (Melillo *et al.*, 1995) might shrink (or increase) if this process were included in addition to C removal by harvest.

If we consider NBP in our global carbon balance (Fig. 6.7), we may find that 100% photosynthetic CO_2 assimilation (gross primary productivity, GPP) is reduced to about 50% by respiration in plants, resulting in NPP. Heterotrophic respiration and disturbance by directional forcing will decrease NPP to less than 5% of GPP. NEP was 3.5% of NPP in the model of Melillo *et al.* (1995). We are aware that the relation between NPP and NEP will be highly variable, and thus a fixed portion can only illustrate the magnitude of this relation. If we consider periodic fires as an additional factor, then we may find that NBP is less than 0.5% of NPP (Kahlbusch & Crutzen, 1995). Despite very high unidirectional fluxes, terrestrial ecosystems appear not to be not a strong permanent sink for carbon. However, on decadal time scales, and due to human activity, we may have a mosaic of very different responses in a patchy landcover. Although the equilibrium situation would make NEP and NBP very small, the transient situation of various forcings will lead to the establishment of various pools along this pathway of C degradation, which will contribute very well to a temporary terrestrial carbon sink.

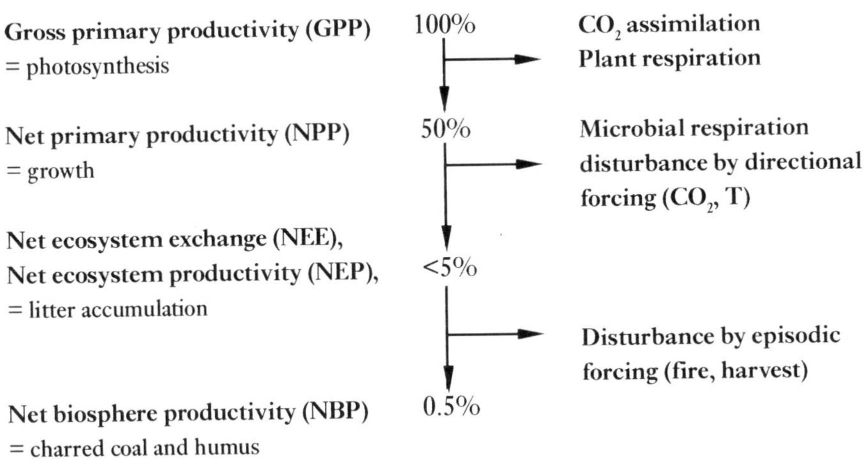

Figure 6.7 *Schematic explanation and estimates of productivity at the leaf (PPP), the whole plant (NPP), the ecosystem (NEP) and the biome (NBP) level.*

References

Amthor, J. S. (1995). Terrestrial higher-plant response to increasing atmospheric CO_2 in relation to the global carbon cycle. *Global Change Biology*, **1**, 243–303.

Arneth, A., Kelliher, F. M., Bauer, G., Hollinger, D. Y., Byers, J. N., Hunt, J. E., McSeveny, T. M., Ziegler, W., Vygodskaya, N. N. & Schulze, E-D. (1995). Environmental regulation of xylem sap flow and total conductance of *Larix gmelinii* (Rupr.) Rupr. trees in Eastern Siberia. *Tree Physiology*, **16**, 247–56.

Cherkatuk, A. C., Rusakova, L. V., Suvorova, G. G. & Uncova, L. S. (1991). *CO_2 Gas Exchange of Forests near Baikal Lake.* Siberian Branch of NAUK, Novosibirsk, 133 pp.

Field, C. & Mooney, H. A. (1983). The photosynthesis nitrogen relationship in wild plants. In *On the Economy of Plant Form and Function*, ed. T. J. Givnish, pp. 28–56. Cambridge: Cambridge University Press.

Gifford, R. M. (1974). A comparison of potential photosynthesis, productivity and yield of plant species with differing photosynthetic metabolism. *Australian Journal of Plant Physiology*, **1**, 107–17.

Harmon, M. E., Ferrel, W. K. & Franklin, J. F. (1990). Effects on carbon storage of the conversion of old-growth forests to young forests. *Science*, **247**, 699–702.

Heywood, V. H. & Watson, R. T. (1995). *Global Biodiversity Assessment*, 1120 pp. Cambridge: Cambridge University Press.

Kahlbusch, T. A. J. & Crutzen, P. J. (1995). Towards a global estimate of black carbon residues of vegetation fires representing a sink of atmospheric CO_2 and a source of O_2. *Global Biogeochemical Cycles*, **9**, 491–501.

Kelliher, F. M., Hollinger, D. Y., Schulze, E. -D., Vygodskaya, N. N., Byers, J. N., Hunt, J. E. & McSeveny, T. M. (1995). Evaporation from an eastern Siberian larch forest. In *Proceedings of the International Symposium on Forest Hydrology*, ed. T. Ohta, pp. 123–30. Tokyo, Japan October 24–28, 1994.

Kolchugina, T. P. & Vinson, T. S. (1995). Role of Russian forests in the global carbon balance. *Ambio*, **24**, 258–64.

Körner, Ch. (1994). Scaling from species to vegetation: the usefulness of functional groups. *Ecological Studies*, **99**, 117–42.

Körner, Ch. (1995). Leaf diffusive conductance in the major vegetation types of the globe. *Ecological Studies*, **100**, 463–90.

Kurz, W. A., Apps, M. J., Webb, T. & MacNamee, P. (1992). The carbon budget of the Canadian Forest Sector. Phase 1: ENFOR Information report NOR-X-326. Forestry Canada Northwest Region, Edmonton, Alberta, 93 pp.

Larcher, W. (1994). *Ôkophysiologie der Pflanzen.* 5th Edition. Ulmer Verlag: Eugen.

Marschner, H. (1995). *Mineral Nutrition of Higher Plants*, 889 pp. London: Academic Press.

Melillo, J. M., McGuire, A. D., Kicklighter, D. W., Moore III, B., Vörösmarty, C .J. & Schloss, A. L. (1993). Global change and terrestrial net primary production. *Nature*, **363**, 234–40.

Melillo, J., Prentice, C., Farquhar, G. & Schulze, E-D. (1995). *Terrestrial Ecosystems: Responses to Global Environmental Change and Feedback to Climate.* IPPC, WG 2, Chapter 9, 445–82.

Pearcy, R. W., Schulze, E-D. & Zimmermann, R. (1989). Measurement of transpiration and leaf conductance. In *Plant Physiological Ecology, Field Methods and Instrumentation*, ed. R. W. Pearcy, J. R. Ehleringer, H. A. Mooney & P. W. Rundel, pp. 137–60. London, New York: Chapman and Hall.

Raschke, K. (1979). Movements of stomata. *Encyclopedia of Plant Physiology*, **7**, 383–441.

Schulze, E-D. (1982). Plant life forms as related to plant carbon, water and nutrient relations. In *Encyclopedia of Plant Physiology. Physiological Plant Ecology.* Vol. 12B, ed. O. L. Lange, P. S. Nobel, C. B. Osmond & H. Ziegler, pp. 615–76. Water relations and photosynthetic productivity, Berlin, Heidelberg,

Schulze, E-D. (1994). *Flux Control in Biological Systems*, 594 pp. San Diego: Academic Press.

Schulze, E-D. & Chapin, F. S. III. (1987). Plant

specialization to environments of different resource availability. *Ecological Studies*, **61**, 120–48.

Schulze, E-D., Kelliher, F. M., Körner, C., Lloyd, J. & Leuning, R. (1994). Relationships among maximum stomatal conductance, ecosystem surface conductance, carbon assimilation rate and plant nitrogen nutrition: A global ecology scaling exercise. *Annual Review of Ecological Systems*, **25**, 629–60.

Schulze, E-D., Schulze, W., Kelliher, F. M., Vygodskaya, N. N., Ziegler, W., Kobak, K. I., Hoch, H., Arneth, A., Kusnetsova, W. A., Sogachev, A., Issajev, A., Bauer, G. & Hollinger, D. Y. (1995). Above-ground biomass and nitrogen nutrition in a chronosequence of pristine Dahurian *Larix* stands in Eastern Siberia. *Canadian Journal of Forest Research*, **25**, 943–60.

Shvidenko, A. & Nilsson, S. (1994). What do we know about the Siberian forest. *Ambio*, **23**, 396–404.

Seiler, W. & Crutzen, P. G. (1980). Estimates of gross and net fluxes of carbon between the biosphere and the atmosphere from biomass burning. *Climate Change*, **2**, 207–47.

Stitt, M. & Schulze, E-D. (1994). Plant growth, storage, and resource allocation: from flux control in a metabolic chain to the whole plant level. In *Flux Control in Biological Systems. From Enzymes to Populations and Ecosystems*, ed. E-D. Schulze, pp. 57–118. San Diego: Academic Press.

Westoby, M. (1984). The self-thinning rule. *Advances in Ecological Research*, **14**, 167–225.

Wilson, M. F. & Henderson-Sellers, A. (1985). A global archive of land cover and soils data for use in general circulation climate models. *Journal of Climatology*, **5**, 119–43.

Wofsy, S. C., Goulden, M. L., Munger, J. W., Fan, S. M., Bakwin, P. S., Daube, B. C., Bassow, S. L. & Bazzaz, F. A. (1993). Net exchange of CO_2 in a mid-latitude forest. *Science*, **260**, 1314–17.

Woodward, F. I. & Smith, T. M. (1995). Predictions and measurements of the maximum photosynthetic rate, A_{max}, at the global scale. *Ecological Studies*, **100**, 491–510.

7 Interannual variation in the terrestrial carbon cycle: significance of Asian tropical forest conversion to imbalances in the global carbon budget

D. L. Skole, W. A. Salas and C. Silapathong

The contemporary global carbon cycle

Carbon is an essential element for life on this planet; approximately half of the mass of plant and animal life is composed of carbon. Carbon exists in several forms and is cycled between several biotic and abiotic reservoirs, including the oceans, terrestrial plant biota and atmosphere (Fig. 7.1). The largest pool of carbon is the lithosphere, but it is relatively inactive. On the other hand, the carbon content of the atmosphere has varied considerably over the past several thousand years, and this variation has been closely coupled to changes in global climate. Variations in atmospheric carbon dioxide are related to changing fluxes of carbon between the atmosphere and the biota, oceans and fossil fuels.

Since atmospheric carbon dioxide measurements were first systematically made in the late 1950s (Keeling, 1960; Keeling *et al.*, 1976), the concentration has steadily increased (Bacastow & Keeling, 1981; Keeling & Whorf, 1994). This increase in atmospheric carbon is attributed to two sources: land cover change, and combustion of fossil fuels (Houghton & Skole, 1990; Schimel *et al.*, 1995). Fossil fuels have been injected into the atmosphere at a steadily increasing rate since the beginning of the industrial revolution, around 1860 (Marland & Rotty, 1984; Keeling, 1994; Marland *et al.*, 1994). Carbon has been released from deforestation and other land cover changes over the last 300 years (Houghton & Skole, 1990; Houghton, 1994).

While the annual increase in carbon content of the atmosphere and the amount of carbon released from fossil fuel combustion are relatively well known, the exact magnitude of fluxes between the atmosphere and the biota, and between the atmosphere and oceans, is uncertain. These uncertainties confound our ability to construct a balanced global budget for carbon.

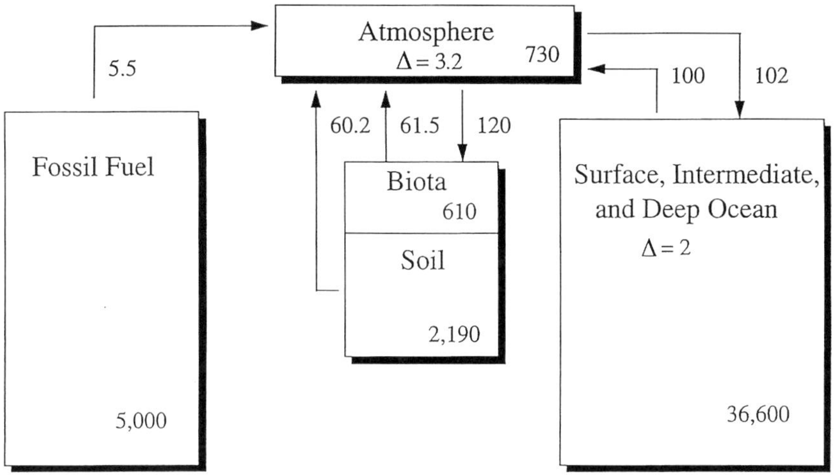

Figure 7.1 *The contemporary global carbon cycle derived from estimates provided in Schimel* et al. *(1995) and Houghton and Skole (1990). Units are 10^{15} g C year^{-1}.*

Imbalances in the global carbon cycle

Reviews of the global carbon cycle can be found in Houghton and Skole (1990) and Schimel *et al.* (1995), and here we shall cover only the more salient issues of current uncertainties in the terrestrial carbon cycle. Atmospheric carbon dioxide has increased from 315 ppmv to 356 ppmv in 1992 (Keeling *et al.*, 1989; Thoning *et al.*, 1994; Keeling & Whorf, 1994). With the development of techniques to analyze the concentration of CO_2 in bubbles of air trapped in deep ice cores, it is also possible to estimate the historical atmospheric concentration of CO_2. Since 1750, there has been a steady increase from 279 ppmv (Neftel *et al.*, 1985). Human activities are largely responsible for the observed increases. Fossil fuel burning is currently the most important source of carbon dioxide. Yet, evidence from ice cores suggests that atmospheric concentrations of CO_2 began to rise prior to 1850, even before there were significant inputs from fossil fuel combustion. Approximately 30% of the current net flux of carbon dioxide is biogenic and related to anthropogenic land cover changes. Moreover, the long-term historical release of carbon dioxide from human alterations of land cover has been approximately equal to the release from fossil fuel burning (Houghton & Skole, 1990).

Net additions to the atmosphere from fossil fuel combustion and biomass oxidation are partly removed by oceanic uptake. However, oceanic uptake from the atmosphere is lower than additions to it, and the atmospheric concentration has risen as a result of this imbalance. Measurements of the atmospheric increase when compared to estimates of the net flux of biogenic carbon, fossil fuel emissions and ocean uptake, cannot be accommodated in a balanced carbon budget.

Houghton *et al.* (1983) and Houghton and Skole (1990) have described this situation using a simple equation of the form:

$$\Delta A_t = B + F - O \quad (1)$$

where ΔA_t is the net increase in the atmospheric carbon at time t, B is the net flux from land clearing and biomass burning, F is the net flux from fossil fuel burning, and O is the net oceanic flux. The units are 10^{15} g C $year^{-1}$. When average annual values for individual terms from the period 1980 to 1990 are substituted into the equation (Schimel *et al.*, 1995), it is possible to describe the current situation:

$$3.2 \neq 1.7 + 5.5 - 2.0 \quad (2)$$

There are important uncertainties associated with each of these terms. Nonetheless, the equation is not balanced. Explanations for this imbalance in the budget center on two possibilities. First, the individual terms of the equation may be in error, or improperly modeled – particularly B and O. The emission from fossil fuels is well known and certain, as are the atmospheric measurements. However, the oceanic and biogenic terms are not. Second, the equation may not be complete, and another term – the so-called missing sink – may be important, and hence:

$$3.2 = 1.7 + 5.5 - 2.0 - b \quad (3)$$

The recent report of the Intergovernmental Panel on Climate Change (Houghton *et al.*, 1995) suggests several possible mechanisms to explain b, including increased forest C storage due to climate-induced changes in ecosystem metabolism, fertilization by carbon dioxide or nitrogen (see Schulze & Heimann, this volume), or unaccounted land use changes such as forest regrowth. At any rate, the value of this term is rather large (*c*. 2) and probably cannot be attributed to errors in any one of the other terms alone.

Recent analyses by Tans *et al.* (1990) and others have suggested a large, unaccounted net sink in undisturbed, predominantly northern, forest ecosystems. The analysis of Tans *et al.* (1990) utilized a constrained approach, coupling measurements of the north–south gradient of atmospheric [CO_2], measurements of oceanic uptake and pCO_2, estimates of land surface fluxes from fossil fuels and deforestation, and a general circulation model. The approach provides a geographically specific assessment of sources and sinks, which is highly dependent on the magnitude of the uncertain biogenic tropical source term.

In another constrained analysis, Houghton and Skole (1990) compared a model estimate of the historic net flux of carbon from land cover change with estimates of the total biospheric net flux derived from a deconvolution of ice core data and ocean models (Siegenthaler & Oeschgher, 1987). This analysis, reproduced in Fig. 7.2, shows general agreement between the estimates prior to about 1930, after which the net flux from land cover change increases, while the total biotic flux declines. This suggests that either the land use or deconvolution analyses are wrong, or that the difference reflects a net sink in undisturbed ecosystems. If a sink in undisturbed ecosystems exists, and is somewhat evenly distributed geographically, the amount per unit area would be too small to measure in the field. Thus, a combination of constrained analyses of this kind and ecosystem models would be required to further elucidate the nature of the sink.

An important consideration in analyses such as Tans *et al.* (1990) and Houghton and Skole (1990) is knowing the biogenic flux from land cover change, particularly the tropical deforestation component. Improved definition of the imbalance in the global carbon budget will come with better definition of the individual terms of the equation above, with the biogenic term being particularly important. Moreover, the use of constrained analyses, in which additional terms to the equation are inferred as residuals from known terms, will require increased geographical and temporal specificity and confidence in the estimate of biogenic fluxes.

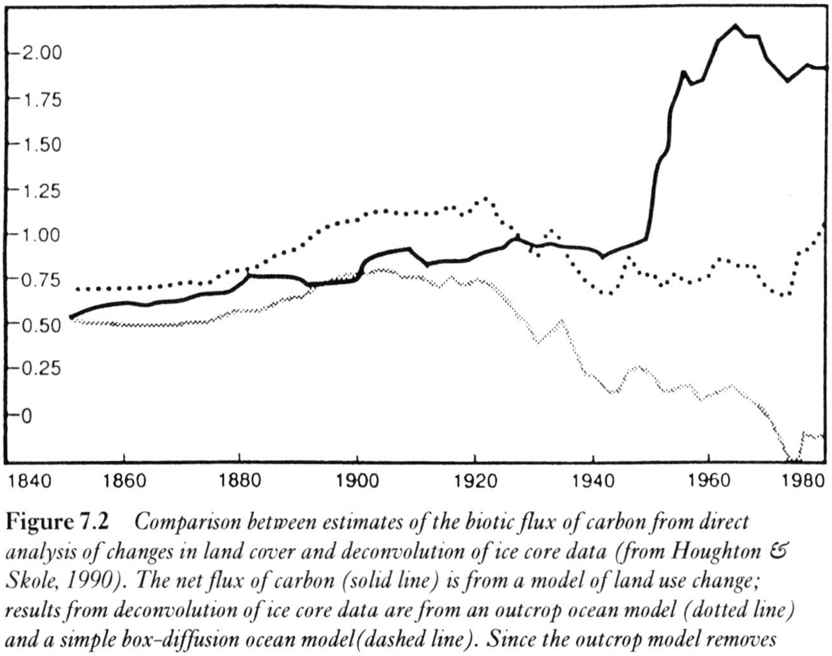

Figure 7.2 *Comparison between estimates of the biotic flux of carbon from direct analysis of changes in land cover and deconvolution of ice core data (from Houghton & Skole, 1990). The net flux of carbon (solid line) is from a model of land use change; results from deconvolution of ice core data are from an outcrop ocean model (dotted line) and a simple box-diffusion ocean model(dashed line). Since the outcrop model removes atmospheric carbon more efficiently than a diffusion model, a larger biotic flux can be accommodated.*

Net fluxes of terrestrial carbon in Asia

Of the combined global net flux of carbon from fossil fuel combustion and land use change in 1990, approximately 30% is from temperate and tropical Asia (Table 7.1). The global net flux from fossil fuels is *c*. 6.1×10^{15} g C year^{-1} (Marland *et al.*, 1994), while the global net flux from land cover conversion is *c*. 1.7×10^{15} g C year^{-1} (Houghton & Hackler, 1995). The net flux from fossil fuel combustion in Asia is *c*. 1.5×10^{15} g C year^{-1}, and the net flux from biogenic sources is 0.8×10^{15} g C year^{-1}. The fossil fuel emissions from Asia represent 25% of the global total fossil fuel inputs to the atmosphere, while the biogenic emissions from Asia represent almost half of the global total inputs from land cover change. The contribution of China's fossil fuel emissions is approximately 43% of the regional total fossil fuel emissions, 10% of the regional biogenic flux, and 32% of the combined fossil fuel and biogenic emissions. When we exclude the Pacific developed countries, which include Japan, Australia and New Zealand, the fossil fuel emissions from Asia are only 18% of the global total fossil fuel inputs, of which China's contribution is almost 60%.

Currently,this is an important region in terms of its biogenic contribution to the global totals, probably more so than in terms of its contribution from fossil fuels. The Southeast Asia subregion is significant in its contribution of biogenic emissions, contributing almost half of the global total. However, it should be noted that growth rates of fossil fuel emissions since 1970 have been significant in this area of the world, and future emissions of fossil fuel carbon will likely be very high in rapidly developing China and the so-called "Asian Tiger" countries of Southeast Asia. The growth rates in fossil fuel emissions in the last 20 years for China and Southeast Asia have exceeded the overall global rate of increase several-fold (213% and 240% vs. 49%).

Table 7.1 *Terrestrial carbon emissions from fossil fuel consumption and land cover conversion in Asia, 1990*

Region or country	Fossil fuel flux	% of global fossil fuel (combined)	% increase since 1970	Biotic flux	% of global biotic (combined)	% increase since 1970
Global	6.10	NA	49	1.70	NA	28
Asia	1.53	25 (20)	156	0.80	47 (10)	52
China	0.66	11 (9)	213	0.08	5 (1)	-42
South and SE Asia	0.50	8 (7)	240	0.70	41 (9)	139
Pacific developed	0.38	6 (5)	52	0.01	1 (<1)	-88

Units: 10^{15} g C year^{-1}.

Biogenic fluxes of carbon from land cover change in Asian and other equitorial tropical forests

Human activities have transformed the natural landscape over the past 300 years in a variety of ways and to varying degrees (Turner *et al.*, 1990; Turner & Meyer, 1994). Quantitatively, the most important type of change in Asia and elsewhere has been the direct conversion of natural land cover to agriculture (Richards, 1984, 1990). Large areas of Asia and other parts of the world are undergoing extensive land transformation as a result of population growth and economic development (see Mackenzie *et al.*, this volume; Malingreau & Zhuang, this volume). Such changes in land cover influence the net exchange of carbon between the land and atmosphere (Detwiler & Hall, 1988; Houghton *et al.*, 1983, 1985, 1987; Moore *et al.*, 1981; Bolin, 1983).

Although large releases of carbon are historically attributed to land cover conversion in the temperate zone, tropical deforestation appears to be the major biogenic source of carbon today. The long-term cumulative release of carbon from land cover change between 1850 and 1990 is *c.*122×10^{15} g C (Schimel *et al.*, 1995; Houghton, 1994). The current annual net flux of carbon from land cover conversion in the tropics is between 0.4 and 2.5×10^{15} g C $year^{-1}$ (Houghton *et al.*, 1985, 1987). Ten countries comprise two-thirds of the net release: Brazil, Columbia, Indonesia, Ivory Coast, Laos, Malaysia, Mexico, Peru, Thailand, and Zaire.

The long-term historical net flux of carbon due to land conversion has been documented by analyses based on direct estimates of land use data and ecosystem response characteristics (Houghton & Skole, 1990). These studies suggest that, while some land transformations increase the amount of carbon stored on land, the overall historical trend has been one of increasing forest conversion to agriculture. This has reduced the amount of carbon stored on land and has resulted in a net flux of carbon to the atmosphere. However, as we point out below, most of these analyses have focused on the long-term, inter-decadal historical trend, rather than on fine temporal analysis of particular years.

The amount of carbon held in terrestrial ecosystems is changed as a result of (1) direct human effects of land use, such as deforestation (see Malingreau & Zhuang, this volume), and (2) indirect human effects on ecosystems, such as increased concentrations of atmospheric CO_2 or climate change, which influence changes in ecosystem metabolism (Schulze & Heimann, this volume). At the present time, the net flux of carbon between the biota and atmosphere due to land use change (deforestation) can be calculated with more confidence than changes in carbon storage associated with large-scale changes in ecosystem metabolism. This is because of the large difference in biomass between forests and the agricultural systems which replace them, and because deforestation and reforestation can be readily documented and quantified, particularly if remote sensing data are utilized.

The current net flux of carbon between the biota and the atmosphere due to land use change is uncertain. Three factors contribute to the range of uncertainty: (1) rates of deforestation, (2) the fate of deforested land (i.e. the amount of secondary forest regrowth and re-clearing), and (3) the stock of biomass and soil organic matter and their response to disturbance, including anthropogenic reductions of carbon stocks within forests due to thinning or degradation. Models developed with improved geographic and temporal data on deforestation rates, better parameterization of the dynamic nature of deforestation and reforestation, and improved data on above- and below-ground carbon response characteristics are needed.

In this chapter we wish to distinguish between changes in carbon storage due to forest *growth* and changes in carbon storage due to forest *regrowth* and secondary succession (see Schulze & Heimann, this volume). The former has been postulated as a component of the so-called missing sink, resulting from changes in ecosystem metabolism in undisturbed forests. The latter is related to land use changes in areas where significant abandonment of agriculture is taking place. One of the central tenents of this paper is that this latter term has been poorly documented, and that, in fact, this could be a significant factor in the global carbon budget. Moreover, asynchronies in clearings in one place and time with abandonment and regrowth in other places and times could result in dampened net biogenic sources during certain periods of time. Most estimates of land cover changes are *net* changes; there have been few studies which document both deforestation and regrowth simultaneously, and virtually no studies document both processes on an annual basis. Should the relative magnitude of these two processes change significantly from one year to the next, there exists a potential for large temporary variations in the magnitude of the biogenic source term, and possibly a significantly dampened source term, even while long-term trends suggest otherwise.

Interannual variations in terrestrial carbon fluxes

Tans *et al.* (1990) used a global circulation model to redistribute atmospheric CO_2 based on surface fluxes of CO_2 from fossil fuel consumption, tropical deforestation, and oceanic uptake. Their analysis predicted that the atmospheric $[CO_2]$ should exhibit a strong north–south gradient of 5.7 to 7.3 ppm. This estimate was significantly higher that the observed meridianal gradient of 3 ppm based on the NOAA flask measurements (Tans *et al.*, 1990). They discussed that the source of this discrepancy could either be their atmospheric transport model or an unaccounted extra-tropical terrestrial sink in the Northern Hemisphere. Analysis of ^{85}Kr and CFC tracer data indicates that inaccuracies in their transport model could account for differences of only 10% and not the two-fold differences measured. Therefore, they assumed that the "missing sink" was a terrestrial sink of 2.0 to 3.4 × 10^{15} g C $year^{-1}$ in the Northern Hemisphere temperate zone.

Siegenthaler and Sarmiento (1993) updated the Tans *et al.* (1990) oceanic uptake estimates by accounting for differences in ocean skin temperature and bulk water temperature, horizontal transport of carbon from the terrestrial biosphere to the ocean via rivers, and the north–south transport of carbon monoxide and its subsequent uptake in the southern oceans. They conclude that there is a terrestrial sink of about 1.8 ± 1.3 × 10^{15} g C $year^{-1}$, or that the tropical source term of 1.6 ± 1.0 × 10^{15} g C $year^{-1}$ is too high, or a combination of both.

Quay *et al.* (1992) estimated the net flux of CO_2 in the ocean and biosphere using isotopic measurements of $d^{13}C$ (essentially the ratio of $^{13}C/^{12}C$) in the atmosphere and dissolved inorganic carbon from ocean surface waters and atmospheric [CO_2]. $d^{13}C$ varies depending on the source of the carbon and changes depending on the uptake mechanism due to isotopic fractionation, therefore changes in this ratio provide an indication of the uptake mechanisms (Quay *et al.*, 1992; Ciais *et al.*, 1995*a*). Because terrestrial photosynthesis discriminates against the heavier isotope (^{13}C), atmospheric uptake can be partitioned between the ocean and the biosphere (Quay *et al.*, 1992; Sarmiento, 1993; Houghton *et al.*, 1995; Ciais *et al.*, 1995*a*). Quay *et al.* (1992) used this technique to estimate that, from 1970 to 1990, the average annual oceanic uptake was 2.1 × 10^{15} g C $year^{-1}$, while the average annual terrestrial uptake was negligible (0.1 × 10^{15} g C $year^{-1}$).

More recently Ciais *et al.* (1995*a*,*b*) applied a hybrid approach using a two-dimensional atmospheric transport model (cf. Tans *et al.*, 1990) with [CO_2] and $d^{13}C$ measurements from NOAA's Climate Monitoring and Diagnostics Laboratory (CMDL) global flask network (cf. Quay *et al.*, 1992) to estimate CO_2 partitioning as a function of latitude and time from 1990 to 1993. These results suggest that the northern temperate zone was a large terrestrial sink (3.5 × 10^{15} g C $year^{-1}$) in 1992 and 1993, whereas the northern tropics (from Equator to 30° N) are a large source (2.0 ± 1.3 × 10^{15} g C $year^{-1}$). This estimate for the northern tropics is considerably higher than the estimate of 0.9 × 10^{15} g C $year^{-1}$ by Houghton *et al.* (1987). Ciais *et al.* (1995*a*) attributed this difference to either a problem with their model for intrahemispheric transport between the tropical and temperate latitudes or that previous deforestation estimates, particularly for tropical regions such as Southeast Asia, have been underestimated.

These results also suggest that the southern tropics were a small terrestrial sink in 1992 and 1993. This is surprising because this latitudinal band contains the Brazilian Amazon and Indonesia. Estimates of CO_2 emission from tropical land use rank these two countries as the top two in terms of total emissions due to deforestation (Houghton *et al.*, 1987). The weak terrestrial sink in the southern tropical zone suggests that either the forests have a positive net ecosystem production (due to fertilization of undisturbed forest and/or secondary growth formation) or previous estimates of deforestation have been overestimated (Ciais *et al.*, 1995*a*).

In another approach, Keeling and Shertz (1992) and Keeling *et al.* (1996) relied on new methods for measuring atmospheric $[O_2]/[N_2]$. Photosynthesis converts CO_2 and water into organic carbon and molecular oxygen (O_2). Combustion and respiration pathways work in the opposite direction, with the oxidation of organic compounds producing CO_2. Thus, atmospheric $[O_2]$ is a function of photosynthesis, respiration and combustion and varies inversely with atmospheric $[CO_2]$. However, changes in atmospheric $[O_2]$ are difficult to measure because these changes are very small compared with the overall atmospheric concentrations. Nonetheless, changes in $[O_2]$ can still be estimated by changes in the ratio of O_2/N_2 because $[N_2]$ is relatively stable. Since O_2 is not chemically active in the ocean owing to its low solubility, changes in atmospheric $[O_2]$ can then be related to fluxes of CO_2 from the biosphere. The net terrestrial exchange of CO_2 can be used to constrain the oceanic CO_2 uptake based on atmospheric measures of CO_2.

Keeling and Shertz (1992) used this technique to estimate an annual oceanic uptake of $3.0 \pm 2.0 \times 10^{15}$ g C year^{-1} and a small annual net terrestrial sink between 1989 and 1991. Keeling *et al.* (1996) used O_2/N_2 data from 1991–1994 to measure hemispheric gradients in CO_2 uptake. They estimated that, overall, on an annual basis from 1991 to 1994 the terrestrial biosphere was a net sink of 2.0 ($\pm$ 0.9) × 10^{15} g C year^{-1}, tropical land biota was not a strong source or sink, and the oceans were a net sink of 1.7 ($\pm$ 0.9) × 10^{15} g C year^{-1}. Using longer-term trends in O_2/N_2 they estimated that the measured slow down in the rate of increase in atmospheric CO_2 was due to either increase CO_2 uptake in the temperate land biota or the removal of a net tropical source of CO_2 due to a decrease of deforestation rates or increase in the area in forest regrowth.

These analyses suggest that the tropics are neither a strong source nor a strong sink (Keeling *et al.*, 1996). Ciais *et al.* (1995*a*,*b*) estimate that the northern tropics are a source, but the southern tropics are a small sink. They propose two explanations for the sink-to-neutral magnitude: (1) an increase in tropical Net Ecosystem Production to offset deforestation sources, and (2) evidence for a combination of reduced rates of deforestation and increased regrowth of previously cleared land "which underscore the uptake of carbon by recovering forests on land abandoned by shifting agriculture and also suggest reduced clearing rates based on satellite analyses".

Interdecadal and interannual land cover change in Asia

The current global rate of deforestation is relatively unknown. Problems documenting it center on: (1) the lack of geographically referenced data at high spatial resolution derived from a single consistent method for the whole of the tropics –

preferably from satellite data, (2) a paucity of quantitative information on the relative rates of new deforestation and abandonment to regrowth, and (3) detailed, annual time-series for specific periods of time from which a complete global carbon budget could be compiled in conjunction with annual atmospheric measurements.

The most recent estimates of the annual rate of tropical deforestation over the last decade by the FAO is 15.4×10^6 ha year^{-1} (FAO, 1993). The regional contributions of this global annual rate are 4.1×10^6 ha year^{-1} for Africa, 3.9×10^6 ha year^{-1} for Asia, and 7.4×10^6 ha year^{-1} for Latin America & Caribbean. Within the Latin America & Caribbean region FAO estimated that the rate for tropical South America was 6.2×10^6 ha year^{-1}, with Brazil alone at 3.7×10^6 ha year^{-1}.

Satellite remote sensing is the only means for resolving discrepancies or quantifying temporal and spatial variations in deforestation rates (see Malingreau & Zhuang, this volume). There are no reasons why satellite-based techniques cannot be applied to a large area like the Southeast Asian subcontinent, or to a significant portion of the tropical forest belt, to resolve the aforementioned controversies and uncertainty, and thereby provide vastly improved forcing functions for global carbon models.

In the two sections which follow, we report new results of analysis of satellite data for Southeast Asia, an important region in the northern tropics, and the Brazilian Amazon, an important region of the southern tropics.

Methods for satellite data analysis

To analyze what is happening in Asia, Landsat MSS and TM data were acquired for the entire Indochina region, including Thailand, Laos, Burma (Myanmar), Vietnam, and Kampuchea. These data were acquired to compile a complete mosaic of satellite data in the mid-1970s, the mid-1980s and the early 1990s in order to compute synoptic average annual deforestation rates over the last 20 years. At selected test sites we compiled multi-year data for analysis of interannual dynamics.

Digital data are selected from the national or foreign archives and from new acquisitions. These data are pre-processed to standard analysis products. This includes conversion to a common format (CCT-P) from other formats (e.g. CCT-X, CCT-A, and others) and the addition of map coordinates. Each scene is registered to a Universal Transverse Mercator (UTM) map projection using satellite navigation data (includes only system corrections), and the digital image imbedded with coordinate information as UTM labeled tic marks. A 1 : 250 000 scale photographic three-band color composite print and a 1 : 1 000 000 color transparency are also produced and distributed.

These pre-processed data are analyzed for deforestation rates. We use a hybrid methodology based on digital data, combined with visual interpretation. Landsat scenes are first georeferenced to "north-up" by using the four corner points provided in the data descriptor record on the data tape as ground control points (GCPs) to define the linear transformation. The rectification is performed using nearest neighbor resampling technique to minimize convolution of the data.

An iterative, self-organizing (ISO), unsupervised clustering technique is applied to the data with a knowledge-based classifier to produce five output classes for our analyses in these regions: forest, non-forest, water, cloud, and cloud shadow. For the ISO analysis in Southeast Asia, we first create a new band which is the ratio of band 4 (NIR) to band 2 (RED). We run the unsupervised classification using only band 2 (RED) and this new band. We apply the ISO algorithm to assign 45 output clusters. This algorithm is based on initially using the statistics of the raw data to arbitrarily determine the first clusters. The statistical mean spectral values of clusters are shifted to locate the actual spectral statistics for each cluster. This step is repeated until 95% of the pixel are not reassigned to new clusters during the iterations that shift the cluster means. Assignment of individual pixels to a cluster is accomplished using a minimum spectral distance decision rule such that all pixels are assigned to one of the 45 output clusters. These output clusters are then recoded into a new classification containing up to five classes. This is done by overlaying the 45 clusters on the original imagery to aggregate the 45 spectral clusters into the final five output classes.

This classification product from is plotted at 1 : 250 000 scale on vellum and then overlaid on a 1 : 250 000 scale photo product of the Landsat scene. Polygons that are misclassified are identified and recoded in the digital geographic information system. The classification product is replotted and checked for further editing. These editing steps are repeated until the classification is completed.

To produce databases covering the entire region of analysis, individual scenes are edge-matched and merged to form a seamless mosaic. The mosaic represents a complete inventory of the forest area as it existed in 1973, 1985, and 1992 for all of continental Southeast Asia. Average annual rates of deforestation are computed by difference between dates. It is not possible to compile complete regional coverage each year in order to estimate the actual annual deforestation and regrowth dynamics. In this case we establish several test sites in each region. In the test sites we co-register annual or semiannual Landsat data and compute change detection analysis for deforested areas and areas in regrowth. For this chapter we only report on the regional analysis of continental Southeast Asia for two dates, 1973 and 1985, and our multi-date time-series analysis from 1989 to 1994 for the test site in Chiang Mai, Thailand (Fig. 7.3).

Results from Asia

Results for Southeast Asia are shown in Fig. 7.4 and Fig. 7.5, which show the mapped extent of forest areas at two dates: *c.* 1973 and 1985. The numerical

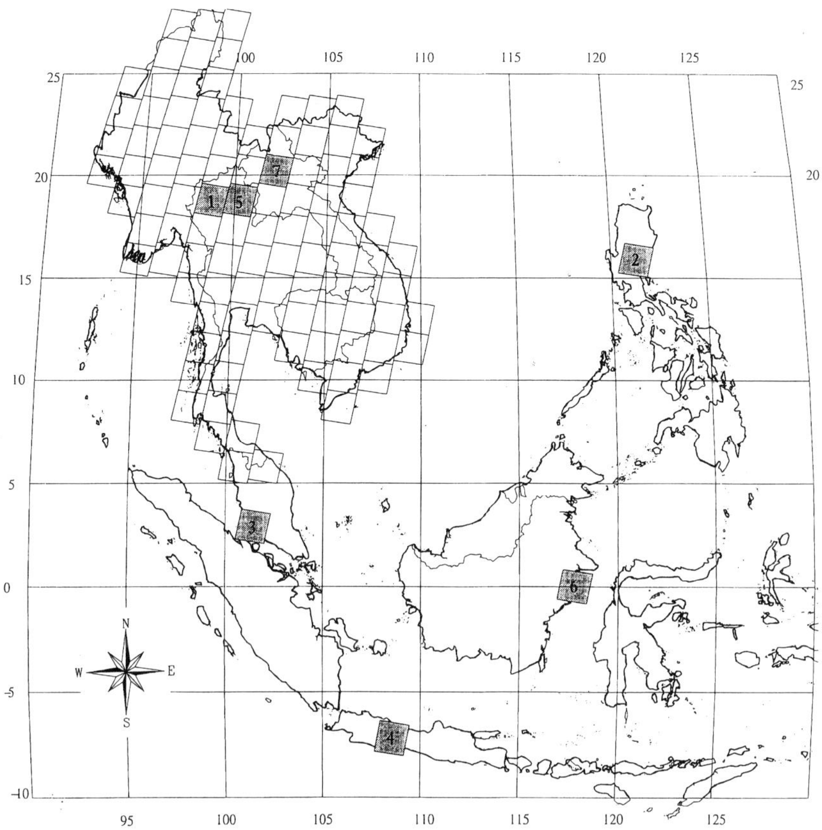

Figure 7.3 *Study area in Southeast Asia, also showing: (1) Chiang Mai, Thailand; (2) Megat Watershed, Phillippines; (3) Klang Langat Watershed, Malaysia; (4) Citarum Watershed, Indonesia; (5) Nan, Thailand; (6) East Kalimantan, Indonesia; and (7) Central Laos. The area covered by Landsat data for continental Southeast Asia reported in this chapter is shown by gridded tiles. Each grid cell represents the location of a Landsat scene. Several test sites are shown, but only the Chiang Mai results are reported here.*

summary of results is shown in Table 7.2. This is the first analysis we know of which reports forest areas and rates of forest loss for the entire region using a single consistent method from satellite data which completely cover the region. It is also the first analysis for the entire region of which we are aware, which has been compiled and mapped in a digital geographic information system at a spatial resolution of less than 100 meters.

The rate of deforestation from the mid-1970s to the mid-1980s was *c.* 1.4×10^6 ha year^{-1}, which is a rate close to that reported for the much larger Brazilian Amazon during the same period (Skole & Tucker, 1993). The highest rate of deforestation occurred in Thailand at 0.48×10^6 ha year^{-1}, representing approximately 35% of the total for the region. It also has the highest percentage loss rate in the region; approximately 26% of the forest area in 1973 had been lost by 1985.

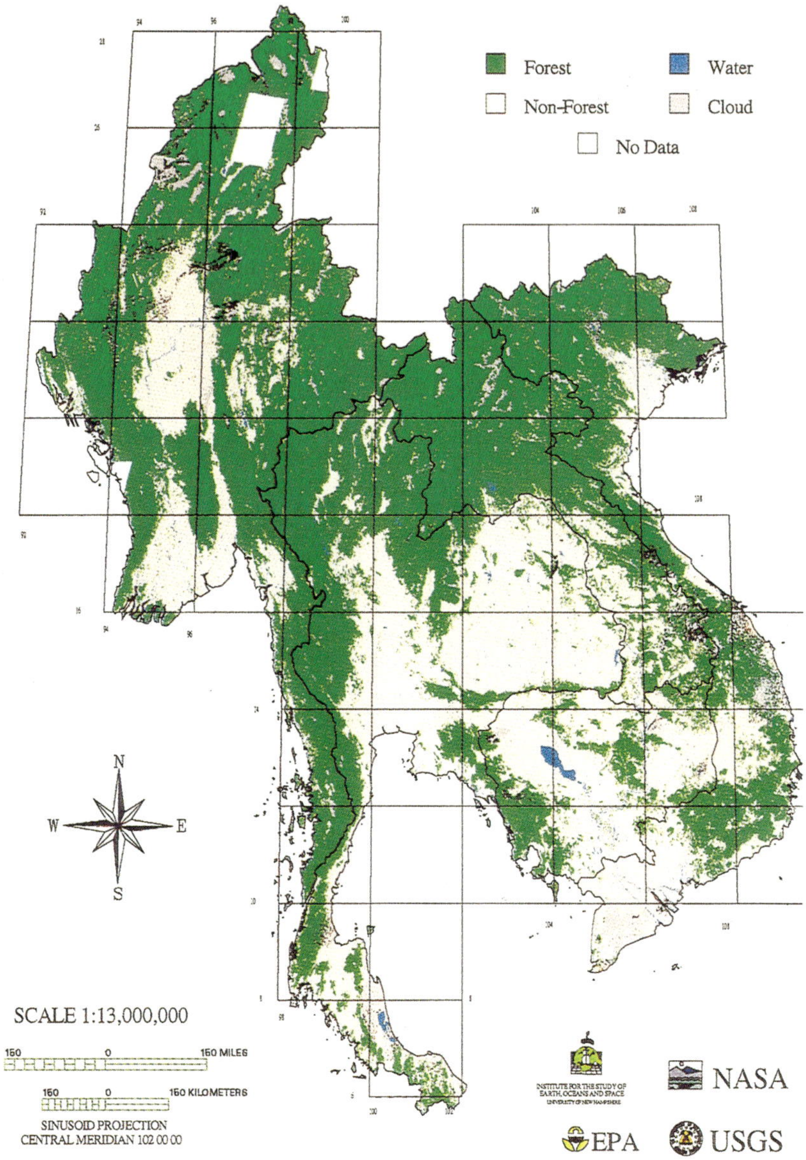

Figure 7.4 *Total forest area in continental Southeast Asia, as derived from analysis of Landsat satellite data for 1973. The areas shaded in green are intact forests.*

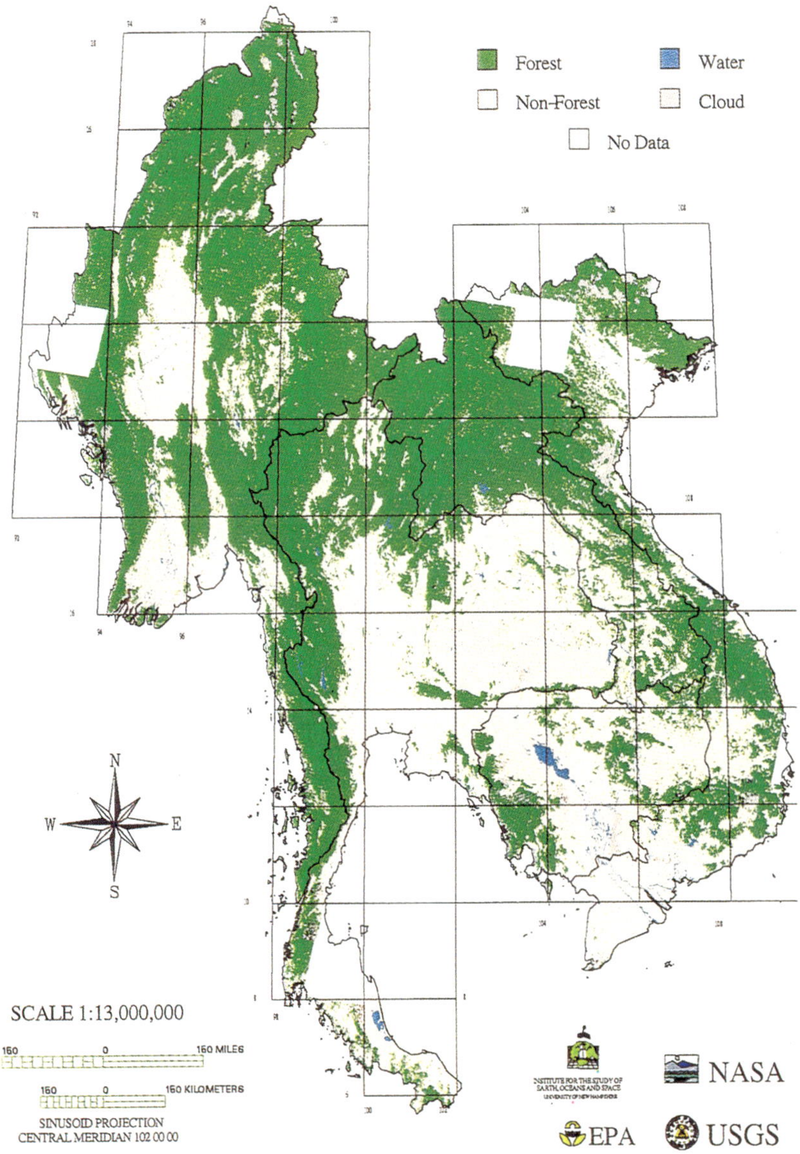

Figure 7.5 *Total forest area in continental Southeast Asia, as derived from analysis of Landsat satellite data for 1985. The areas shaded in green are intact forests.*

Table 7.2 *Summary of results from satellite data analysis of forest areas and their rate of change in continental Southeast Asia*

Country in 1973	Forest area in 1985	Forest area change	Forest area	% Change rate per year	Deforestation
Cambodia	5.25	3.98	1.27	24	0.11
Laos	18.28	16.52	1.76	10	0.15
Thailand	22.56	16.74	5.81	26	0.49
Vietnam	19.92	16.15	3.77	19	0.31
Myanmar	48.71	44.82	3.88	8	0.32
Total	114.70	98.21	16.49	14	1.37

Units: 10^6 ha or 10^6 ha year^{-1}.

Most of the deforestation in Thailand took place in the northern mountainous region around the province of Chiang Mai and in eastern Thailand along the Cambodian border where shifting cultivation, expansion of commercial agriculture and logging are very active. The total area of Northern Thailand was169 000 km^2 with 84 000 km^2 (or 49% of the total area) in forest in 1985. By 1993 the Royal Forestry Department of Thailand (RFD) reported that there were just over 75 000 km^2 (or 44%) of forest remaining. This loss of forest corresponds to an annual clearing rate of over 0.11×10^6 ha year^{-1}. During this period from 1985 through 1993, all of Thailand had an average annual deforestation rate of just under 0.2×10^6 ha year^{-1}, hence almost 60% of the deforestation occurs in an area less than one-third the size of the entire country.

These region-wide results are interesting in that they are higher than that reported by the World Resources Institute (*c.* 0.95×10^6 ha year^{-1}) for the period 1981–1985 (WRI, 1992) and FAO (1.3×10^6 ha year^{-1}) for a later period, 1981 to 1990 (FAO, 1993). This latter comparison implies that the current estimates from FAO, upon which some carbon flux studies have been based (Houghton, 1994), are either too low or the rate has dropped since 1985. Our analysis of Thailand is higher than that reported by the World Resources Institute (*c.* 0.25×10^6 ha year^{-1}) for closed forests for a later period, 1985–1988 and approximately comparable to those reported by FAO for the later period 1981–1990. At the time of writing we do not have results from satellite analysis for the entire region for the period 1985–1993, but this analysis will be forthcoming soon. Nonetheless, the indications from these data for the entire region and for Thailand suggest that rates of deforestation peaked in the late 1980s and have declined since.

Shifting cultivation is a main land use agent of forest loss in this part of Southeast Asia. United Nations' estimates suggest that shifting cultivation supports about 28 million people in insular and continental Southeast Asia. During the early 1980s FAO estimated that shifting cultivation contributed to 49% of all the deforestation in Southeast Asia (Lanly, 1983). FAO estimated that 1×10^6 ha year^{-1} were in shifting cultivation in Thailand alone at this time. Kammerer (1989) estimated that 50 000 members of hill tribes in Northern Thailand were practising integral shifting cultivation. In the mid-1980s it was estimated that 0.5×10^6 ha year^{-1} of land was under some sort of shifting cultivation cycle in Northern Thailand.

The test site in Chiang Mai (Fig. 7.3) is in the center of this heavy deforestation pressure. Our results of multi-temporal analysis of co-registered annual data from 1986 to 1994 reveals that the area in cultivation increased by 38% from 4.5×10^3 ha in 1989 to 6.2×10^3 ha in 1994 (Fig. 7.6). From 1989 until 1992 the cultivation area increased annually until it reached 7.4×10^3 ha in 1992. The next two years saw a drop in the total cultivated area. During the 6-year period, 1.1×10^3 ha of land remained under cultivation, hence the percentage of area in permanent cultivation or in long cultivation cycles (greater than 6 years) varied from 14% to 23% of the total area cultivated at any given time. During this time period 66.2×10^3 ha of the over 81×10^3 ha of study area was undisturbed, hence 81% of the area was in long fallow, secondary forest or undisturbed primary forest (not likely since many of the hill tribes prefer to clear primary forest over secondary forest since it leads to higher crop yields).

Fig. 7.7 shows the annual extent of new clearings and abandonment areas. Over the time period of observation, the area of new clearings in any given year varied from 2.3×10^3 ha in 1990 to a peak of 3.7×10^3 ha in 1992 with a decline to 2.4×10^3 ha cleared for 1994. In 1990 and 1991, over 50% of the area in cultivation was on areas that had just been cleared. This percentage decreased every year until 1994 when only 39% of the cultivated area was cleared that year. The extent of the

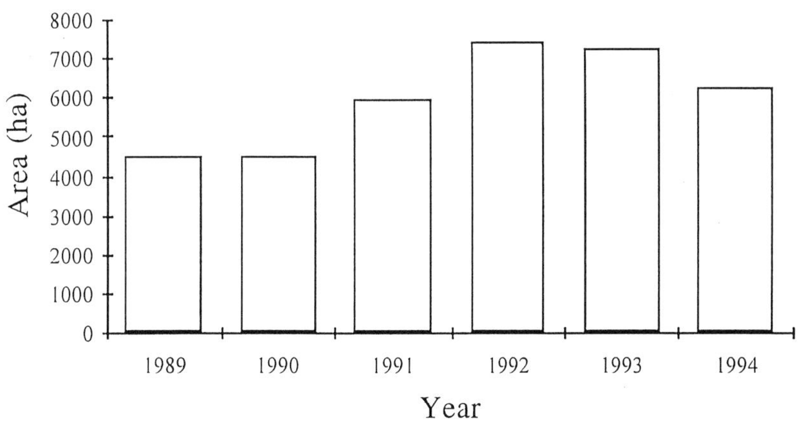

Figure 7.6 *Change in area in active agriculture over time as derived from multi-temporal analysis of Landsat data for the test site in Chiang Mai, northern Thailand.*

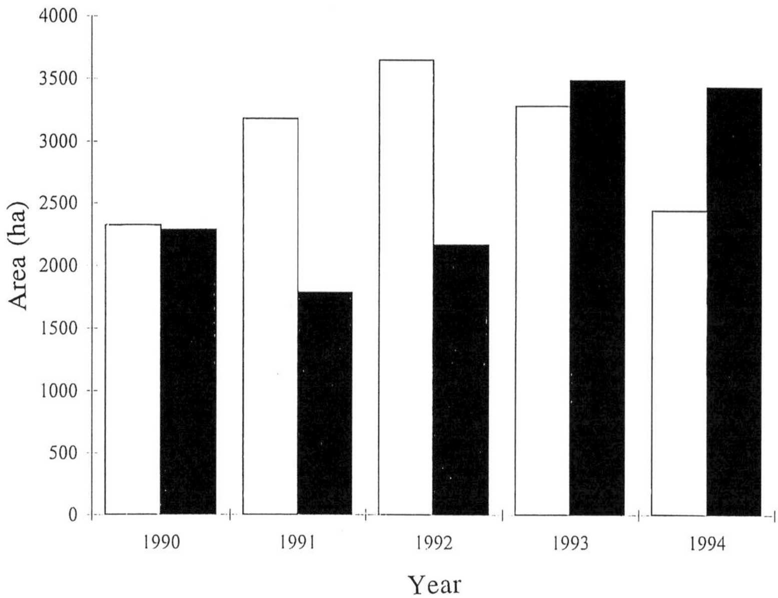

Figure 7.7 *Annual rates of formation of new clearings (empty column) and new areas of abandoned fallow or secondary growth (solid column) in Chiang Mai, northern Thailand.*

abandonment area also fluctuated significantly over this time period. The smallest abandonment occurred from 1990 to 1991 with only 1.8×10^3 ha abandoned, while the extent of abandonment almost doubled to 3.5×10^3 ha and 3.4×10^3 ha in 1993 and 1994, respectively.

These results of the multi-temporal analysis suggest that deforestation rates are highly variable from one year to the next and that the process of deforestation involves considerable abandonment to fallow, or secondary forests. The rate of abandonment is also highly variable and sometimes not synchronous with clearing (i.e. one could have years with large amounts of fallow formation and very little new deforestation). Carbon-accumulating regrowth is a predominant feature of the Asian landscape. The overall trend in the region appears to be characterized as having rates of deforestation peaking in the late 1980s and declining somewhat in the 1990s. As this "pulse" of deforestation passes through the system as a general *interdecadal* trend which would suggest increased carbon emissions, the high variation in rates of clearing and abandonment (which are not necessarily occurring simultaneously) could create *interannual* periods in which the annual carbon flux is characteristically different than the *decadal* trend.

More work is under way at this time to test these ideas in other locations throughout the tropics on an interannual basis and to develop region-wide analyses for more recent time periods, such as 1985–1993 and 1993–1996.

Comparison with results from other tropical regions

Regional analysis in the Brazilian Amazon

Analyses of satellite data for the Brazilian Amazon indicate that, from 1978 through 1988 the average annual rate of deforestation was between 1.6×10^6– 2.2×10^6 ha year^{-1} (INPE, 1992; Skole & Tucker, 1993). Our own work-in-progress for the same period suggests that the annual rate of deforestation in the Brazilian Amazon peaked during the period 1978–1986 at around 1.8×10^6 ha year^{-1} and then declined to *c*. 1.4×10^6 ha year^{-1} from 1986 to 1993.

The Brazilian Space Agency, INPE, reports annual deforestation rates from 1988–1991 for the Amazon of *c*. 1.9, 1.4, and 1.1×10^6 ha year^{-1} , respectively, for this 3-year time period (D. S. Alves, personal communication). The 1990–1991 rate was 67% less than the average rate for this 3-year census. Apparently, the deforestation rate from 1990–1991 was half of the average annual rate from 1978 through 1986.

Other data from the Brazilian Amazon suggest large differences in deforestation rates and the existence of large areas of abandonment each year (Alves & Skole, 1996; Skole *et al.*, 1994). Approximately 30% of the deforested areas in the Amazon are in secondary growth. From a few site studies in which interannual time series data have been acquired, it has been shown that the abandonment rate can be as high as 82% of the deforestation rate (1991–1992 in Alves & Skole, 1996). At the same time, the deforestation rate varied annually by as much as 100% with an average over 6 years of 50%.

These results bolster those reported here from Southeast Asia and give strong indications that there are indeed significant influences of interannual changes in the deforestation rate on net fluxes of carbon, but without further analyses in more sites it is not possible to draw final conclusions.

The significance of secondary growth

Results reported here for Southeast Asia indicate extensive areas of forest fallow and secondary growth in Continental Southeast Asia exist. Similar results exist for the Legal Amazon. For example, 30% of the deforested areas in the Legal Amazon were in some form of secondary growth in 1986 and in 1992. This percentage represents areas over 77 000 km^2 and 98 000 km^2 in secondary growth in 1986 and 1992, respectively.

Satellite analysis by Alves and Skole (1996) indicate that, for a site in Rondonia, Brazil, secondary growth is a large (over 40% of the deforested areas) and a rapidly changing pool. Almost 60% of the areas in secondary growth in 1986 were re-cleared at least once by 1992, and over 55% of the deforested areas in 1986 were abandoned into secondary growth for some period of time by 1992 (Alves & Skole, 1996). This analysis and other similar satellite-based analyses (Steininger, 1996; Foody *et al.*, 1996; Brondizio *et al.*, 1994; Moran *et al.*, 1994; Lucas *et al.*, 1993; Mausel *et al.*, 1993) are useful for characterizing the dynamics of the secondary growth resulting from tropical land use, and support similar evidence reported here for Southeast Asia.

These studies of deforestation from satellite data show it is a highly dynamic process of clearing, abandonment and re-clearing, and the rates at which land is cleared or abandoned are related to the land use and management system the forest farmers employ. In some cases, deforestation and secondary succession exist in tandem as a tightly coupled system in which secondary growth is continually recycled back into farmland. In other cases, active land management maintains the land in agriculture, or the lack of active land management or population displacement results in long-term succession.

Thus land *use* change influences land *cover* change and is integrated with dynamics of ecosystem structure, function, and response. Within a focused research program, it will be important to frame such questions as: what human land use and land management strategies are employed in different situations and how do they control, or interact with, the dynamics of ecosystem response to disturbance? Because deforestation and abandonment have opposite effects on atmospheric carbon dioxide (cf. uptake vs. release), we can re-state the aforementioned general question in terms specific to the carbon cycle: over time, what land use strategies determine the abundance and spatial distribution of secondary growth, how do they determine the balance between clearing and regrowth rates, and how are these land use strategies in turn influenced by ecological conditions?

Implications of recent carbon cycle analyses for tropical deforestation rates

Analysis for specific recent years (cf. 1992–1994) reported by Keeling *et al.* (1996) and Ciais *et al.* (1995*a*,*b*) which suggest that the tropics as a whole, and the southern tropics in particular, are not large sources of biogenic carbon, are somewhat inconsistent with interdecadal results from previous terrestrial carbon models based on land use change by Houghton and Skole (1990), Houghton (1994) and others, and recent observations from satellite analyses of deforestation rates for Southeast Asia and the Amazon (Skole & Tucker, 1993), which suggest that tropical deforestation has contributed to a significant net source of carbon.

Based on results presented above, an explanation worth exploring further is that pervious estimates of the tropical source term were based on decadal mean values, rather than explicitly on any specific year or set of years when atmospheric measurements are made (cf. 1992–1994). This would be especially relevant if tropical land use change is characterized by large areas and rates of abandonment with significant inter-annual variation as appears to be the case in both Southeast Asia and the Amazon. Significant interannual differences in the long-term trend in deforestation rates and the relative rates of deforestation and abandonment could result in periods of asynchrony where clearing rates, which relate to carbon emissions, and abandonment rates, which relate to carbon uptake, are very different. Such dynamics would result in wide interannual variations in the relative contribution to the net flux from clearing vs. regrowth, and, potentially, a dampening in the decadal mean values for particular years.

In this scenario, two important conditions must be met: (1) large *interannual* departures in rates of deforestation from the *decadal mean* values reported by previous analyses, and (2) large areas of forest regrowth as a result of land abandonment. In years in which deforestation rates are greatly less than the decadal mean, and with the large abundance of areas of secondary succession and forest regrowth, due to high rates of abandonment in previous years, we would expect to see a significant dampening of the carbon source computed from the decadal mean data alone.

These issues then beg a very interesting suite of questions concerning the factors which *cause* deforestation rates to vary from year to year, and what factors determine or control the balance between clearing and abandonment. The resolution of such questions depends completely on: (1) making fine temporal resolution measurements (annually) of deforestation rates over large areas using satellite remote sensing, (2) making fine spatial and temporal scale analyses of the dynamics of land use and cover change in order to document if regrowth is a significant factor, and (3) developing an improved quantitative and diagnostic capability to determine the factors (i.e. the so-called human dimensions) which control relative rates of clearing and regrowth from one year to the next.

An approach to IGBP and IHDP research in tropical Asia

To better quantify and understand the significance of interannual changes in the rate of deforestation and abandonment, the IGBP could undertake a research effort in Southeast Asia which is described below. It would be composed of four central activities, which could be carried out through collaboration of several Program Elements of the IGBP and IHDP, particularly LUCC, GCTE, IGAC and GAIM. Additional participation of DIS and START would be essential framework elements.

Activity 1

Determine the decadal mean rate of forest loss for the region from 1986 through 1996 using a two-date complete inventory of satellite data.

This work would be similar to that described in this paper but updated to include the addition of one more date *c.* 1996. This could readily be achieved through collaboration with satellite observation programs already in place, such as the NASA Landsat Pathfinder project whose results are reported in this paper, Japanese efforts now under way, and several European Union projects such as the TREES project. This effort provides an additional inventory year in 1996 to determine the average rate of change from 1986 to 1996.

Activity 2

Determine if *annual* rates of deforestation have been significantly different from the decadal mean rate over large areas and over the region as whole.

This activity would be developed to define the interannual rate of deforestation for the region as a whole, using stratified sampling based on the decadal complete inventories, using the results described above as a starting point. A stratified sampling scheme in Southeast Asia could be easily developed and then implemented to estimate the annual change in forest area from 1986 to 1994, the time frame relevant to atmospheric CO_2 and O_2 measurements. This analysis would focus only on forest to non-forest, and provide a single region-wide estimate which can be used to interpolate annually between the average rates of forest loss estimated by Activity 1.

Activity 3

Develop case studies aimed at the following question: is secondary growth important and does land cover change dynamically on an annual basis; what are the land use transition probabilities?

In five to eight selected case study sites, the interprogram-element project would analyze annual time series of satellite data from 1986 to the present to calculate transition sequences and their associated probabilities for deforestation and secondary growth. Each case study would be an area as large as a single Landsat scene and could be located in the following countries of the region: Thailand, Laos, Vietnam, Indonesia, Malaysia and the Philippines, since there exists already an ongoing precursor project being developed under START and LUCC. It would be desirable to evaluate the addition of two other sites in Myanmar and Kampuchea. Each case study would focus on estimating the transition matrix for the following classes: forest, secondary growth, paddy, upland crops, tree crops, and urban/developed.

To carry out the case studies, it will be necessary to solicit the collaboration of a well-established team of regional scientists through the START programs and in collaboration with established groups, which have been working since 1993 as part of an IGBP/IHDP Land Use and Cover Change (LUCC) project in Southeast Asia.

Activity 4

Develop diagnostic models of the deforestation process to better understand and quantify the differential controls on rates of deforestation and abandonment.

The causes of deforestation have often been attributed to such things as population growth, road building, and a host of other similar monotonically varying variables (Panayotou, 1991). Although population and infrastructure are important long-term determinants of general trends over decades, interannual variability is not tightly linked to such parameters. Econometric variables such as variations in price structures and local market demands are likely to be better predictors of deforestation and abandonment rates over short time cycles. This activity would couple the satellite-based deforestation dataset derived from the aforementioned activities to socio-economic data to build diagnostic models of the deforestation process.

Such models can provide: (1) an improved understanding of the factors which control variations in deforestation rates annually, (2) an improved understanding of the factors which determine the balance between creation of new farmland through deforestation and the creation of new farmland by clearing secondary fallow forests, (3) an improved understanding of the factors which control the balance between deforestation (the bringing of land into production) and abandonment (the taking of land out of production), and (4) a method for spatial and temporal extrapolation of the results from the activities described above.

These models should be developed at two levels. The first would be focused on each of the case study sites with an emphasis on establishing the critical village-level variables, dynamics and proximate determinants. The second level of model analysis would be at the scale of countries or the region. The case studies will be used to calibrate the regional model and provide a basis and location for validation of regional results.

References

Alves, D. S. & Skole, D. L. (1996). Characterizing land cover dynamics using multi-temporal imagery. *International Journal of Remote Sensing*, 17(4), 835–9.

Bacastow, R. B. & Keeling, C. D. (1981). Atmospheric carbon dioxide concentration and the observed airborne fraction. In *Carbon Cycle Modeling*, ed. B. Bolin. SCOPE 16. New York: John Wiley.

Bolin, B. (1983). The carbon cycle. In *The Major Biogeochemical Cycles and Their Interactions*, ed. B. Bolin & R.B. Cook. SCOPE 21. New York: John Wiley.

Brondizio, E. S., Moran, E. F., Mausel, P. & Wu, Y. (1994). Land use change in the Amazon estuary: patterns of Cabocolo settlement and landscape management. *Human Ecology*, **22**, 249–78.

Ciais, P., Tans, P. P., White, J. W. C., Trolier, M., Francey, R. J., Berry, J. A., Randall, D. A., Sellers, P. J., Collatz, J. & Schimel, D. S. (1995*a*). Partitioning of ocean and land uptake of CO_2 as inferred by ^{13}C measurements from the NOAA Climate Monitoring and Diagnostics Laboratory Global Air Sampling Network. *Journal of Geophysical Research*, **100(D3)**, 5051–70.

Ciais, P., Tans, P. P., Trolier, M., White, J. W. C. & Francey, R. J. (1995*b*). A large northern hemisphere terrestrial CO_2 sink indicated by the $^{13}C/^{12}C$ ratio of atmospheric CO_2. *Science*, **269**, 1098–102.

Detwiler, R. P. & Hall, C. A. S. (1988). Tropical forests and global carbon budget. *Science*, **239**, 42–7.

FAO, (1993). *Forest Resources Assessment 1990 – Tropical Countries*. Rome: United Nations Food and Agriculture Organization.

Foody, G. M., Palubinskas, G., Lucas, R. M., Curran, P. J. & Honzak, M. (1996). Identifying terrestrial carbon sinks: classification of successional stages in regenerating tropical forest from Landsat TM data. *Remote Sensing of the Environment*, **55**, 205–16.

Houghton, R. A. (1994). Emissions of carbon from land use change. In *The Carbon Cycle*. ed. T. M. L Wigley & D. Schimel. New York: Cambridge University Press.

Houghton, R. A. & Hackler, J. L. (1995). *Continental Scale Estimates of the Biotic Carbon Flux from Land Cover Change: 1850 to 1980*. ORNL/CDIAC-79, NDP-050, 144 pp. Oak Ridge, TN: Oak Ridge National Laboratory.

Houghton, R. A. & Skole, D. L. (1990). Carbon. In *The Earth as Transformed by Human Action*, ed. B. L. Turner II, W. C. Clark, R. W. Kates, J. F. Richards, J. T. Mathews, W. B. Meyer. New York: Cambridge University Press.

Houghton, R. A., Hobbie, J. E., Melillo, J. M., Moore, B., Peterson, B. J., Shaver, G. R. & Woodwell, G. M. (1983). Changes in the carbon content of terrestrial biota and soils between 1860 and 1980: a net release of CO_2 to the atmosphere. *Ecological Monographs*, **53**, 235–62.

Houghton, R. A., Boone, R. D., Melillo, J. M., Myers, N., Palm, C. A., Skole, D. L., Moore, B. & Woodwell, G. M. (1985). Net flux of carbon from tropical forests in 1980. *Nature*, **316(6029)**, 617–20.

Houghton, R. A., Boone, R. D., Fruci, J. R., Hobbie, J. E., Melillo, J. M., Palm, C. A., Peterson, B. J., Shaver, G. R., Woodwell, G. M., Moore, B., Skole, D. L. & Myers, N. (1987). The flux of carbon from terrestrial ecosystems to the atmosphere in 1980 due to changes in land use: geographic distribution of the global flux. *Tellus*, **39(B)**, 122–39.

Houghton, J. T., Meira Fihlo, L. G., Bruce, J., Lee, H., Chandler, B. A., Haites, E., Harris, N. & Maskell, K. (eds.). (1995). *Climate Change 1994. Radiativeorcing of Climate Change and an Evaluation of the IPCC IS92 Emission Scenarios*. Cambridge: Cambridge University Press.

INPE. (1992). *Deforestation in Brazilian Amazonia*. Sao Paulo, Brazil: Instituto Nacional de Pesquisas Espaciais.

Kammerer, C. A. (1989). Of labels and laws: Thailand's resettlement and repatriation policies. *Cultural Survival Quarterly*, **12(4)**, 7–12.

Keeling, C. D. (1960). The concentration and isotopic abundance of carbon dioxide in the atmosphere. *Tellus*, **12**, 200–3.

Keeling, C. D. (1994). Global historical CO_2 emissions. pp. 501–504. In *Trends '93: A Compendium of Data on Global Change*, ed. T. A. Boden, D. P. Kaiser, R. J. Sepanski, & F. W. Stoss, ORNL/CDIAC-65. Oak Ridge, TN: Carbon Dioxide Information Analysis Center, Oak Ridge National Laboratory.

Keeling, R. F. & Shertz, S. R. (1992). Seasonal and interannual variations in atmospheric oxygen and implications for the global carbon cycle. *Nature*, **358**, 723–27.

Keeling, C. D. & Whorf, T. P. (1994). Atmospheric CO_2 records from sites in the SIO air sampling network. In *Trends '93: A Compendium of Data on Global Change*, ed. T. A. Boden, D. P. Kaiser, R. J. Sepanski & F. W. Stoss, pp. 16–26. ORNL/CDIAC-65. Oak Ridge, TN: Carbon Dioxide Information Analysis Center, Oak Ridge National Laboratory.

Keeling, C. D., Bacastow, R. B., Bainbridge, A. E., Ekdahl, C. A., Guenther, P. R., Waterman, L. S. & Chin, J. F. S. (1976). Atmospheric carbon dioxide variations at Mauna Loa Observatory, Hawaii. *Tellus*, **28(6)**, 538–51.

Keeling, C. D., Bacastow, R. B., Carter, A. F., Piper, S. C., Whorf, T. P., Heimannn, M., Mook, W. G. & Roeloffzen, H. (1989). A three-dimensional model of atmoshperic CO_2 transport based on observed winds: 1. Analysis of observational data. In *Aspects of Climate Varianbility in the Pacific and the Westerns Americas*, ed. D. H. Peterson, pp. 165–236. *Geophysical Monograph*, **55**, AGU, Washington, USA.

Keeling, R. F., Piper, S. C. & Heimann, M. (1996). Global and hemispheric CO_2 sinks deduced from changes in atmospheric O_2 concentration. *Nature*, **382**, 218–21.

Lanly, J. P. (1983). Assessment of the forest resources of the tropics. *FAO Forestry Abstracts: Review Article*, **44**, 287–317.

Lucas, R. M., Honzak, M., Foody, G. M., Curran, P. J., Corves, C. (1993). Characterizing tropical secondary forests using multitemporal Landsat sensor imagery. *International Journal of Remote Sensing*, **14**, 3061–7.

Marland, G. & Rotty, R. M. (1984). Carbon dioxide emissions from fossil fuels: a procedure for estimation of results for 1950–82. *Tellus*, **36(B)**, 232–61.

Marland, G., Andres, R. J. & Boden, T. A. (1994). Global, regional, and national CO_2 emissions. pp. 505-584. In *Trends '93: A Compendium of Data on Global Change*. ed. T. A. Boden, D. P. Kaiser, R. J. Sepanski & F. W. Stoss, ORNL/CDIAC-65. Oak Ridge, TN: Carbon Dioxide Information Analysis Center, Oak Ridge National Laboratory.

Mausel, P., Wu, Y., Li, Y., Moran, E. F. & Brondizio, E. S. (1993). Spectral identification of successional stages following deforestation in the Amazon. *Geocarto International*, **4**, 61–71.

Moore, B., Boone, R. D., Hobbie, J. E., Houghton, R. A., Melillo, J. M., Peterson, B. J., Shaver, G. R., Vorosmarty, C. J. & Woodwell, G. M. (1981). A simple model for analysis for the role of terrestrial ecosystems in the global carbon budget. In *Carbon Cycle Modelling*, ed. B. Bolin, pp. 365 - 86. Chichester: John Wiley.

Moran, E. F., Brondizio, E., Mausel, P., Wu, Y. (1994). Integrating Amazonain vegetation, land-use, and satellite data. *Bioscience*, **44**, 329–38.

Neftel, A., Moor, E., Oeschger, H. & Stauffer, B. (1985). Evidence from polar ice cores for the increase in atmospheric CO_2 in the past two centuries, *Nature*, **315**, 45–7.

Panayotou, T. (1991). Population Change and Land Use in Developing Countries: The Case of Thailand. HIID Working Paper. Cambridge, Massachusetts.

Quay, P. D., Tilbrook, B., Wong, C. S. (1992). Oceanic uptake of fossil fuel CO_2: carbon-13 evidence. *Science*, **256**, 74–9.

Richards, J. F. (1984). Global patterns of land conversion. *Environment*, **26(9)**, 6–38.

Richards, J. F. (1990). Land transformation. In *The Earth as transformed by human action: Global and regional changes in the biosphere over the past 300 years*, ed. B. L. Turner III, W. C. Clark, R. W. Kates, J. F. Richards, J. T. Mathews & W. B. Meyer, pp. 163–79. Cambridge: Cambridge University Press.

Sarmiento, J. (1993). Atmospheric CO_2 stalled. *Nature*, **365**, 697–8.

Schimel, D., Enting, I. G., Heimann, M., Wigley, T. M. L., Raynaud, D., Alves, D. & Seigenthaler, U. (1995). CO_2 and the carbon cycle. In *Climate Change 1994. Radiative Forcing of Climate Change and an Evaluation of the IPCC IS92 Emission Scenarios*, ed. J. T. Houghton, L. G. Meira Fihlo, J. Bruce, H. Lee, B. A. Chandler, E. Haites, N. Harris & K. Maskell. Cambridge: Cambridge University Press.

Siegenthaler, U. & Oeschger, H. (1987). Biospheric CO_2 emissions during the past 200 years reconstructed by deconvolution of ice core data. *Tellus*, **39(B)**, 140–54.

Siegenthaler, U. & Sarmiento, J. L. (1993). Atmospheric carbon dioxide and the ocean. *Nature*, **365**, 119–25.

Skole, D. S. & Tucker, C. J. (1993). Tropical deforestation and habitat fragmentation in the Amazon: Satellite data from 1978 to 1988. *Science*, **260**, 1905–10.

Skole, D. L., Chomentowski, W. H., Salas, W. A. & Nobre, A. D. (1994). Physical and human dimensions of deforestation in Amazonia. *BioScience*, **44(5)**, 314–22.

Steininger, M. K. (1996). Tropical secondary forest growth in the Amazon: age, area and change estimation with Thematic Mapper data. *International Journal of Remote Sensing*, **17**, 9–17.

Tans, P. P., Fung, I. Y. & Takahashi, T. (1990). Latitudinal distribution of sources and sinks of atmospheric carbon dioxide. *Science*, **247**, 1431–8.

Thoning, K. W., Tans, P. P. & Waterman, L. S. (1994). Atmospheric CO2 records from sites in the NOAA/CMDL continuous monitoring network. In *Trends '93: A Compendium of Data on Global Change*, ed. T. A. Boden, D. P. Kaiser, R. J. Sepanski & F. W. Stoss, pp. 28–40. ORNL/CDIAC-65. Oak Ridge, TN: Carbon Dioxide Information Analysis Center, Oak Ridge National Laboratory.

Turner, B. L. III & Meyer, W. B. (1994). Global land-use and land-cover change: an overview. In *Changes in Land Use and Land Cover: A Global Perspective*, ed. W. B. Meyer & B. L. Turner III, pp. 3–11. Cambridge: Cambridge University Press.

Turner, B. L. III, Clark, W. C., Kates, R. W., Richards, J. F., Mathews, J. T. & Meyer, W. B. (eds.). (1990). The Earth as transformed by human action: global and regional changes in the biosphere over the past 300 years. Cambridge: Cambridge University Press.

WRI. (1992). *World Resources 1992-1993*. A Report by the World Resources Institute in collabroation with the United Nations Environment Programme and the United Nations Development Programme, 385 pp. New York: Oxford University Press.

Methane emissions from rice fields

H. U. Neue and J. Boonjawat

Introduction

Methane (CH_4) is an important greenhouse gas. The present atmospheric CH_4 concentration is more than double its pre-industrial level of about 0.8 ppmv. The current burden of CH_4 in the atmosphere is approximately 4700 Tg (1 Tg = 1 million tonnes), and the global annual emission is about 500 Tg with an apparent net flux of 40 Tg year^{-1} (Cicerone & Oremland, 1988). Isotopic measurements of atmospheric CH_4 show that 70–80% is of biogenic origin (Wahlen *et al.*, 1989). Flooded rice fields are a significant source of atmospheric CH_4. Estimates of annual CH_4 emission from rice fields to the atmosphere range from less than 20 to 100 million tonnes with best estimates of 50 ± 25 Tg. CH_4 emission from rice fields to the atmosphere is the net result of opposing bacterial processes, production in anaerobic microenvironments, and consumption and oxidation in aerobic microenvironments, both of which can be found side by side in flooded rice soils. Flooding a rice field cuts off the soils oxygen supply causing anaerobic fermentation of organic matter and concomitant CH_4 formation. CH_4 transport to the atmosphere takes place through ebullition of gas bubbles from the soil and through the gas-conduit system of rice plants that enable wetland plants to supply atmospheric oxygen to the submerged roots. Because rice was originally grown on soils that were naturally flooded, respective CH_4 emission can be classified as a natural source. However, bunding of rice fields, concomitant water harvesting, and artifical flooding have added to that natural source and are considered an anthropogenic CH_4 source. The expansion of irrigation and increased production of wetland rice has increased CH_4 fluxes from rice fields to the atmosphere. Given the required increase in rice production from 520 million tonnes today up to 1 billion tonnes in the next century, further increase of CH_4 emission will occur if current technologies are continued.

Comprehensive reviews on CH_4 emission from rice fields have been published recently (Neue & Roger, 1993; Neue, 1993; Neue & Sass, 1994) and various technical papers in national and international journals have contributed to the understanding of the relationships between CH_4 fluxes, agronomic practices and biogeochemical processes and factors. The objective of this chapter is to update the state of research on CH_4 emission from rice fields, focusing on Asia where more than 90% of the world's rice production is harvested and consumed.

Asian case studies

Various national and international studies on CH_4 emission from rice fields have been conducted in Asia, especially in China, India, Indonesia, Japan, the Philippines, and Thailand. Most of the studies have become an integral part of the Rice Activity within the International Global Atmospheric Chemistry (IGAC) project of the International Geosphere Biosphere Programme (IGBP) and have been linked to respective Global Change and Terrestrial Ecosystems (GCTE) efforts and Global Change System for Analysis, Research and Training (START) activities. The Rice Activity of IGAC enhanced communication and interaction, developed an overall research agenda, assisted in integration and interpretation of research results, and established standard methods and protocols for CH_4 flux measurements from rice fields (IGAC-CPO, 1994). The most comprehensive study is an interregional research program with participation of national institutions in China, India, Indonesia, the Philippines, and Thailand. The program is funded by the Global Environmental Facility (GEF) of the United Nation Development Programme (UNDP) and coordinated by the International Rice Research Institute (IRRI) and aims to determine the CH_4 source strength of major rice ecologies, its controlling factors and processes, and mitigation options. National and international studies on CH_4 emission from rice fields in Asia have resulted in a better understanding and quantifying of major processes controlling CH_4 fluxes and have identified feasible mitigation options. The variation of natural and cultural components affecting CH_4 fluxes makes it extremely difficult to accurately calculate the regional and global source strength of atmospheric CH_4 from rice fields at present, but uncertainties have been defined more clearly.

Methane emission pattern

CH_4 fluxes in rice fields showing distinct diel and seasonal variations which are expressions of changing balances among CH_4 formation, CH_4 oxidation, and CH_4 transport mechanisms. All of these processes are controlled by various interactive biotic and abiotic factors such as temperature, carbon source, soil redox potential, soil pH, soil microbes, and the properties of the rice plant itself.

Seasonal pattern of methane emission

Generally, one or two distinct seasonal maxima are observed in tropical lowland ricefields, such as in irrigated rice, that are kept flooded during the entire growth season. The first maxima may occur within 4 weeks after flooding and are governed by CH_4 production from soil organic matter and organic amendments (Neue *et al.*, 1994). This early maximum is pronounced in fields that received organic manure, whereas fields with low organic inputs generally show a gradual increase of emission rates in the early stage of the growing season. Ebullition is the dominant transport pathway after the initial flooding, during the following land preparation, and when rice plants are still small (Wassmann *et al.*, 1996). Rice fields are mostly flooded for land preparation 2–4 weeks before transplanting, and the initial CH_4 emission may be missed when monitoring CH_4 fluxes only during crop growth. After the first flush, CH_4 emission may decrease when degradable soil carbon becomes limiting. A second maxima is generally observed during heading and flowering about 9 weeks after transplanting. The stable low soil redox potential and neutral soil pH, the increased release of root-borne carbon sources, and the increasing capacity of plant-mediated CH_4 emission boost CH_4 formation and release (Neue & Sass, 1994; Kludze *et al.*, 1993, 1997). High CH_4 fluxes may continue or even increase until the end of the season. CH_4 emission may decline during the ripening and maturity stage if root porosity and root transport capacities decline because of root aging and degradation (Wang B., 1995). The balance between root exudation, root porosity, and root oxidation power controls the CH_4 emission rate during the reproductive and ripening stages (Wang, B. 1995). If fields are drained for harvest, CH_4 production declines. CH_4 emission increases sharply during the initial drying phase, releasing most of the soil entrapped CH_4, when macropores become aerated (Neue *et al.*, 1994). CH_4 emission ceases when previously flooded soils become fully aerated and the whole soil matrix shows positive redox potentials.

Temperature changes, flooding regimes and cultural practices modify this general seasonal pattern of CH_4 emission. CH_4 emission from irrigated rice fields are generally higher in the dry season than in the wet season because of higher solar radiation and resulting higher temperatures and plant biomass production (Wassmann *et al.*, 1994). Above emission pattern change and total CH_4 emission are reduced from irrigated rice if mid-season drainage is practised. A 6-to-10-day aeration period may reduce CH_4 emission up to 80% (Wassmann *et al.*, 1995). A great deal of rainfed rice, dependent on precipitation, is prone to drought periods. Deepwater rice, grown in areas that become flooded at water depths from 50 to 100 cm shows similar emission patterns to irrigated rice as long as rice plants do not get submerged for a significant period of time. No information is available on CH_4 emission rates and pattern from areas grown to floating rice where water levels may rise up to some meters. CH_4 emission of floating rice is probably low and erratic. Its root system consists mainly of nodal roots in the floodwater, and its crown roots at the production sites of CH_4 in the soil degrade after water levels rise above 1 m.

Diel pattern of methane emission

CH_4 fluxes from ricefields show distinct diel patterns (Schütz *et al.*, 1989; Wang M. *et al.*, 1990; Yagi & Minami, 1993; Wassmann *et al.*, 1994; Wang, B., 1995). Though initial ebullition of CH_4 tends to show diel variations, such patterns become distinct as the rice plants act increasingly as a carbon source and emission conduit during the season. Emission rates usually increase rapidly after sunrise, peak in the early afternoon, then decline rapidly and level off at night. Diel changes in light intensity (photosynthesis, respiration) and humidity (transpiration) do not affect CH_4 emission (Wang, B., 1995). Diel changes of soil solution temperature is the major controlling factor and the partial pressure of CH_4 in the soil is the major modifier (Wang, B., 1995). Mariko *et al.* (1991) showed that CH_4 concentration in the growth medium is correlated positively with CH_4 emission rates. The partial pressure of CH_4 in the soil is not only a function of CH_4 formation (carbon source) but also a function of CH_4 oxidation and CH_4 emission. Diel changes in root oxidation power, and of the redox potential, and pH in the rhizosphere may also affect the diel emission pattern. The effect of temperature on CH_4 emission is non-linear, differs between cultivars, and changes with growth stages (Wang, B. 1995). He found only small effects below 25 °C. Increasing temperature above 25 °C (up to 40 °C) dramatically enhanced CH_4 emission, especially at later growth stages. Fitting diel CH_4 emissions with the Fickian equation to a function of soil temperature and CH_4 concentrations in the soil solution results in correlation coefficients higher than 0.9 (Wang, B., 1995). Fitting diel emission rates with the Arrhenius equation only to a function of soil temperature results in delayed increases and decreases (Wang, B., 1995).

Soils

According to Neue (1989, 1991) wetland rice soils have free water at the surface for the major part of the rice growing season. Water may be retained by levelling, by establishing a dense pan below the plow layer to reduce percolation, or by building levees or dikes. Flooding drastically reduces the diffusion of atmospheric oxygen into the soil, and facultative and anaerobic organisms proliferate and sequentially reduce soil substrates. The redox potential, measured as the electric potential in volts, is a quantitative indicator measuring the tendencies of different oxidations and reductions (net flow of electrons) that occur. The higher the value of the redox potential, the greater the presence of strong oxidizing agents in a soil. The reduction of soils upon flooding proceeds stepwise in a thermodynamic sequence. Free oxygen (O_2) is reduced first, followed, in order, by nitrate (NO_3^-), manganic manganese (Mn^{4+})compounds, ferric iron (Fe^{3+})compounds, sulfate (SO_4^{2-}), and last, carbon dioxide (CO_2). Consequently, the redox potential falls sharply and the partial pressure of CO_2 increases considerably. As a result, the pH of all soils, irrespective if acid or alkaline when dry, stabilizes at about neutral, and availability of most nutrients increases. The anaerobic fermentation of organic matter produces and accumulates an array of substances, many of them transitory and not found in well-aerated soils. These substances include various gases, hydrocarbon, alcohols, carbonyls, volatile and nonvolatile fatty acids, phenolic acids, and volatile sulfur compounds. CO_2 and CH_4 are the end products of organic matter decomposition in flooded rice soils.

Soils can be classified into four groups according to the total amount of CH_4 produced during anaerobic incubation (Neue *et al.*, 1994; Wang, Z. *et al.*, 1993*b*). Each group has a distinct pattern of CH_4 formation and differs by magnitudes in the total amount of CH_4 produced. Sufficiently reduced differences in CH_4 production are more related to substrate (like hydrogen, acetate) limitation rather than to soil conditions. Soil types show a wide range of CH_4 production potential. Only a few correlations between stable soil properties, used in classification systems to discriminate soil types, and CH_4 production have been found. Total organic carbon is correlated with CH_4 production if other limiting factors are mitigated (Neue & Roger 1993; Wang, Z. *et al.*, 1993*b*). Sandy soils produce more CH_4 than do clayey soils with similar carbon contents (Neue & Roger, 1993). Sandy soils show lower entrapped CH_4 (Wang, Z. *et al.*, 1993*b*) because their pore size distribution enhances ebullition and diffusion (Neue *et al.*, 1994). The negative impact of clayey texture on CH_4 production may also be due to formation of organo-mineral complexes which protect organic compounds from being degraded.

Dynamic soil properties, typically not well correlated to stable soil properties, are important in controlling CH_4 formation. Initiation of CH_4 production requires a redox potential of less than -150 mV, low molecular fatty acids derived from easily degradable organic matter, a pH between 6 and 8, low salt, especially sulfate concentrations, and temperatures above 10 °C (Wang, Z. *et al.*, 1992, 1993*a*). The supply of molecular oxygen rather than the redox potential controls CH_4 formation (Fetzer & Conrad, 1993). To achieve a redox potential lower than -150 mV, NO_3^-, Mn^{4+}, Fe^{3+}, and SO_4^{2-} in the soil must be reduced (Patrick & Reddy, 1978). The rate of reduction correlates significantly with CH_4 production (Neue & Roger, 1993). Puddling (stirring) flooded rice soils at the end of field preparation disperses soil colloids and increases the water to soil ratio, resulting in very low soil bulk densities, which enhance reduction and ease transplanting. High soil bulk density, resulting from less intense field preparation or drying of flooded soils retards organic matter decomposition, increases the concentration and residue time of organic acids, and reduces the speed of Eh and pH changes as well as CH_4 formation. As long as a soil contains significant amounts of active oxidants, CH_4 formation is restricted, and more decomposable carbon may be mineralized to CO_2. In calcareous and alkaline soils, CH_4 production may occur within hours after flooding air-dried soils, whereas in acidic soils it may take weeks before CH_4

is formed. In highly acidic soils, CH_4 may not be formed at any time. Higher soil temperature speeds up CH_4 formation, but total amounts may not differ when comparing weeks or month rather than days. While high sulfate concentrations reduce CH_4 fluxes, high salinity levels based on NaCl may not result in lower emissions because such conditions reduce CH_4 oxidation much more than CH_4 production (Denier van der Gon & Neue, 1995*a*). Most likely, due to enlarging rooting systems and the resulting increased oxygen released into the rhizisphere, CH_4 oxidation increases as the season progresses, and may account for as much as 81% of the CH_4 produced (Sass & Fisher, 1995).

Neue *et al.* (1990) amalgamated conditions for high CH_4 production in wetland rice soils into six crucial parameters: water regime, Eh/pH buffer, carbon supply, temperature, texture and mineralogy, and salinity. Linking these criteria to soil types according to soil taxonomy, (SMSS, 1987) they suggested that all of the Oxisols, most of the Ultisols, and some of the Aridisols, Entisols, and Inceptisols are less favorable to CH_4 production when flooded. Rice soils that are prone to CH_4 production mainly belong to the soil orders of Entisols, Inceptisols, Alfisols, Vertisols, and Mollisols. Neue and Roger (1993) verified the impact of the above soil parameters, on methane emission by multiple correlation analysis and Bachelet and Neue (1993) quantified the ranking of soil types and included regional soil information when modeling CH_4 emission of countries in Asia. The use of soil classifications and soil maps to extrapolate CH_4 emissions to regional and global scales remains highly limited since most criteria used for soil classifications do not affect CH_4 production directly.

Organic matter

Submergence is often equated with retarded decomposition and accumulation of organic matter, but wetland rice soils in the tropics with high temperature in all seasons show rapid mineralization and weak humification (Bonneau, 1982; Neue & Scharpenseel, 1987; Neue, 1991), both of which favor CH_4 formation. Experiments in tropical rice fields revealed that submergence only slightly retarded decomposition of ^{14}C-labeled rice straw during the initial 3 months compared with respective upland soils (Neue & Scharpenseel, 1987; Neue, 1991). The rate of decomposition decreases with soil depth (Neue, 1985). Decomposition of the remaining, more resistant, metabolites and residues was also very rapid in all but the very acidic soils, irrespective of water regimes. Half-lives of about 2 years indicate that total fermentation in a cultivated rice soil, even when continuously flooded, is by no means exclusively anoxic. Anaerobic fermentation produces an array of organic substances, many of them transitory and not found in well-aerated soils. Ponnamperuma (1984) listed various gases, hydrocarbons, alcohols, carbonyls, volatile fatty acids, non-volatile fatty acids, phenolic acids, and volatile S compounds. Methanogens constitute the last step in the electron transfer chain generated by the anaerobic degradation of organic matter. Readily mineralizable organic soil matter is the main source for the fermentation products that finally drive the CH_4 formation in wetland rice soils. Generally, there is a strong correlation between water-soluble carbon and CH_4 production (Vermoesen *et al.*, 1991).

Organic amendments to flooded soils increase CH_4 production and emission (Schütz *et al.*, 1989; Yagi & Minami, 1990; Sass *et al.*, 1991; Cicerone *et al.*, 1992; Neue *et al.*, 1994) by providing carbon sources and apparently enhancing the reductive capacity of soils. Changes are more pronounced when organic substrates are added to soils low in organic matter. Based on the content of readily mineralizable carbon, rice straw or green manures produce more CH_4 per unit carbon, while humified substrates like compost produce less. Higher increases in CH_4 emission are observed when stubbles or straw are tilled into flooded rather than into aerated soil. Wang, Z. *et al.* (1992) showed that CH_4 production increased in proportion to the application rate of rice straw, indicating that CH_4 production may be carbon limited in most soils. The highest increases in emission rates was observed when applied straw was partially burned (Sass & Fisher, 1995). The application of rice straw increased CH_4 emission remarkably, irrespective of soil type, while the addition of compost enhanced CH_4 emission only slightly (Yagi & Minami, 1990; Minami & Neue, 1994). Depending on the amount of fresh organic matter added, CH_4 emission may increase only during few weeks following incorporation, and flooding or may be increased throughout the flooded season (Sass *et al.*, 1991). CH_4 emission was about four times higher throughout the growing season from plots fertilized with green manure than from urea-fertilized plots (Denier van der Gon & Neue, 1995*a*). Quality and quantity of organic materials influence CH_4 formation. Small differences in the carbon balance among fields and seasons can result in large differences of CH_4 emissions.

Water management and fertilization

Water control (irrigation and drainage) is one of the most important factors controlling rice production and CH_4 fluxes. In many rice fields, crops suffer from either too much or too little water because of rainfall patterns and topography. During the monsoonal rainy season in tropical Asia, ricelands are flooded naturally. In field operation, the average water requirement for an irrigated transplanted rice crop in Asia is 1240 mm (Yoshida, 1981), while upland crops may need less than half that amount. In the United States, the shortest flooding period of dry seeded rice consistent with high grain yields is from 75 to 105 days after seeding (Sass *et al.*, 1992). Moisture stress of 50 kPa (slightly above field capacity) may reduce grain yield to 20–25% of the yield of continually flooded treatments (De Datta, 1981). The rice plant is most sensitive to water deficit during the reproductive stage causing a high percentage of sterility (Yoshida, 1981). Water deficits during the vegetative stage may reduce plant height, tiller number, and leaf area that may also highly reduce yields if plants do not recover before flowering.

CH_4 emission rates vary markedly with water regimes. A single mid-season drainage may reduce seasonal emission rates by about 50% (Sass *et al.*, 1992; Kimura *et al.*, 1991). Multiple aeration for 2–3 days at 3, 6, and 9 weeks after initial flooding reduced CH_4 emission by 88% and did not reduce rice yields compared to the normal irrigation in the southern US, where dry-seeded rice is flooded 4–5 weeks after seedling emergence until 10–14 days before harvest (Sass *et al.*, 1992). Multiple aeration also required 2.7 times more water than the normal floodwater treatment. Fields flooded 4 weeks later than normal showed the same initial time course of CH_4 fluxes observed in other fields but at a later phase in plant development. Daily emissions increased to much higher values thereafter than in the other water regimes, leading to the highest total emission. Upon drying, most of the entrapped CH_4 is released to the atmosphere, when macropores become aerated (Neue *et al.*, 1994). Short aeration periods release entrapped CH_4 to the atmosphere without oxidizing the soil sufficiently to interrupt CH_4 production, resulting in significantly higher emissions rather than lower emissions. The release of entrapped CH_4 upon drying at the end of the flooded growing season may account for about 10% of the total seasonal emission (Byrnes *et al.*, 1995; Denier van der Gon *et al.*, 1996). Sass *et al.* (1992) may have overestimated the reduction due to mid-season and multiple aeration, missing the release of entrapped CH_4 at the onset of soil aeration because of low time resolution of measurements. Soil-entrapped CH_4 is released to the atmosphere as a result of physical disturbances of flooded soils by cultural practices such as puddling, harrowing, transplanting, weeding, fertilizer/pesticide application, and harvesting that may account for up to 20% of total emission. Increasing percolating rates of water may supply sufficient oxygen to the soil to raise the Eh and decrease CH_4 production and increase CH_4 oxidation. Percolating water also transports organic solutes and dissolved gases into the subsoil or groundwater, where leached CH_4 may be oxidized or be released to the atmosphere elsewhere (Kimura *et al.*, 1992). In soils with high CH_4 production, Inubushi *et al.* (1992) found that a percolation rate of about 4 mm per day reduced CH_4 emission up to 58%.

Mineral fertilizer (source, mode, and rate) influences CH_4 production and emission in various ways. Generally, fertilizer applications enhance plant and root growth which increases CH_4 emission presumably due to higher root exudates and biomass for CH_4 transfer. The addition of sulfate-containing fertilizer decreases CH_4 production. In the presence of exogenous Fe^{3+} or SO_4^{2-}, methanogenic bacteria are outcompeted for hydrogen (H_2) by bacteria reducing Fe^{3+} and SO_4^{2-} (Achtnich *et al.*, 1995). CH_4 emission is low in reduced soils if sulfate concentrations are high (Gambrell & Patrick, 1978; DeLaune *et al.*, 1983). Successful outcompetition of methanogens by sulfate reducers occurred in freshwater sediments at *in situ* concentrations as low as 60 mM (Lovely & Klug, 1983).

Gypsum ($CaSO_4$) applications (6.7 t ha^{-1}) increased concentrations of SO_4^{2-} in the soil solution up to about 10 mM and reduced CH_4 mission by 55–70% (Denier van der Gon & Neue, 1994). Fertilization with ammonium sulfate [$(NH_4)_2SO_4$] at rates of 50–200 kg ha^{-1}, as used in most field experiments, results in *in situ* concentrations of SO_4^{2-} at about the critical level. But, plant uptake and reduction to S^{2-} quickly deplete this pool except that SO_4^{2-} is recycled through oxidation, for example, in the rice rhizosphere. Because of the various ways sulfate-containing fertilizer can affect CH_4 emission from rice fields, it is easy to understand why contradicting impacts have been reported from field experiments.

Rice cultivars

Rice plants play an important role in the flux of CH_4. Up to 90% of the CH_4 released from ricefields to the atmosphere is emitted through the rice plant (Seiler *et al.*, 1984; Holzapfel-Pschorn *et al.*, 1986). Well-developed intracellular air spaces (aerenchyma) in leaf blades, leaf sheaths, culm, and roots provide a transport system for the conduction of CH_4 from the bulk soil into the atmosphere (Nouchi *et al.*, 1990; Kludze *et al.*, 1993; Schütz *et al.*, 1991). Openings of the aerenchyma to the atmosphere are found on leaves and nodes, but not on internodes of stems (Wang, B., 1995). Before shoot elongation, about 50% of the CH_4 is released from leaf blades. At later growth stages, major release sites are pores connecting medullar cavities and air in the junction of leaf sheath and culm. The number of tillers per m^2 were found positively correlated to CH_4 emission (Mariko *et al.*, 1991). However, this relationship was verified only for the vegetative growth stage (Wang, B., 1995). CH_4 concentrations in and emission from rice plants show a clear gradient. The highest concentrations are found in the arenchyma below the water level and highest emission occurs through openings immediately above the water level. The flux of gases in the aerenchyma depends on concentration gradients, and diffusion coefficients of roots and the internal structure of the aerenchyma. The diffusion coefficient of the transition between root and shoot aerenchyma seems to control the gas transport through the rice plant (Butterbach-Bahl, 1992). Increasing the floodwater depth decreases CH_4 emission only briefly as long as part of the shoot remains above the water. Cutting shoots above the floodwater does not affect CH_4 fluxes through rice plants within one or two days verifying that the closed chamber technique is a reliable method to measure CH_4 fluxes.

Schütz *et al.* (1989) suggested that a second and higher peak of CH_4 emission observed during the reproductive stage of rice might be due to the rice plant's organic root exudates and/or litter. The observed very close positive correlation between CH_4 production and live root biomass in wetlands (Wilson *et al.*, 1989) and in a Texas rice field (Sass *et al.*, 1990) support this hypothesis. Root exudates refer to organic materials released partially by roots into the surrounding soil volume and consist mainly of carbohydrates, organic acids, amino acids and phenolic compounds (Andal *et al.*, 1956; MacRae & Castro, 1966; Marschner, 1985) that provide an easily available carbon and energy source to the microbial community. A variety of factors, including soil water stress, temperature, light intensity, age and species of plants, mineral nutrition, soil microorganisms and foliar applications of chemicals have been reported to affect root exudation (Hale & Moore, 1979).

Kludze *et al.* (1997) found a significant dependence of exudation rates on plant parameters, especially root weight and root density. The effect of these plant characteristics were also dependent on cultivar type and the physiological age of the plant. The highest amounts of exudates were released during the flowering stage and the least during the seedling stage of plant growth. Ladha *et al.* (1986) demonstrated the wide variation among rice cultivars in the amount of carbohydrate material lost from rice roots, and Vancura and Stotzky (1976) reported that the amount of substances liberated from plant roots differs both in plant species and among cultivars of the same species. The studies by Kludze *et al.* (1997) also demonstrated a strong positive effect of rice root exudates on CH_4 production. The amount of exudate carbon utilized in CH_4 formation varied among rice cultivars, suggesting that not only exudate quantity but also the quality (chemical composition) influence the rates of CH_4 generation. Comparison of *de novo* synthesized proteins in the root with that secreted in the root exudates among nine rice cultivars showed that about 40–80% of total protein synthesized in the root were secreted in the 4-day seedling stage, which varied among different cultivars (Boonjawat *et al.*, 1995). Large cultivar differences in root oxidation power (Neue & Roger, 1993) and in emission rates (Parashar *et al.*, 1990) have been reported. Modern cultivars emit generally less CH_4 than traditional cultivars when compared on a single plant basis (Neue *et al.*, 1997). Under field conditions, a comparison of cultivars is more complex because farmers adjust management practices to cultivars in order to achieve an optimum productivity. Higher fertilizer rates apply to modern cultivars because they are more responsive at increasing both plant biomass (grain yield) and CH_4 emission. Rising atmospheric CO_2 concentrations increase plant biomass (see Schulze & Heimann, this volume; Neue *et al.*, 1995*a*) and probably root exudates, both of which should result in higher CH_4 fluxes. Anyway, the observed differences in tiller number, root exudation, root oxidation power and probably gas transfer of rices between and within types of cultivars may provide the possibility to breed rices both for reduced CH_4 emission and higher grain yield.

Source strength

Wetland rice fields contribute between 20–100 Tg to the total annual CH_4 source strength of 515 Tg (IPCC, 1992). The large uncertainty of CH_4 emission from rice fields is the consequence of simple extrapolations from a limited number of flux measurements, often obtained only for short periods, to areas and time spans for which they may not necessarily be representative. All present estimates are based on area-weighted summations of periodic CH_4 emission measurements at a few sites between transplanting and harvest. Key controlling factors like water regimes, soil properties, agronomic practices and plant characteristics are mostly not well documented.

A major impediment to the development of reliable estimates is the insufficient time resolution of the flux measurements used to estimate seasonal or annual fluxes. Evaluating bi-hourly CH_4 emission measurements over three seasons from 18 experiments with three or four replications at four irrigated sites in Indonesia, the Philippines and Thailand, Buendia *et al.* (1995) determined strategies for sporadic sampling. Weekly measurements are only appropriate for the period from panicle initiation to early flowering, while daily measurements at 6, 12, and 18 o'clock are required at all other growth stages (from pre-planting to post-harvest) to accurately quantify CH_4 emissions in continously flooded rice fields. The 12 and 18 o'clock schedule may be changed if the highest diurnal temperature occurs later than 14 o'clock. The measurement schedule was found by minimizing the coefficient of variation and reflects diurnal measurement periods with minimal flux changes over time. The measurement resolution has to be increased in rainfed rice and when midseason aerations are practised in irrigated rice.

Despite all methodological shortcomings, the median of reported CH_4 fluxes can be discriminated according to rice ecosystems and countries (Table 8.1). Excluding the very low fluxes for irrigated rice in India reported by Mitra (1992), irrigated rice reveals the highest seasonal CH_4 flux with a median ranging from 250 to 480 kg ha^{-1}, followed by deepwater rice with 190 kg ha^{-1} and rainfed rice with 70–170 kg ha^{-1}. Rice fields representing irrigated rice in an Indian measurement campaign on 34 fields/plots at 14 sites (irrigated rice 4, deepwater rice 3, and rainfed rice 7 sites) during the 1991 Kharif (wet) season emitted less than 10 kg ha^{-1} (Mitra, 1992). These fields became sometimes even a sink for atmospheric CH_4, indicating fully aerated soils for most parts of the growing season which is also supported by the low total dry plant biomass (3 to 9 tonnes ha^{-1}) at N fertilizer rates of up to 120 kg ha^{-1}. Extrapolating the results, Mitra (1992) concluded that irrigated rice in India is an insignificant source of CH_4, releasing an annual maximum of 0.335 Tg. This extrapolation does not comprise a consistent and realistic estimate of CH_4 emission from irrigated rice fields in India. Such low emission may demonstrate the potential to mitigate CH_4 emission in irrigated rice fields through modified water regimes, but its consistency with required higher production levels in the future would still have to be established.

Irrigated rice is defined as rice grown on fields with assured irrigation and water control to maintain shallow floodwater during the season (Khush, 1984). Irrigated

Table 8.1 *Minimum, maximum and median of reported CH_4 fluxes from rice fields from seeding or transplanting to harvest*

Country	Number of observations	Emission rate ($gCH_4\ m^{-2}\ day^{-1}$)			Total emission ($gCH_4\ m^{-2}$)			
		min.	median	max.	min.	median	max.	
China								Wang, M. X. *et al.*, 1993, Wang, 1995; Lu *et al.*, 1995
irrigated	74	0.06	0.34	1.41	5	34	155	
India								
irrigated	7			0.02	0.06	0.74	2	Mitra, 1992
irrigated	3	0.21	0.26	0.41	20	25	39	Adhya *et al.*, 1994
rainfed	8				5	17	60	Mitra, 1992
deepwater	7				14	19	24	Mitra, 1992
Indonesia								Nugroho *et al.*, 1994; Makarim *et al.*, 1995
irrigated	10	0.16	0.52	0.78	14	31	47	
rainfed	4	0.01	0.06	0.10	4	8	10	
Italy								Schütz *et al.*, 1989
irrigated	22	0.10	0.29	0.68	12		77	
Japan								Yagi & Minami, 1990, 1991; Kimura & Minami, 1995
irrigated	28	0.01		0.39	1		45	
Korea (ROK)								
irrigated	4	0.07	0.24	0.46	9	33	63	Shin *et al.*, 1995

Philippines								IRRI, 1996; Metra-Corton *et al.*, 1995
irrigated	56	0.09	0.25	0.79	10	27	87	
rainfed	1	0.09			7			
Spain								Seiler *et al.*, 1984
irrigated	1	0.10			12			
Thailand								Jermsawatdipong *et al.*, 1994*a*,*b*; Chairoj, 1994; Kimura & Minami, 1995
irrigated	27	0.38	0.48	0.72	34	48	86	
rainfed	4	0.02	0.15	0.45	1	15	68	
deepwater	2	0.09		0.17	12		32	Chareonsilp *et al.*, 1995 Siriratpiriya, 1994;
USA								Lindau *et al.*, 1991; Cicerone *et al.* 1992; Sass & Fisher, 1995
irrigated	41	0.05	0.27	0.55	1	25	48	

wet season areas are those where irrigation water may be added to the rice fields during the local wet season as a supplement to rainfall in assuring the success of a crop threatened by erratic precipitation. Irrigated dry season areas are those where no rice crop can be grown without irrigation. In the dry season, evapotranspiration is high and water needs are considerably greater than during the wet season. In most irrigated dry season areas the combination of assured water throughout the season, high solar radiation, low pest incidence, and high input levels result in the year's highest yield. Expanding irrigation systems have increasingly assured water supply in irrigated rice in India, raising the harvested irrigated rice area to 19 million ha (45% of the total) in the past two decades and contributing 61% of the total rice yield harvested (IRRI, 1993). Assured water control has also led to higher inputs and input efficiency raising grain yields by 55–98% in some areas, such as Punjab, Haryana, and Tamil Nadu. The average yield level of irrigated rice in India is 3.6 t ha^{-1} similar to that of Thailand (4.0 t ha^{-1}), Sri Lanka (3.7 t ha^{-1}), the Philippines (3.4 t ha^{-1}), and Malaysia (3.0 t ha^{-1}) (IRRI, 1993), indicating that average water regimes, input levels, cultural practices and resulting CH_4 fluxes should not be expected to be very different. Indeed, Adhya *et al.* (1994) reported emission rates from irrigated rice fields in India that are similar to that found in other countries.

Combining all reported CH_4 flux measurements with information on rice ecosystems, its source strength follows the order: irrigated rice ≥ favorable rainfed rice > rice deepwater rice ≥ flood-prone rainfed > drought-prone rainfed rice > tidal wetland rice. Upland rice is not a source of CH_4, since upland rice is grown like wheat in aerated soils that never become flooded for a significant period of time.

The irrigated rice ecosystem has the highest CH_4 source strength considering the area planted (74 million ha in Asia, 79 million ha worldwide; IRRI, 1993), the assured water supply and the high yield (biomass) levels. China, Japan, Indonesia and Korea typify high-yielding irrigated rice areas in Asia with grain yields of > 5 t ha^{-1}. Medium-yielding areas with yields of 4–5 t ha^{-1} include Bangladesh, northwestern and southern India, Laos, Malaysia, Myanmar, the Philippines, Sri Lanka, Thailand and Vietnam. Cambodia, eastern India, Madagascar, Nepal, and Pakistan are representative of low-yielding areas of the irrigated rice ecosystem with yields of < 4 t ha^{-1}. Main yield-limiting factors which also reduce CH_4 emission increase from high- to low-yielding areas: inadequate water supply, inefficient use of scarce irrigation water, and inadequate drainage leading to the buildup of salinity and alkalinity (IRRI, 1993). Variations of CH_4 emission between sites of comparable yield levels are due to differences in residue recycling and organic amendments, timing and number of short aeration periods, soils (mineralizable carbon content and texture), fertilization, and rice cultivars. The highest CH_4 fluxes are observed in fields receiving organic amendments and being continously flooded during the growing season. The lowest CH_4 fluxes are recorded in fields with low residue recycling, multiple aeration periods, poor soils and low fertilization with resulting poor rice growth and low yields.

The source strength of rainfed rice, planted on 39 million ha in Asia (41 million ha worldwide), is most uncertain because of its high variability in all factors controlling CH_4 emission. CH_4 fluxes are generally lower from rainfed rice because of drought periods and low rice yields. Having basically only one rice crop per year and harvesting not only lower grain, but also lower straw, yields residue recycling and organic amendments that are generally less in rainfed rice compared to irrigated rice. CH_4 fluxes may match that of irrigated rice in some favorable, as well as flood-prone, rainfed rice areas. Future improvements of rice production in rainfed areas, through irrigation, higher fertilizer input, increasing organic recycling and manuring, will likely enhance CH_4 emission if not combined with mitigation technologies.

CH_4 emission in deepwater rice ecosystems (10 million ha in Asia, 11 million worldwide) may vary according to floodwater heights and rice yields. Deepwater rice planted in areas with water levels between 50 and 100 cm only and yielding up to 4 t ha^{-1} may even have a higher source strength than irrigated rice because of long flooded growing periods (up to 7 months) and high rice biomass production. Floating rice grown in areas that are flooded up to as much as 8 m and yielding 0.5–2.5 t ha^{-1} should emit less because the lower part of the stem and roots in the soil may be dead at later growth stages.

The source strength of tidal wetland rice is small because of its area (about 5 million ha) and the fact that it is affected by saltwater high in sulfate. Salinity in the form of NaCl may not reduce CH_4 fluxes because it reduces CH_4 oxidation more than production (Denier van der Gon & Neue, 1995*b*).

The best estimate of current annual global CH_4 emission from rice fields remains at about 40 ± 20 Tg as indicated by various model calculations (Aselmann & Crutzen, 1989; Neue *et al.*, 1989; Yagi & Minami, 1990; Bachelet & Neue, 1993). Various earlier estimates did use high default values and ignored variations according to rice ecosystems and major controlling factors. On the other hand, almost no measurements recorded emissions during field preparation nor did they catch the increased fluxes due to soil disturbances during the growing season and at the beginning of aeration periods. These unaccounted fluxes may comprise up to 25% of the total seasonal emission. The global and regional source strength of rice fields remains uncertain as long as mechanistic information on CH_4 fluxes and respective geographic information of controlling factors are not sufficiently established for reliable regional modeling. Understanding processes has progressed well and large-scale information on rice growing areas, growing seasons, temperature regimes, and soil types is available. Essential geographic information on water regimes, organic recycling and amendments, relevant soil properties, rice cultivars, and cultural practices are still insufficient or not available at all. The high spatial and temporal variability, even within small scales, of each of these factors as well as their interrelations and feedback limits further fast improvements in defining the source strength. Merely increasing the sites of CH_4 flux measurements will not reduce uncertainties. Sound upscaling and verification methods have to be established. Upscaling from the plot level to 1° grids with micrometerological methods, in combination with empirical relations based on mechanistic understanding and downscaling by inverse modeling from regional atmospheric background levels to the same grid with dominant CH_4 sources from rice, seems to be a feasible, promising approach.

Abatement strategies

Abatement strategies can only be developed if the source strength of rice fields is reliably identified and discriminated according to the various rice ecosystems and production systems. They will only be accepted if mitigation technologies are in accord with increased rice production and productivity at least at the farm level. Promising mitigation candidates are water management, organic amendments, fertilization, cultural practices and rice cultivars. While present knowledge of processes controlling CH_4 fluxes should allow the development of mitigation technologies, information is still lacking on trade-offs and socio-economic feasibilities.

CH_4 emission has the greatest potential to be reduced in irrigated rice. Increasing water use efficiency, dry seeding and keeping soils only flooded during the critical growth stage from maximum tillering until early maturity is a feasible option in dry season irrigated rice if high water control systems are available. Flooding for land preparation and during most of the season will probably remain common practice at least on heavy textured rice soils which become very hard when dry. In Asia, 40% of rice soils have at least 45% clay (Kawaguchi & Kyuma, 1977). The advantages of wet tillage are lower draft requirements, reduced water percolation, ease of transplanting, and improved weed control. Defined mid-season aeration and drying before harvest reduces CH_4 fluxes without increasing nitrous oxid (N_2O) emissions, and without reducing nitrogen efficiency as well as rice yields, should be feasible in all irrigated rice during dry seasons. Precipitation may prevent such defined aeration periods in the wet season.

Developing rice cultivars with lower CH_4 emission potentials, optimizing residue recycling and organic amendment, and using sulfate-containing fertilizer would effectively supplement modifications of water regimes. In areas and seasons where precipitation does flood rice fields anyway and drainage is not possible, above-mentioned supplementary technologies would become major options to reduce CH_4 emission. Crop diversification (i.e. sequential cropping of an upland crop before or after one or two crops of rice) is a feasible option to reduce CH_4

emission in line with economic benefits in rice-growing areas with year-round irrigation and good access to markets (Neue *et al.*, 1995*b*; Ranganathan *et al.*, 1995). Common upland crops that are widely grown in rotation with wetland rice are wheat, mungbean, soybean, maize, and vegetables.

Abatement strategies for the favorable rainfed rice ecosystem are similar to that of the wetseason irrigated rice. Water management is hardly a feasible mitigation strategy in the flood-prone rainfed ecosystem which is characterized by severe flooding. Mitigating CH_4 emission in drought-prone rainfed areas may not be worthwhile because of its low source strength. Developing rice cultivars with low emission potentials may be the only feasible mitigation option for deepwater rice ecosystems where flooding regimes are hard to change or to manage. Some of the area grown to deepwater rice may be left idle in future because of economic reason. Others may be turned into fish farms during the flooding periods and grown to irrigated rice or upland crops during the dry season when irrigation water and short duration crops are available. Tidal wetland rice areas will only emit significant amounts of CH_4 when fully reclaimed into irrigated rice.

Conclusions

Rice production systems have been productive, stable and environmentally sound agricultural systems with immanent features (leveling, bunding, flooding) to conserve water and soil and to increase fertility of cultivated land. Traditional rice systems remained relatively unchanged for more than 2000 years. In contrast, the magnitude of changes in a time span of one human generation since the 1960s has indeed been a revolution in the means of rice production, especially in irrigated rice. The development of semidwarf rice cultivars with high yield potential (10–11 t ha^{-1}), shorter growth duration (around 100 days) and associated management practices led to double- and triple-crop rice systems with an increase in time soils remained submerged and planted, with more soil tillage and puddling operations, and greater amounts of applied nutrients from irrigation water and fertilization. Given these dramatic changes in a relatively short period of time, we are now only beginning to understand and quantify the impact of intensification on the natural resource base and the environment one of which is increased CH_4 emission.

Looking into the next century, one of the greatest challenges is to close the gap between the supply and demand for food and non-renewable resources upon which our life support systems depend. Today, rice is the staple food for about 2.4 billion people and provides more than 20% of their daily calorie intake. This number will increase to 4.6 billion people by 2050. Just to meet the projected demand for rice in the 21st century, the world's rough rice production must increase from 520 million tonnes today to at least 880 million tonnes by 2025, an increase of almost 70%. That requirement may rise to one billion tonnes by the year 2050 (Lampe, 1995). The doubling of production will have to be achieved on decreasing arable land, with reduced water resources, with less labor input, and with fewer agrochemical inputs.

Rice research has never before been confronted with a more demanding, complex, and difficult agenda than today. In the 1960s, rice research focused mainly on a single objective: increasing yield. Today, a multidisciplinary set of targets have to be met :

- increasing yield, total production, and productivity
- protecting the environment and sustaining the non-renewable resources on which rice is grown
- enhancing working conditions for farmers in order to compete with alternative job opportunities outside rice and agricultural production
- improving income and living conditions of rice farming families
- achieving price levels of rice acceptable to urban and rural low-income families

Food comes first and will remain the first priority. Efforts to reduce CH_4 emission and other environmental consequences must go hand in hand with it.

References

Achtnich, C., Bak, F. & Conrad, R. (1995). Competition for electron donors among nitrate reducers, ferric iron reducers, sulfate reducers, and methanogens in anoxic paddy soil. *Biology and Fertility of Soils*, **19(1)**, 65–72.

Adhya, T. K., Rath, A. K., Gupta, P. K., Rao, V. R., Das, S. N., Parida, K. M., Parashar, D. C. & Sethunathan, N. (1994). Methane emission from flooded rice fields under irrigated conditions. *Biology and Fertility of Soils*, **18**, 243–8.

Andal, R., Bhuvanesware, K. & Subba-Rao, N. (1956). Root exudates of paddy. *Nature*, **178**, 1063.

Aselmann, I. & Crutzen, P. (1989). Global distribution of natural freshwater wetlands and rice paddies: Their net primary productivity, seasonality and possible methane emission. *Journal of Atmospheric Chemistry*, **8**, 307–58.

Bachelet, D. & Neue, H. U. (1993). Methane emission from wetland rice areas of Asia. *Chemosphere*, **26**, 219–38.

Bonneau, M. (1982). Soil temperature. In *Constituents and Properties of Soils*, ed. M. Bonneau & B. Souchier, pp. 366–71. London: Academic Press.

Boonjawat, J., Chaopongpang, S. & Pitaksutheepong, C. (1995). Biochemical significance of lectins in plant–microbial interaction. In *Biopolymers and Bioproducts: Structure, Function and Applications*, ed. J. Svasti, V. Rimphanitchyakit, S. Soontaros, A. Tasanakajorn, T. Limpaseni, P. Pongsawasdi, P. Wilairat, B. Sonthayanon, J. Boonjawat, K. Packdibumrung & S. Kamolsiripichaiporn, pp. 190–202. Bangkok: Samakkisan (Dokya) Public Co. Ltd.

Buendia, L. V., Neue, H. U., Wassmann, R., Lantin, R. S., Corton, T., Makarim, A., Chareonsilp, N., Javellana, A. M. & Lignes, V. (1995). Sporadic sampling strategies for accurate determination of methane emission by closed chamber method. Manila, Philippines: International Rice Research Institute.

Butterbach-Bahl, K. (1992). Mechanismen der

Produktion und Emission von Methan in Reisfeldern: Abhängigkeit von Felddüngung und angebauter Varietät. Diss. Techn. Univ. München. Schriftenreihe des Fraunhofer Instituts für Atmospherische Umweltforschung Bd. 14. Wiss. Verl. Mauraun, Frankfurt/M.

Byrnes, B. H., Austin, E. R. & Tays, B. K. (1995). Methane emission from flooded rice soils and plants under controlled conditions. *Soil Biology and Biochemistry*, **27**, 331–9.

Chairoj, P. (1994). Research Report: a cooperative project on methane emission fron rice paddy fields. Department of Agriculture, Minister of Agriculture and Cooperatives, Thailand and Department of Envirnomental Science, Japan.

Charoensilp, N., Buddhaboon, B., Charoendham, P. (1995). Methane emission from deepwater rice fields. Prachinburi Rice Research Center, RRI, Department of Agriculture, Bangkok, Thailand.

Cicerone, R. J. & Oremland, R. S. (1988). Biogeochemical aspects of atmospheric methane. *Global Biogeochemical Cycles*, **2**, 299–327.

Cicerone, R. J., Delwiche, C. C., Tyler, T. C. & Zimmermann, P. R. (1992). Methane emission from Californian rice paddies with varied treatment. *Global Biogeochemical Cycles*, **6**, 233–48.

De Datta, S. K. (1981). Principles and practices of rice production. New York: John Wiley.

DeLaune, R. D., Smith, C. J. & Patrick, W. H. (1983). Methane releases from Gulf coast wetlands. *Tellus*, **35B**, 8–15.

Denier van der Gon, H. A. C. & Neue, H. U. (1994). Impact of gypsum application on the methane emission from a wetland rice field. *Global Biogeochemical Cycles*, **8**, 127–34.

Denier van der Gon, H. A. C. & Neue, H. U. (1995*a*). Influence of organic matter incorporation on the methane emission from a wetland rice field. *Global Biogeochemical Cycles*, **9**, 11–22.

Denier van der Gon, H. A. C. & Neue, H. U. (1995*b*). Methane emission from a wetland rice field as affected by salinity. *Plant and Soil*, **170**, 307–13.

Denier van der Gon, H. A. C., van Bremen, N., Neue, H. U., Lantin, R. S., Aduna, J. B., Alberto, M. C. R. & Wassmann, R. (1996). Release of entrapped methane from wetland rice fields upon drying. *Global Biogeochemical Cycles*, **10**, 1–9.

Fetzer, S. & Conrad, R. (1993). Effect of redox potential on methanogenesis by *Methanosarcina bakeri*. *Archives of Microbiology*, **160**, 108–13.

Gambrell, R. P. & Pattrick, W. H. (1978). Chemical and microbiological properties of anaerobic soils and sediments. In *Plant life in anaerobic soils and sediments*, ed. D. D. Hook & R. M. M. Crawford, pp. 375–423. Ann Arbor, MI: Ann Arbor Science Publications, Inc.

Hale, M. G. & Moore, L. D. (1979). Factors affecting root exudition. *Advances in Agronomy*, **31**, 93–123.

Holzapfel-Pschorn, A., Conrad, R. & Seiler, W. (1986). Effects of vegetation on the emission of methane from submerged paddy soil. *Plant Soil*, **92**, 223–33.

IGAC-CPO (1994). Global measurement standardization of methane emission from irrigated rice cultivation. IGAC Core Project Office. Massachusetts Institute of Technology, Cambridge, AA 02139, USA. 6pp.

Inubushi, K., Muramatsu, Y. & Umebayashi, M. (1992). Influence of percolation on methane emission from flooded paddy soil. *Japanese Journal of Soil Science and Plant Nutrition*, **63**, 184–9.

IPPC - Intergovernmental Panel on Climate Change (1992). Climate change: the supplementary report to the IPCC scientific assessment. Cambridge: Cambridge University Press.

IRRI (1993). *IRRI Rice Almanac.* Manila, Philippines: International Rice Research Institute.

IRRI (1996). *Program Report 1995.* Manila, Philippines: International Rice Research Institute.

Jermsawatdipong, P., Prabuddham, P., Hasathon, Y., Khomthong, N., Naklang, K. & Chinda, M. (1994*a*). Methane emission paddy fields in Thailand. In *Thailand's Present and*

Projected Future Contribution to Global Emissions of GHG, pp. 3–35. Report to Asian Development Bank.

Jermsawatdipong, P., Murase, P.,Prabuddham, P., Hasathon, Y., Chinda , M., Chaiwatana, B., Khomthong, N., Sattawatananon, S., Naklang, K., Watanabe, A., Haraguchi, H. & Kimura, M. (1994*b*). Methane emission from plots with differences in fertilizer application in Thai paddy. *Soil Science and Plant Nutrition*, **40**, 63–71.

Kawaguchi, K. & Kyuma, K. (1977). Paddy soils in tropical Asia: their material nature and fertility. University Press of Hawaii, Honolulu, USA.

Khush, G. S. (1984). Terminology for rice growing environments. In *Terminology for rice growing environments*, pp. 5–10. Manila, Philippines: International Rice Research Institute.

Kimura, M. & Minami, K. (1995). Dynamics of methane in rice fields: Emissions to the atmosphere in Japan and Thailand. In *Climate Change and Rice*, ed. S. Peng, K. T. Ingram, H. U. Neue & L. H. Ziska, pp. 30–45. Berlin: Springer-Verlag.

Kimura, M., Miura, Y., Watanabe, A., Katoh, A. & Haraguchi, H. (1991). Methane emission from paddy field (part 1). Effect of fertilization, growth stage and midsummer drainage: pot experiment. *Environmental Science*, **4**, 265–71.

Kimura, M., Miura, Y., Watanabe, A., Murase, J. & Kuwatsuka, S. (1992). Methane production and its fate in paddies. I. Effect of rice straw application and percolation rate on the leaching into subsoil of methane and other soil components. *Soil Science and Plant Nutrition*, **38**, 665–72.

Kludze, H. K., DeLaune, R. D. & Patrick, W. H. Jr. (1993). Aerenchyma formation and methane and oxygen exchange in rice. *Soil Science Society of America Journal*, **57**, 386–91.

Kludze, H. K., Neue, H. U., Llenarezas, D. & Lantin, R. S. (1997). Rice root exudation and its impact on methane production. *Soil Science Society of America Journal*, (in press).

Ladha, J. K., Tirol, A. C., Darby, L. G., Caldo, G., Ventura, W. Watanabe, I. (1986). Plant associated N_2 fixation (C_2H_2-reduction) by five rice varieties and relationship with plant growth characters as affected by straw incorporation. *Soil Science and Plant Nutrition*, **32**, 91–106.

Lampe, K. (1995). Rice research: Food for 4 billion people. *GeoJournal*, **35** (3), 253–9.

Lindau, C. W., Bollich, P. K., DeLaune, R. D., Pattrick, W. H. & Law, V. J. (1991). Effect of urea fertilizer and environmental factors on methane emissions from a Louisiana, USA rice field. *Plant and Soil*, **136**, 195–203.

Lovely, D. R. & Klug, M. J. (1983). Sulfate reducers can outcompete methanogens at freshwater sulfate concentrations. *Applied Environmental Microbiology*, **45**, 187–92.

Lu, W., Chen, W., Guo, W. & Duan, B. (1995). Effect of water regimes on methane emission from irrigated rice. China National Rice Research Institute, Hangzhou 310006, Zhejiang, China.

MacRae, I. C., Castro, T. F. (1966). Carbohydrates and amino acids in the root exudates of rice seedlings. *Phyton*, **23**, 95–100.

Makarim, A. K., Bey, A., Boer, R. & Setyanto, P. (1995). Methane gas from lowland rice fields in Indonesia: Development and research directions (in Indonesian). Central Research Institute for Food Crops, Bogor, Indonesia.

Mariko, S., Harazono, Y., Owa, N. & Nouchi, I. (1991). Methane in flooded soil, water and the emission through rice plants to the atmosphere. *Environmental and Experimental Botany*, **31(3)**, 343–50.

Marschner, H. (1985). Mineral nutrition in higher plants. New York: Academic Press.

Metra-Corton, T., Bajita, J. B., Asis, C. A., Pamplona, R. R. & Salamanca (1995). Methane emission from an irrigated rice field. Munoz, Nueva Ecija, Philippines: Philippine Rice Research Institute.

Minami, K. & Neue, H. U. (1994). Rice paddies as a methane source. *Climate Change*, **27**, 13–26.

Mitra, A. P. (1992). Greenhouse gas emission in India: 1991 Methane Campaign. Sci. Rep. no 2. New Delhi: Council of Scientific and Industrial Research and Ministry of Environment and Forests.

Neue, H. U. (1985). Organic matter dynamica in wetland soils. In *Wetland Soils: Characterisation, Classification and Utilization*, pp. 109–122. Manila, Philippines: International Rice Research Institute.

Neue, H. U. (1989). Rice growing soils: constraints, utilization and research needs. FFTC Book Series No. 39: 1–14. Taipei, Taiwan: Food and Fertilizer Technology Center for the Asian Pacific Region.

Neue, H. U. (1991). Holistic view of chemistry of flooded soil. In *Soil Management for Sustainable Rice Production in the Tropics*, IBSRAM, Monogragh No.2. pp. 5–32. International Board for Soil Research and Management.

Neue, H. U. (1993). Methane emission from rice fields. *BioScience*, **43(7)**, 466–75.

Neue, H. U. & Roger, P. A. (1993). Rice agriculture: factors controlling emissions. In *Atmospheric Methane: Sources, Sinks, and Role in Global Change*, ed. M. A. K. Khalil, pp. 254–98. NATO Advanced Science Institute Series, Series I: Global Environmental Change, Vol. 13. Berlin: Springer Verlag.

Neue, H. U. & Sass, R. (1994). Trace gas emissions from rice fields. In *Global Atmospheric-Biospheric Chemistry*, ed. R. G. Prinn, pp. 119–48. Environmental Science Research 48. New York: Plenum Press.

Neue, H. U. & Scharpenseel, H. W. (1987). Decomposition pattern of ^{14}C-labeled rice straw in aerobic and submerged rice soils of the Philippines. *The Science of the Total Environment*, **62**, 431–4.

Neue, H. U., Becker-Heidmann, P. & Scharpenseel, H. W. (1989). Organic matter dynamics, soil properties, and cultural practices in rice lands and their relationship to methane production. In *Soils and the Geenhouse Effect*, ed. A. F. Bouwman. New York: John Wiley.

Neue, H. U., P. Becker-Heidmann & Scharpenseel, H. W. (1990). Organic matter dynamics soil properties and cultural practices in ricelands and their relationship to methane production. In *Soil and the Greenhouse Effect*, ed. A. F. Bouwman, pp. 457–66. Chichester, UK: John Wiley.

Neue, H. U., Lantin, R. S., Wassmann, R., Aduna, J. B., Alberto, M. C. R. & Andales, M. J. F. (1994). Methane emission from rice soils of the Philippines. In *CH_4 and N_2O - Global Emissions and Controls from Rice Fields and Other Agricultural and Industrial Sources*, ed. K. Minami, A. R. Mosier & R. L. Sass, pp. 55–77. NIAES Series 2, Tokyo, Japan.

Neue, H. U., Ziska, L. H., Matthews, R. B., Dai, Q. (1995*a*). Reducing global warming – the role of rice. *GeoJournal*, **35(3)**, 351–62.

Neue, H. U., Wassmann, R. & Lantin, R. S. (1995*b*). Mitigation options for methane emission from rice fields. In *Climate Change and Rice*, ed. S. Peng, K. T. Ingram, H. U. Neue & L. H. Ziska, pp. 136–44. Berlin: Springer-Verlag.

Neue, H. U., Wassmann, R., Kludze, H. K., Wang B. & Lantin, R. S. (1997). Factors and processes controlling methane emissions from ricefields. *Fertilizer Research*. (in press).

Nouchi, I., Mariko, S. & Aoki, K. (1990). Mechanism of methane transport from the rhizosphere to the atmosphere through rice plants. *Plant Physiology*, **94**, 59–66.

Nugroho, S. G., Lumbanraja, L., Suprapto, H., Sunyoto, Ardjasa, W. S., Haraguchi, H. & Kimura, M. (1994). Methane emission from Indonesian paddy field subjected to several fertilizer treatments. *Soil Science and Plant Nutrition*, **40(2)**, 275–81.

Parashar, D. C., Rai, J., Gupta, P. K. & Singh, N. (1990). Parameters affecting methane emission from paddy fields. *Indian Journal of Radio & Space Physics*, **20**, 12–17.

Patrick Jr., W. H. & Reddy, C. N. (1978). Chemical changes in rice soils. In *Soils and Rice*, pp. 361–80. Manila, Philippines: International Rice Research Institute.

Ponnamperuma, F.N. (1984). Effects of flooding on soils. In *Flooding and Plant Growth*, ed. T. T. Kozlowski, pp. 9–45. New York: Academic Press.

Ranganathan, R., Neue, H. U. & Pingali, P. L. (1995). Global climate change: role of rice in methane emission and prospects for mitigation. In *Climate Change and Rice*, ed. S. Peng, K. T. Ingram, H. U. Neue & L. H. Ziska, pp. 122–135. Berlin: Springer-Verlag.

Sass, R. L. & Fisher, F. M. (1995). Methane emissions from Texax rice fields: a five year study. In *Climate Change and Rice*, ed. S. Peng, K. T. Ingram, H. U. Neue & L. H. Ziska, pp. 46–59. Berlin: Springer-Verlag.

Sass, R. L., Fisher, F. M., Harcombe, P. A. & Turner, F. T. (1990). Methane production and emission in a Texas rice field. *Global Biogeochemical Cycles*, **4(1)**, 47–68.

Sass, R. L., Fisher, F. M., Harcombe, P. A. & Turner, F. T. (1991). Mitigation of methane emission from rice fields: effect of incorporated rice straw. *Global Biogeochemical Cycles*, **5**, 275–88.

Sass, R. L., Fisher, F. M., Wang, Y. B., Turner, F. T. & Jund, F. T. (1992). Methane emission from rice fields: the effect of flood water management. *Global Biogeochemical Cycles*, **6(3)**, 249–62.

Schütz, H., Holzapfel-Pschorn, A., Conrad, R., Rennenberg, H., & Seiler, W. (1989). A three-year continuous record on the influence of daytime season and fertilizer treatment on methane emission rates from an Italian rice paddy field. *Journal of Geophysical Research*, **94**, 16405–16.

Schütz, H., Schröder, P. & Rennenberg, H. (1991). Role of plants in regulating the methane flux to the atmosphere. In *Trace Gas Emission from Plants*, ed. T. D. Sharkey, E. A. Holland & H. A. Mooney, pp. 29–61. New York: Academic Press.

Seiler, W., Holzapfel-Pschorn, A., Conrad, R. & Scharfe, D. (1984). Methane emission from rice paddies. *Journal of Atmospheric Chemistry*, **1**, 241–68.

Shin, Y. K., Yun, S. H. & Park, M. E. (1995). Estimation of methane emission by water management and rice straw application in paddy soil in Korea. *Journal of Korean Society of Soil Science and Fertility*, **28(3)**, 261–5.

Siriratpiriya, O. (1994). Methane emission from rice fields in Thailand. Institute of Environmental Research, Chulalongkorn University, Bangkok, Thailand.

SMSS Soil Survey Staff. (1987). Keys to Soil Taxonomy. SMSS Technical Monograph no. 6. Ithaca, New York.

Vancura, V. & Stotzky, G. (1976). Gaseous and volatile exudates from germinating seeds and seedlings. *Canadian Journal of Botany*, **54**, 518–32.

Vermoesen, A., Ramon, H. & van Cleemput, O. (1991). Composition of the soil gas phase. Permanent gases and hydrocarbons. *Pedology*, **41**, 119–32.

Wahlen, M., Tanaka, N., Henry, R., Deck, B., Zeglen, J., Vogel, J. S., Southon, J., Shemesh, A., Fairbanks, R. & Broecker, W. (1989). Carbon-14 in methane sources and atmospheric methane: the contribution from fossil carbon. *Science*, **245**, 286–90.

Wang, B. (1995). Effects of rice cultivars on diel and seasonal methane emission. PhD thesis, University of the Philippines, Los Banos, Philippines, 130 pp.

Wang, M. X. (1995). Methane emission rates from various rice fields in China. In *China Contribution to Global Change Studies*, ed. D. Ye, & H. Lin, pp.154–60. China Global Change Report No. 2. Beijing, China: Science Press.

Wang, M. X., Dai, A. G. X., Shen, R. X., Schütz, H., Rennenberg, H., Seiler, W. & Wu, H. B. (1990). Methane emission from a Chinese rice paddy field. *Acta Meteorological Sinica*, **4**, 265–75.

Wang, M. X., Dai, A. G., Huang, J., Ren, L. X.,Shen, R. X., Schütz, H., Seiler, W., Rasmussen, R. A. & Khalil, M. A. K. (1993). Methane source from China. *Scientia Atmospherica Sinica*, **17(2)**, 52–64.

Wang, Z. P., Lindau, C. W., Delaune, R. D., & Patrick Jr., W. H. (1992). Methane production from anaerobic soil amended with rice straw and nitrogen fertilizers. *Fertilizer Research*, **33**, 115–21.

Wang, Z. P., Delaune, R. D., Masscheleyn, P. H. & Patrick Jr, W. H. (1993*a*). Soil redox and pH effects on methane production in a flooded rice soil. *Soil Science Society of America Journal*, **57**, 382–85.

Wang, Z. P., Lindau, C. W., De Laune, R. D. & Patrick, W. H. (1993*b*). Methane emission and entrapment in flooded rice soils as affected by soil properties. *Biology and Fertility of Soils*, **16**, 163–8.

Wassmann, R., Neue, H. U., Lantin, R. S., Aduna, J. B., Alberto, M. C. R., Andales, M.

J., Tan, M. J., Denier van der Gon, H. A. C., Hoffmann, H., Papen, H., Rennenberg, H. & Seiler, W. (1994). Temporal patterns of methane emissions from wetland ricefields treated by different modes of N-application. *Journal of Geophysical Research*, **99**, 16457–62.

Wassmann, R., Neue, H. U., Lantin, R. S., Javellana, M. J., Diego, R., Lignes, V. E., Hoffmann, H., Papen, H. & Rennenberg, H. (1995). Methane emission from rainfed rice fields. In *Fragile Lives in Fragile Ecosystems*, pp. 218–25. Manila, Philippines: International Rice Research Institute.

Wassmann, R., Neue, H. U., Alberto, M. C. R., Lantin, R. S., Bueno, C., Llenaresas, D., Arah, J. R. M., Papen, H., Seiler, W. & Rennenberg, H. (1996). Fluxes and pools of methane in wetland rice soils with varying organic inputs. *Environmental Monitoring and Assessment*, **42**, 163–73.

Wilson, J. O., Crill, P. M., Bartlett, K. B., Sebacher, D. I., Harris P. C. & Sass, R. L. (1989). Seasonal variation of methane emission from a temperate swamp. *Biogeochemistry*, **8**, 55–71.

Yagi, K. & Minami, K. (1990). Effects of organic matter application on methane emission from some Japanese paddy fields. *Soil Science and Plant Nutrition*, **36**, 599–610.

Yagi, K. & Minami, K. (1991). Emission and production of methane in the paddy fields of Japan. *Japan Agricultural Research Quarter*, **25**, 165–71.

Yagi, K. & Minami, K. (1993). Spatial and temporal variations of methane flux from a paddy field. In *Biogeochemistry and Global Change*, ed. R. S. Oremland, pp. 353–68. New York: Chapman & Hall.

Yoshida, S. (1981). Fundamentals of rice crop science, 269 pp. Manila, Philippines: International Rice Research Institute.

9 Drainage basins, river systems, and anthropogenic change: the Chinese example

C. J. Vörösmarty, C. Li, J. Sun and Z. Dai

Introduction

An adequate supply of freshwater is of great importance to the sustainability of human society as well as to the terrestrial and aquatic ecosystems upon which we rely. Freshwater at or near the Earth's surface is a key regulator of atmospheric dynamics, primary productivity of the biosphere, chemical and mass wasting of geological systems, and the suitability of habitat for living organisms. Adequate water resources are critical to human health, food production, and our ability to meet the enduring expectation of continued industrial development and rising standards of living.

Despite its importance to humans, the issue of anthropogenic change and how it has influenced rivers has received relatively little attention in comparison to the greenhouse effect and changes to the global carbon cycle. This, in part, can be traced to the absence of a clearly defined "Mauna Loa curve" for river systems, the physics and chemistry of which are highly site specific and temporally variable. As a consequence, global estimates of the impact of excessive water use and pollution are few and the dimensions of the problem have only recently begun to be synthesized (Postel *et al.*, 1996; L'vovich & White, 1990; Meybeck & Helmer, 1989).

The purpose of this contribution is to provide a broad overview of the impacts of anthropogenic change on the terrestrial hydrologic cycle and drainage basins which deliver water and constituents to the coastal zones of the world. The world's river systems, as a major conduit for water, nutrients, and organic matter, serve as an important linkage in regional and global-scale biogeochemical cycles, and there is mounting evidence that they are experiencing unprecedented rates of change due to the growth of human population and associated economic development. A prime example of such change is in East Asia and China. We will focus on emerging patterns in this region as an illustration of larger global trends. Our emphasis is on the recent past (from 1960) and the contemporary condition, a period of rapid economic growth, population increase, dam construction, and expanded use of industrial fertilizers in agriculture. We will also draw appropriately from the long history of China and its close relationship to the behavior and condition of its river basins.

Within this framework, our discussion has three supporting goals. The first is to identify the concept of global change with respect to hydrology and nutrient biogeochemistry. The second goal is to explore these key issues using China as an example of a rapidly developing economy. Finally, we discuss the potential role that IGBP could play in improving our ability to coherently monitor and improve our understanding of this facet of global change. Throughout our discussion we will highlight the human dimension of these issues.

Anthropogenic disturbance, drainage basin hydrology, and nutrient biogeochemistry

We identify three important agents of global change of consequence to land surface hydrology and constituent biogeochemistry: (1) climate variability and climate change, (2) water engineering works, and (3) human settlement, the redistribution of natural landscapes, and industrial development (Fig. 9.1). Each of these changes conveys a characteristic impact on the behavior of water and waterborne materials within drainage basins. However, a more complete picture can be secured by viewing these disturbances synergistically. We present below a series of case studies highlighting both the individual impacts of these effects as well as key interactions among them.

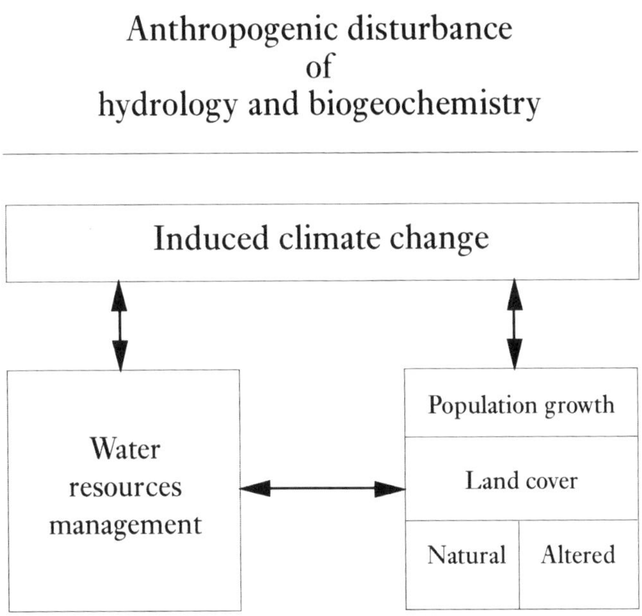

Figure 9.1 *Elements of global anthropogenic change as applied to drainage basin hydrology and nutrient biogeochemistry. These changes have both individual and synergistic effects.*

We have organized this chapter around a drainage basin perspective, in which we highlight interconnections between processes occurring on the landscape with responses of the coupled hydrological system. By their very nature, drainage basins afford an integration over both space and time as rivers focus the end products of numerous distributed hydrological and biogeochemical processes. We will demonstrate how the character of hydrologic "signals" at the drainage basin scale can be transformed both directly and inadvertently by human activities.

Climate variability and climate change

Persistently, and throughout its long history, China and China's major drainage basins have encountered a high degree of natural climatic variability. On a regional basis, two important factors contribute to this effect. The first is a general decrease in precipitation from the humid subtropics in the South to the arid North (Korzoun *et al.*, 1978; Baumgartner & Reichel, 1975), where interannual variability is particularly high due to precipitation patterns typical of arid and semiarid areas (Hartmann, 1994). Second, the region is subject to the East Asian monsoon and interannual variations associated with this event. Strong summer monsoons bring heavy precipitation to the North China Plain and drought to southern China (Guo, 1987). Other work (Diaz & Fu, 1987) has linked droughts in both northern and southern China with the occurrence of El Niño Southern Oscillation (ENSO). Chang and King (1994) analyzed possible relationships between rainfall patterns in the eastern Yangtze River (Changjiang) basin and large-scale atmosphere–ocean interactions. They discovered that Yangtze River (Changjiang) flooding and droughts in northern and southern China were associated with a phenomenon known as the Quasi-biennial Oscillation (QBO) which, in turn, was correlated strongly and positively to the East Asia Circulation Index (sum of pressure at the 50° N/130° E and 50° N/130° E). Major flooding was associated with a combination of characteristic frequencies of 2–3 and 5–6 years, implicating both QBO and ENSO events, respectively.

These variations are of more than simple academic interest. In China such extreme events have held an important and often tragic impact on human society. The Yellow River (Huanghe) is an unpredictable and dangerous river, flooding more than 2500 times and changing its course 26 times (with nine major reroutings) over the last 3000 years (Tregear, 1980). The 1931 flood of the Yellow River was responsible for nearly 4 million deaths (Famighetti,1995; Cotterell & Morgan, 1975), and we know of no single, event-based natural disaster that exceeds these losses as measured in the context of human life. Even the relatively smaller 1887 flood of the Yellow River killed 900 000 (Famighetti, 1995; Matthews, 1992), earning it a place among the four most costly natural disasters in the history of humankind.

Such variability also transfers important effects onto Chinese agriculture and industry. Since the 10% of all Chinese territory that produces over two-thirds of agricultural and economic output is below the flood stage of China's large rivers (Smil, 1992), important strategic resources are at great risk. The 1954 flooding of the Yangtze River, for example, affected 17 million people downstream of the Three-Gorges region, submerging 1 MHa of farmland and effecting a 10 billion Yuan loss (Liu & Zuo, 1987). Such events are not isolated, and a tabulation of those areas experiencing flooding for the 40 years spanning 1950 to 1990 show the presence of several significant events as well as a high year-to-year variability (Fig. 9.2*a*). This situation also characterizes an inventory of drought severity and extent (Fig. 9.2*b*). Annual sums for the area affected by drought have several times exceeded 15 MHa, representing a significant portion of the total available cropland in China, approximately 100 MHa (State Statistical Bureau of China, 1989, 1995). In fact, from these same government statistics describing conditions from 1978 through 1994, we conclude that this high proportion is now likely to be the norm. More than 20 MHa are at risk from drought nearly every year, and would suffer crop losses if not for the ameliorating effects of irrigation. Since the per capita availability of cropland in China (<0.1 ha) places it among the lowest in the world (Smil, 1992) such drought-induced losses have clear implications for the adequacy of future food supplies.

In the context of such natural variability are superimposed current uncertainties, about future climate and the greenhouse effect. To illustrate these uncertainties we applied a Water Balance and Transport Model (WBM/WTM) (Vörösmarty *et al.*, 1996; Vörösmarty & Moore, 1991) to simulate runoff and river flow under contemporary and 2 × CO_2 climates for East Asia. The WBM/WTM requires several biophysical data sets which describe vegetation, soils, topography, and climatic variables. The model operates at a quasi-daily time step and with approximately a 50 km spatial resolution. We contrast the capacity of the WBM/WTM to predict annual streamflow when driven by a station-based climatology for precipitation, temperature, and radiation (Legates & Willmott, 1988; Hahn *et al.*, 1988) and two examples of recent GCM experiments (Geophysical Fluid Dynamics Laboratory; Goddard Institute of Space Studies) as described in Melillo *et al.* (1993). Fig. 9.3 gives computed vs. observed discharge near the mouths of several rivers in the region and the apparent, striking differences yielded by the two approaches. The distributed pattern of runoff derived from the station-based climatology bears a close resemblance to observed runoff fields (Korzoun *et al.*, 1978; Baumgartner & Reichel, 1975) (Fig. 9.4), whereas the GCM-based predictions show a significantly wetter East Asian region. Such difficulties in predicting river flow using GCMs are now clearly documented (Dümenil & Todini, 1992; Miller & Russell, 1992; Kite *et al.*, 1994), and much

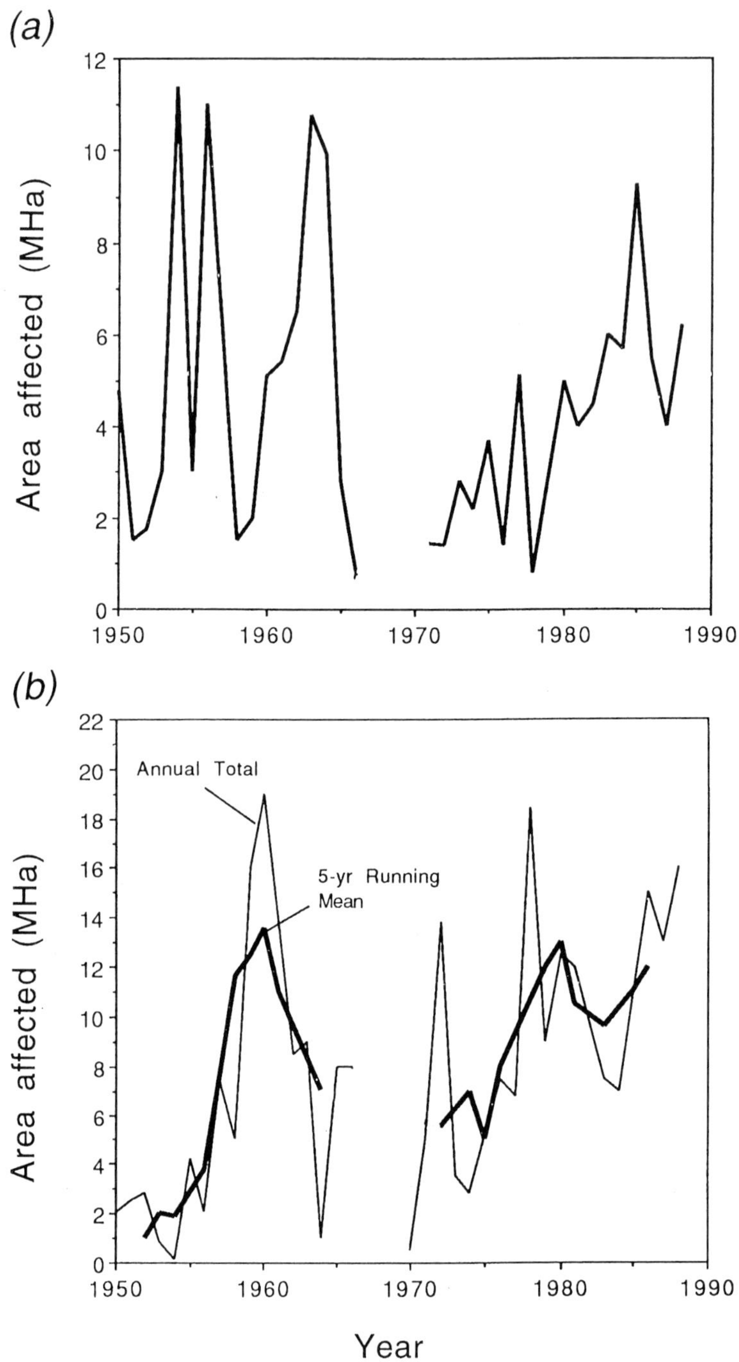

Figure 9.2 *Interannual variations in the area affected by (a) flooding and (b) drought (heavy line, 5-year running mean; light line, annual total). The areas influenced by drought are for croplands and represent a significant fraction of the approximately 100 MHa of cultivated land available in China. (From Smil, 1992.)*

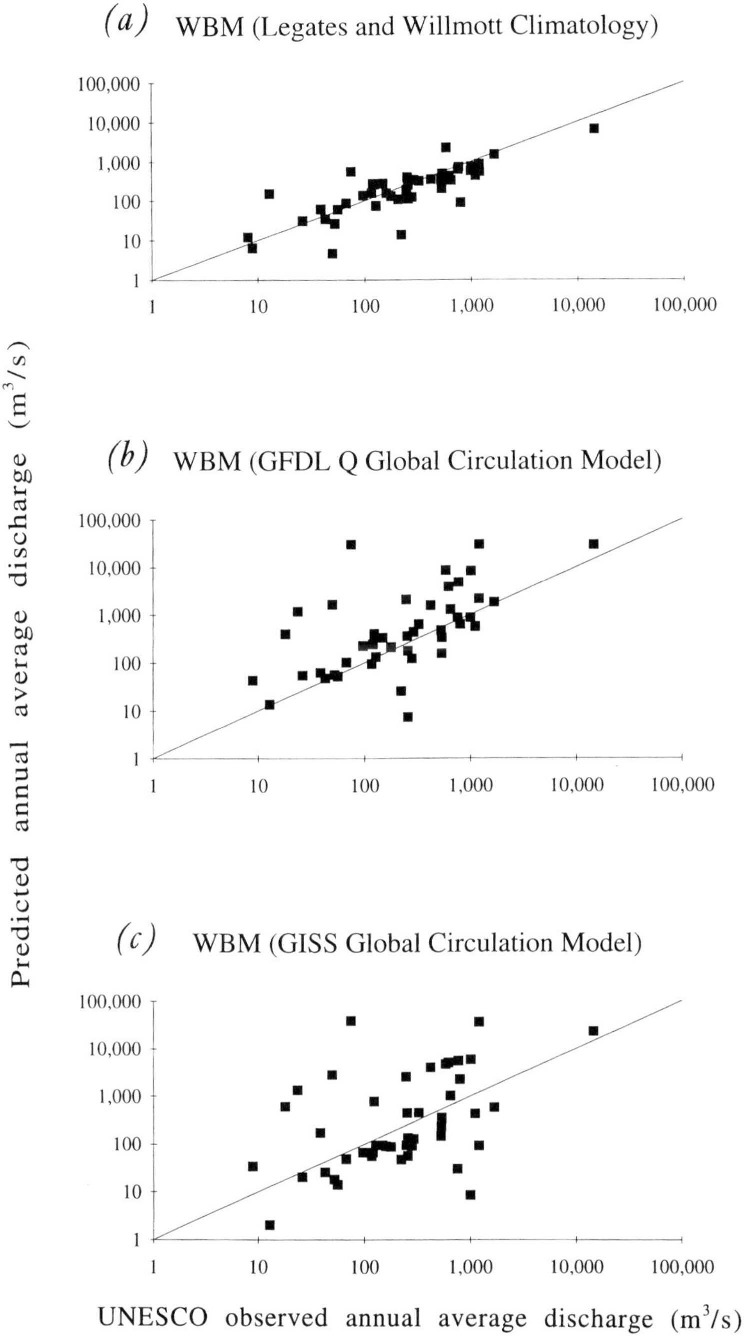

Figure 9.3 *Observed and WBM/WTM - predicted discharges for East Asia. Predictions are by WBM/WTM driven by observational climate fields (top) and GCM outputs (middle/bottom) depicting contemporary climate. Observed flows are from UNESCO (various years). (GFDL, Geophysical Fluid Dynamics Laboratory; GISS, Goddard Institute of Space Studies; WMB/WTM, Water Balance and Transport Model.)*

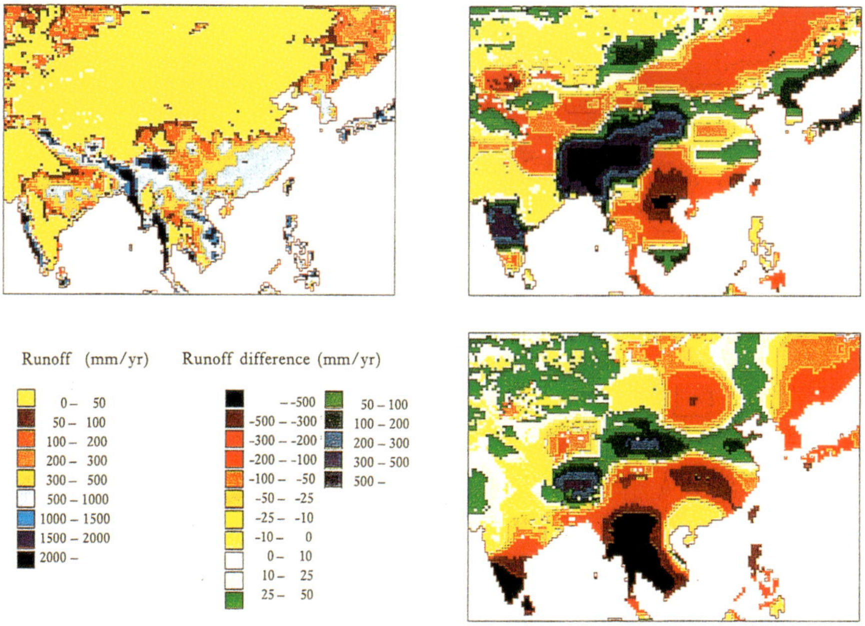

Figure 9.4 *Change in WBM / WTM-predicted runoff (30-minute resolution) for 2 × CO_2 global warming relative to contemporary runoff derived from climatic drivers generated by two GCM models: GFDL-Q Global Circulation Model (top, right) and GISS Global Circulation Model (bottom, right). For reference, contemporary runoff computed by WBM / WTM and based on meteorological station data is also presented (top, left).*

energy is being devoted toward developing an improved representation of land surface hydrology in these models (Sivapalan & Kalma, 1995; Feddes, 1995).

Until suitable refinements are forthcoming, we are left to rely on outputs from the current generation of GCMs to interpret the possible dimension of future water cycle changes associated with the greenhouse effect. Current IPCC (1995) regional assessments, for example, rely on both GCM scenarios as well as hypothetical temperature and precipitation changes to predict impacts on regional hydrology and freshwater ecology. With due caution, we attempt here to interpret at least the broadest of features associated with potential runoff changes predicted by these models for East Asia. We do this with the aim of documenting the degree of improvement necessary to make such models ultimately more useful in a regional water resources context.

Fig. 9.4 gives the change in distributed runoff between the two doubled CO_2 scenarios and their respective contemporary predictions. Contemporary runoff based on station data is shown for reference. Two issues are readily apparent from the complex patterns that emerge. The first is that the spatial distribution of the runoff change bears only modest resemblance across the two scenarios, especially when viewed for individual subregions within East Asia. Second, the magnitude and indeed the direction of these changes are large in absolute terms and indicate that several areas could experience severe drought and/or increased flood potential relative to their current state of water availability.

Discharge at the mouths of the rivers shows this effect as well (Fig. 9.5), and reflects the source area of runoff from which the changes are derived in each scenario. In China, both GCM scenarios under a 2 × CO_2 climate show the majority of rivers south of the North China Plain yielding decreases in discharge. In contrast, those in the north, including the Yellow River, show increases. Notable disparities exist between the models for China's largest river, the Yangtze River. Outside of China, there are significant differences as well (e.g. Vietnam and Cambodia, the Korean peninsula, eastern Russia) pointing again to the great uncertainties with which we currently must deal in assessing these potential changes.

For all of the East Asian rivers shown in Fig. 9.5, the aggregate, gross increase in discharge under greenhouse forcing simulated by the GFDL model is 11 200 m^3

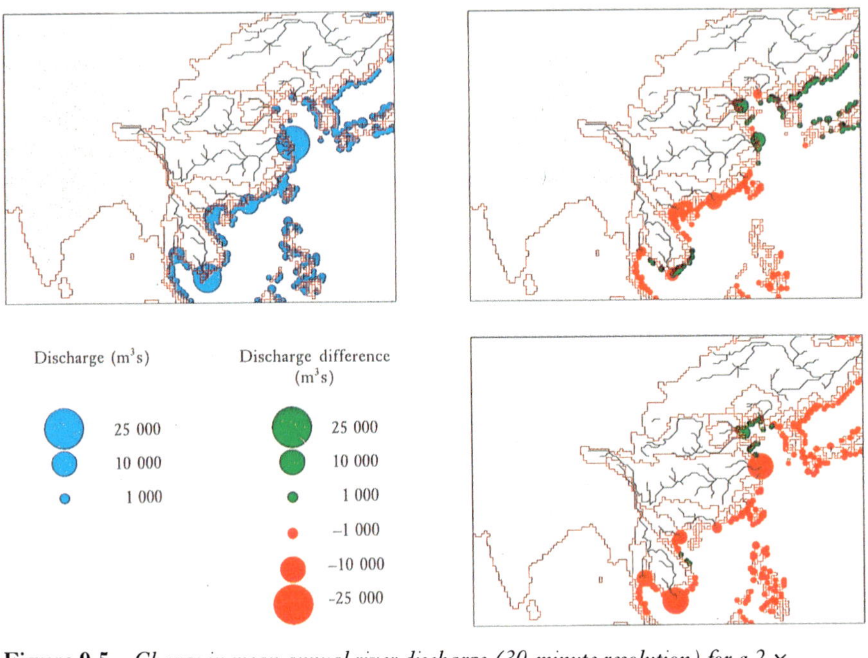

Figure 9.5 *Change in mean annual river discharge (30-minute resolution) for a 2 × CO_2 global warming relative to contemporary discharge derived from climate drivers generated by two GCM models: GFDL-Q Global Circulation Model (top, right) and GISS Global Circulation Model (bottom, right). For reference, contemporary river discharge computed by WBM/WTM and based on meteorological station data is also presented (top, left).*

s^{-1}, while collective decreases total 11 600 $m^3 s^{-1}$. The respective values for the GISS model are 2600 and 34 800 $m^3 s^{-1}$, indicating a much more severe drying overall. These changes are significant, given an accumulated contemporary flux calculated by the WBM/WTM at 78 500 $m^3 s^{-1}$. If either of these scenarios approaches reality, the water management community will face an increasingly urgent challenge in coping with the dual uncertainties of potential water scarcity and water excess at the regional level. In this context, it is interesting to note that the Yellow River has been experiencing a decade-long drought (Milliman & Mei-e, 1995; Smil, 1992), precisely the opposite of what this river basin is expected to sustain under the climate change scenarios that we tested.

Water engineering works

Climatic variability and its attendant, often disastrous, consequences have played a defining role throughout Chinese history. Add to this the potential hydrological effects of deforestation and land cover change, and it is not surprising that river control has served as an important preoccupation for Chinese society over many thousands of years. Examples of water management date back as far as 5000–6000 years ago when Emperor Dai Yu attempted the first massive control of the Yellow River in Henan Province (Tregear, 1980). The original Dujiangyan irrigation project located in the Yangtze River basin, and begun 2200 years ago, serves as the main water supply to more than 700 000 ha of cropland now in use in modern China (Appleton, 1984). Construction of the Grand Canal, the longest artificial waterway in the world, was built in the Sui Dynasty (AD 581–618) to unify the country and to ensure access by drought-prone northern capitals to a reliable source of rice from the southern granary centered in the Changjiang valley (Chang, 1993; Tregear, 1980). It served for hundreds of years as the major transportation route linking northern and southern China and as a means of stabilizing food supplies behind a backdrop of unpredictable climate and water availability.

Once again, the Yellow River provides an important example of the interrelationships between human activity and drainage basin hydrology. Massive dykes were constructed along the Yellow River to selectively protect northern and southern territories from flooding (Chang, 1993). From AD 960–1126 the capital was located at Kaifeng and was protected by the differential strengthening of the southern bank of the river. From the Jin Dynasty starting in 1115 and for 700 years thereafter, the capital at Beijing was protected by improvements on dykes along the north bank, a policy which resulted in major shifts in the river's course and several major floods on the North China Plains. River engineering thus amplified the apparent variability in river discharge. Exacerbating this situation were the likely increase in flooding and river bed instability associated with runoff and erosional changes due to deforestation, amply demonstrating the complex interactions linking climatic variability, land cover change and water resources management (Fig. 9.1).

Water engineering continues in earnest in modern China and is an important focal point for ongoing development efforts. The social investment in such projects is immense. From 1972–1977, more than 100 million people were enlisted during the winter months to work on irrigation, drainage, and flood control projects with the aim of stabilizing agricultural production in the light of unpredictable rainfall, water supply and river discharge (Sigurdson, 1977). More than 85 000 water conservation projects have been instituted since 1949, necessitating the re-settlement of more than 10 million people (Chau, 1995). Irrigation provides an important impetus for harnessing both surface and subsurface water resources. Currently, more than 53% of cultivated land in China is under irrigation, producing in excess of 80% of the national crop output (State Statistical Bureau of China, 1995). The use of groundwater has also been expanding. For example, the number of tube wells increased 20-fold (to two million) in the decade from 1965 to 1975 (Sigurdson, 1977). It has been estimated that 5.5 km^3 or 7% of the 78.6 km^3 of the total groundwater pool is withdrawn annually (Qu *et al.*, 1989).

Dam-building for economic and agricultural development continues to proliferate in China, representing a significant proportion of global river control activities. From 1950 through 1986 Chinese engineers built 17 400 dams greater than 15 m in height, this number representing nearly 50% of the global total for all such structures (ICOLD, 1988, 1984). Construction of the massive Three Gorges Dam on the Yangtze River has begun and is the culmination of feasibility studies first presented in the 1930s (Chau, 1995). Its chief benefits are to be flood control and hydroelectricity, but the project exemplifies the reality that such engineering works are not without their attendant and often significant environmental costs. Several substantial impacts are postulated including: inundation of 632 km^2, about 50% of which is farmland; the displacement of 1.13 million people; inundation of 19 cities with associated economic, municipal, and cultural infrastructure; shifts in commercially important fisheries; siltation of 300 km of navigable channels upstream due to high erosion rates in the surrounding catchment and efficient trapping of sediment by the reservoir (Chau, 1995; Liu & Zuo, 1987). The Three Gorges Dam project demonstrates how a single engineering work can propagate several reciprocal effects through the physical, biological, and socio-economic systems of engineered rivers. The aggregate, global effect of such large impoundments are difficult to assess, but recent inventories on a broad scale show a potentially important imprint of humans (Dynesius & Nielson, 1994; Vörösmarty *et al.*, 1997*a,b*).

Land use, human settlement, and industrial development

A useful heuristic model for predicting the impact of humans on watersheds is one that considers the contrast between natural ecosystems and those disturbed by humans from activities such as forestry, cultivation, grazing, and urbanization. Natural ecosystems process and release biogeochemical constituents based on the interactions among soils, resident vegetation, underlying lithology, and history of weathering. In broad terms, undisturbed terrestrial ecosystems can be considered relatively "closed" and cycle water, carbon, and nutrients through conservative pathways that are self-organizing and self-sustaining (Moore *et al.*, 1989; Vitousek, 1983) (Fig. 9.6). Upon disturbance, these cycles become relatively "open", throughputs become quantitatively more important, and materials are translocated in large quantities from the land into aquatic ecosystems. There is a well-established body of empirical evidence to support this contention at the local scale for both water and nutrients (Oyebande, 1988; Shiklomanov & Krestovsky, 1988; Swank, 1988; Likens *et al.*, 1977; Dillon & Kirchner, 1975). The impact of such changes accumulates across drainage basins and cascades downstream through large river systems, thus helping to define the hydrological response of rivers and providing a link between human activities on the continental land mass and the coastal oceans. We briefly review some of these aggregate effects from both a water quantity and quality perspective.

Impacts on river discharge

Water flows in the Yellow River reflect a long and highly dynamic relationship with human activities occurring inside the basin. The Yellow River provides a useful case study of how land cover change in tandem with climate and water engineering works can potentially alter discharge regime, in particular the occurrence of extreme events. We explore now some of the evidence linking population growth and land cover change to an apparent increase in the incidence of flooding over the last 2000 years.

Through analysis of Chinese chronicles, Fang and Xie (1994) constructed a time series of vegetation redistribution and flood events. In Fig. 9.7 we present their flood frequencies, upon which we have superimposed key events in the history of China (Zhao, 1994; Tregear, 1980; Cotterell & Morgan, 1975). Prior to 200 BC, agriculture within the basin was restricted, a natural forest cover predominated in the loess plateau (Fang & Xie, 1994; Milliman *et al.*, 1987), and reported floods were relatively rare. From 200 BC through the end of the Tang Dynasty at approximately AD 900, population was relatively stable (Cotterell & Morgan, 1975) and the time series shows relatively lower flood frequencies. The lowest values are associated with periods of civil strife and invasion which, according to the authors, restricted sedentary agriculture and facilitated reversion of the landscape back to natural forest or pasture. Starting with the Northern Song dynasty, flood frequen-

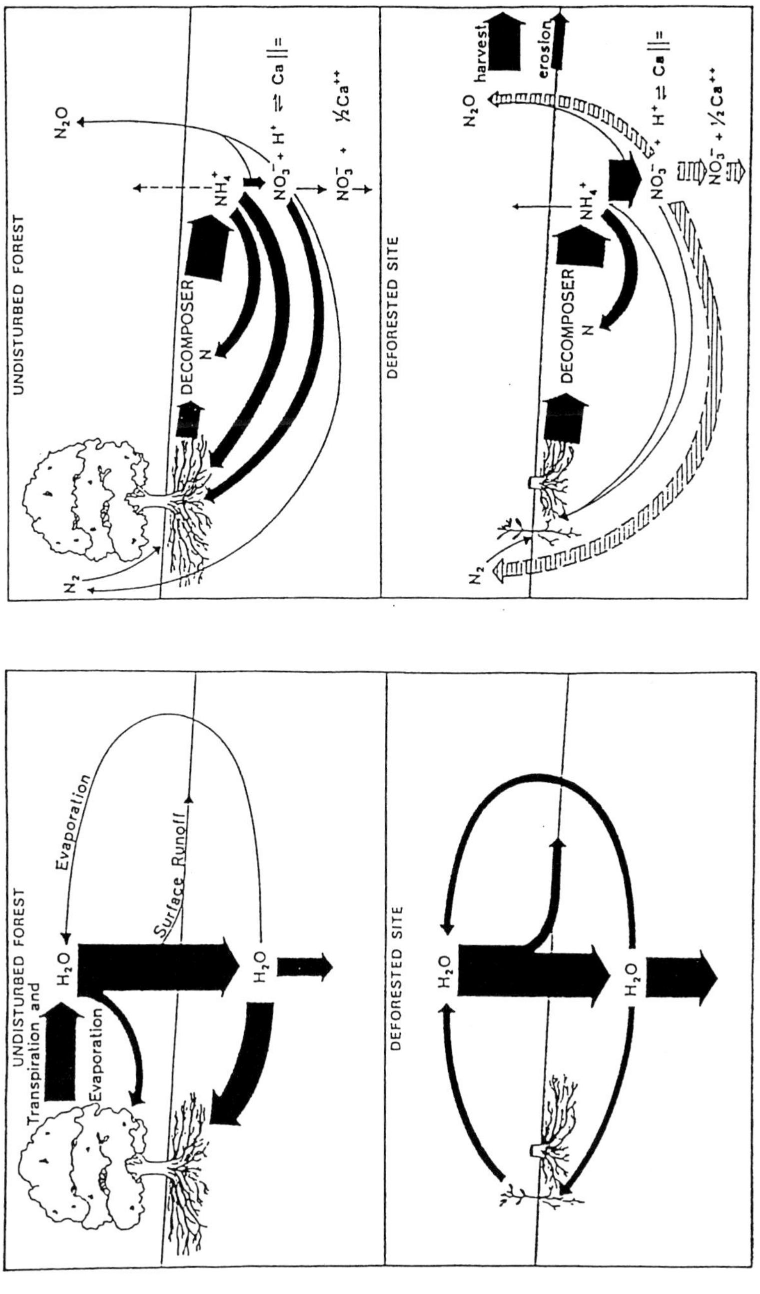

Figure 9.6 *A heuristic model of the effect of deforestation on the water and nitrogen cycles. (From Vitousek, 1983.)*

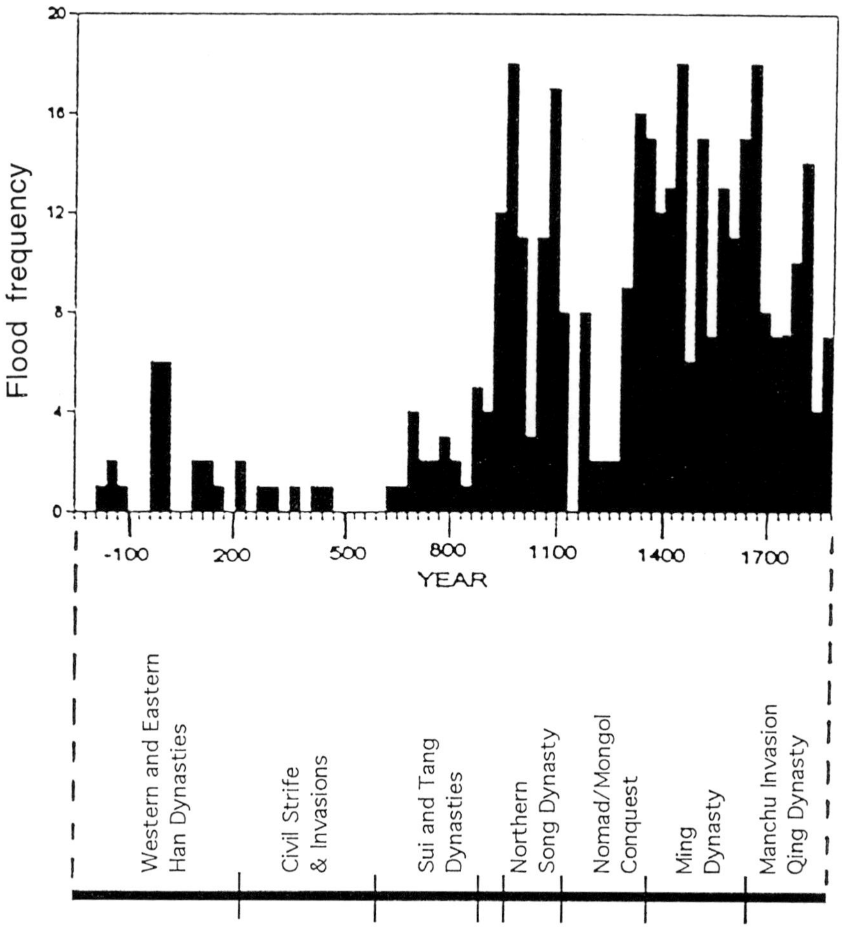

Figure 9.7 *Long-term record of flood frequency (per 30-year interval) in the Yellow River drainage basin and key events in the history of China. (From Fang & Xie, 1994; Zhao, 1994; Tregear, 1980; Cotterell & Morgan, 1975.)*

cies were substantially greater, aided by a rapidly expanding population and punctuated by dramatic periods of high and low flooding frequencies, again associated with periods of civil unrest and/or invasions. Substantially reduced flooding is apparent during the period of the nomad and Mongol conquests from AD 1126 to 1368 as well as during the unrest accompanying the Qing Dynasty, which began in 1644.

Although it can be argued that such archive-based records may be incomplete or inaccurate, when they are corroborated by several independent sources, they can provide reliable proxy information on past climates (Dhérent & Petit-Renaud, 1994; Zhang & Crowley, 1989). In the case of the Yellow River time series (Fig. 9.7) and its interpretation with regard to vegetation change, several factors were analyzed including the historical documents themselves, fossil pollen records, soil information, and archaeological data. Admittedly, some periods of warfare (e.g. AD 420–589) yielded relatively meager records (Chu, 1973), and prior to AD 950 the records for China as a whole are poor. However, those for the central Yellow River are considered to be generally reliable (Wang *et al.*, 1987).

Fang and Xie hypothesize that apparent increases in flooding were brought about by massive deforestation of the loess plateau for fuelwood, charcoal production, agriculture, and brick-making. This, in turn, can be traced to the combined effects of elevated runoff, stream channelization, levee construction, and streambed instability. The instability arises from excessive siltation within the river channel itself. The deposited sediments are derived ultimately from the highly erodible lands cultivated upstream. The Yellow River is well known for this effect, and up to 30% of the sediment load emerging from the loess plateau is deposited before reaching the coast (Martin *et al.*, 1993), causing an aggradation of the river bed up to 10 m above the surrounding floodplain (Ren & Shi, 1986). As the river bed rises, so does its potential for catastrophic flooding. An important and independent confirmation of excessive sediment transport and bed deposition during the period of agricultural expansion is contained in the sedimentary record (Saito & Yang, 1995; Milliman *et al.*, 1987). Starting at approximately 1000–2000 years BP, an order of magnitude increase in particulate transport by the Yellow River was likely.

An opposite set of effects, namely, a lower incidence of river bed siltation and flooding would be expected in the presence of expanded natural forests, afforestation, and/or conversion to pasture. According to Fang and Xie, such land cover changes would be associated with low or migrating populations, foreign invasions, and/or civil strife. Net reductions in streambed siltation and channel stabilization would be expected to take place under such conditions through increased scouring by the relatively underloaded river (Hong *et al.*, 1991).

This interpretation of events, although plausible, ignores the additional influence of climatic variability which may be necessary for correctly interpreting the time series. Two characteristics of the time series are important in this context. The first is the large and broad increase in the general degree of flooding from AD 650 to 1125 and again from AD 1150 to 1880. Chu (1973) presented a 5000-year reconstruction of Chinese climate based on phenological reports derived from historical documents and validated against independent global temperature fluctuations. The period prior to AD 1100 was generally warmer in relation to the present mean, while the opposite was true after AD 1100. Zhang and Crowley (1989) related the Chu time series to several independent measures of wetness/dryness and concluded that relatively warmer temperatures in China are generally associated with wetter but less flood-prone conditions, while cooler temperatures yield the opposite effect. The low flood frequencies from 260 BC to AD 650 as well as the broad expansion of flooding after AD 1125 are both consistent with their conclusion. However, the first broad peak of flood frequencies between AD 650 and 1125 cannot be explained by this conclusion alone, suggesting a role for rapid population growth and land cover change during this period.

The second important feature of the flooding time series are the periodic, sharp reductions within the latter half of the reconstruction. The very large reductions associated with the nomad/Mongol and later Manchu invasions are intertwined with periods of excessive drought as evidenced by widespread dust storms (Zhang, 1984) and increased reports of drought in the historical literature (Wang *et al.*, 1987). So severe were these events in northern China during the latter period that there are reports of a complete absence of rain for the entire growing season, widespread famine, and "men eating men" (Zhao, 1994; Wang & Zhao, 1981). During these times, reduced flooding could simply be a reflection of little or no precipitation.

Thus, we are left with a complex picture of the interplay between population growth, land cover change and climate variability. We speculate that humans have increased the susceptibility of the Yellow River to flooding, and they have done so through the combined effects of land degradation and river control. Accelerated erosion elevates the streambed downstream, reduces its stability, and increases the risk of catastrophic flooding when dykes are breached. The studies we have reviewed here also indicate the importance of climate variability and we see widespread landscape change as setting up the appropriate set of conditions under which climate variations are exacerbated and translated into extreme events. Human management of the landscape within the Yellow River basin is thus implicated as a major factor controlling the flux of water and potentially its variability over a several hundred-year time frame, but it does so in tandem with river engineering and climatic change.

Waterborne constituents

SEDIMENT FLUX

Closely linked to the disturbance of regional runoff and discharge regime is the acceleration of erosion and sediment transport due to land use and land cover change. Table 9.1 summarizes the substantial increase in sediment yields attributable to landscape change in numerous settings around the world. Several-fold increases are common and, in some cases, orders of magnitude greater sediment flux have been observed. Milliman *et al.* (1987), Saito and Yang (1995) and Saito *et al.* (1995) chronicle a time series of sediment transport in the Yellow River based on sedimentological studies (analysis of cores; seismic data; ^{14}C dating; particle size analysis; documentation of coastal progradation) (Table 9.2). According to their interpretation, widespread human use of highly erodible loess soils in what was once predominantly wooded landscape has converted the Yellow River (82nd

Table 9.1 *Observed increases in sediment erosion due to various human activities*

Setting	Magnitude of increase
Cultivation of forest land	
Java	2 times
Trinidad	9
Ivory Coast	18
Tanzania	5
Forest clearance	
Cameron Highlands, Malaysia	5
South Island, New Zealand	Up to 100
Erosion of forest roads	
Idaho (USA)	200 – 500
Urbanization in rainforest	
Malaysia	20
In large rivers, generally	3.5
In small rivers, generally	8

Source: Douglas, 1990; based on several original sources.

Table 9.2 *Relationship between land use in the Yellow River basin and suspended sediment transport.*

Years	Event	Estimated percentage of modern load	Estimated annual load (10^9 MT)
AD 600 – present	Heavy agricultural use; extensive erosion of Loess Plateau	100	1.2
AD 50 – AD 600	Mongol invasion; steppes return to grazed pastures	50	0.6
200 BC – AD 60	Agricultural use with many river channels diverting silt to ocean	80	1.0
340 BC - 200 BC	River channeled; Loess Plateau mostly forested; relatively little sediment lost by channel overflow	35	0.4

Source: Milliman *et al.* (1987).

largest in terms of discharge (Meybeck & Ragu, 1996)) into the world's second largest source of suspended sediment to the ocean.

Such erosional losses are of clear importance to the sustainability and economic viability of Chinese agriculture. A national survey of 45 MHa showed more than 30% of farmland to be suffering from excessive erosion, a proportion that is even greater (50%) for the Yellow River basin (Smil, 1992). There are important repercussions downstream as well, in particular, on the integrity of water engineering schemes. In Shaanxi Province, one-third of reservoir capacity is lost per year relative to new construction due to reservoir siltation (Zeng & Zhou, 1989). The Guanting Reservoir in the Hai He basin (Hebei province) showed a 26% loss of capacity from 1953–1981 (Qu, 1987). Several reservoirs along the Yellow River mainstem have shown 50–80% storage loss within 5 to 7 years and many tributary reservoirs have been abandoned outright (Zeng & Zhou, 1989). In the case of the Sanmenxia Reservoir, a 60% decline in its initial operational capacity within 6 years of construction necessitated costly redesign and reconstruction. The impoundment is reported to operate routinely at one-tenth its maximum design capacity and at 23% of its hydroelectric potential (Volker & Henry, 1988; Zeng & Zhou, 1989). The role of humans in the suspended sediment budget of the Yellow River thus provides a good example of the complex linkages among landscape management, altered hydrology, economics, and the operation of water engineering works.

BIOGEOCHEMICAL FLUXES

The imprint of human activities on river transport processes is not restricted to water and particulate materials alone. Loadings of nutrients are known to have increased dramatically for individual rivers since the beginning of the Industrial Era (Nixon, 1995; Meybeck, 1993; Stanley, 1992; Hinga *et al.*, 1991; Meybeck & Helmer, 1989; Vörösmarty *et al.*, 1986). Cultural eutrophication is also the source of widespread deterioration in lake water quality (e.g. Ryding & Rast, 1989). It has been estimated that only a minority of the world's drainage basins (approximately 20%) have nearly pristine water quality and that the global riverine transport of inorganic N and P have increased several-fold over the last 150 to 200 years (Meybeck, 1982; Wollast, 1983). From numerous individual river basin and coastal zone studies – the Baltic region (Rosenberg *et al.*, 1990); Mississippi River/Gulf of Mexico (Ortner & Dagg, 1995; Turner & Rabalais, 1991, 1994); Northern Adriatic (Justic *et al.*, 1995); the Black Sea (Mee, 1992) – we know that excessive riverborne nutrients, shifts in nutrient limitation, coastal eutrophication, toxic phytoplankton blooms, and bottom-water hypoxia go hand in hand, making the issue of clear relevance to contemporary society.

Accelerated biogeochemical fluxes reflect the collective impact of several anthropogenic activities that transform watersheds. These include deforestation, conversion of natural systems to agriculture and grazing land, wetland destruction, mechanized agriculture, increased application of industrial fertilizer, population and industrial growth, point source pollution, and atmospheric deposition. A useful and composite index of human activities occurring within drainage basins is the *demophoric index*, computed as the product of relative energy use (above subsistence) and population (Vallentyne, 1978). When this index is compared to either inorganic N or P discharged by rivers (Fig. 9.8), it is apparent that high population densities and industrial development yield substantial additional discharge of inorganic nutrients into rivers (Meybeck, 1982).

How does China fit within this pattern of global change? Fig. 9.9 shows some quantitative aspects of China's population and recent exploitation of key strategic resources. In absolute terms, the population of China is about one-sixth of the global total, with a rate of increase not dissimilar to the mean for the rest of the world. Population density is high, however, averaging over 100 people per km^2 of land area for the nation as a whole and rivaling the densities recorded for highly developed western Europe (State Statistical Bureau of China, 1995; Hammond, 1992). And, if we consider that a sizeable fraction of the country is in extremely arid or otherwise inhospitable territory, the effective density is even higher. The Jiang-Su province in the downstream portions of the Yangtze River basin, for example, has 400 inhabitants per km^2 (State Statistical Bureau of China, 1995). Urban areas in China have population densities measured in tens of thousands per km^2 (Zhao, 1994; Houghton-Mifflin, 1992). For the country as a whole, China has one of the largest, contiguous distributions of high population density in the world (Tobler *et al.*, 1995). In terms of both CO_2 emission and industrial fertilizer consumption, China appears to make contributions that are roughly proportional to its share of the global population. However, if expressed in relative terms, there has been a 10-fold increase in CO_2 use from 1955 to 1990 and a nearly 50-fold increase in fertilizer consumption over the same period. Both these changes are remarkable, given the prevailing rate of global increase. China is also an important locus for atmospheric deposition of inorganic N (Galloway *et al.*, 1994), with potential negative impacts on terrestrial ecosystems and enhanced loadings of nutrients into streams (Aber *et al.*, 1996).

With a realization that a significant fraction of the world's population lives in China under conditions of high population density, rapid industrialization, and increasing pressure on land and water resources, we would expect Chinese drainage basins to bear a strong anthropogenic imprint. Table 9.3 lists concentrations and stoichiometric ratios of inorganic nutrients observed in several large rivers of the world. Relative to rivers without significant anthropogenic influence (e.g. Amazon, Yukon), concentrations of inorganic N and P in highly populated drainage basins (e.g. Po, Rhine) show large increases. In the case of N and P in densely populated and highly industrialized Europe, there is an order of magni-

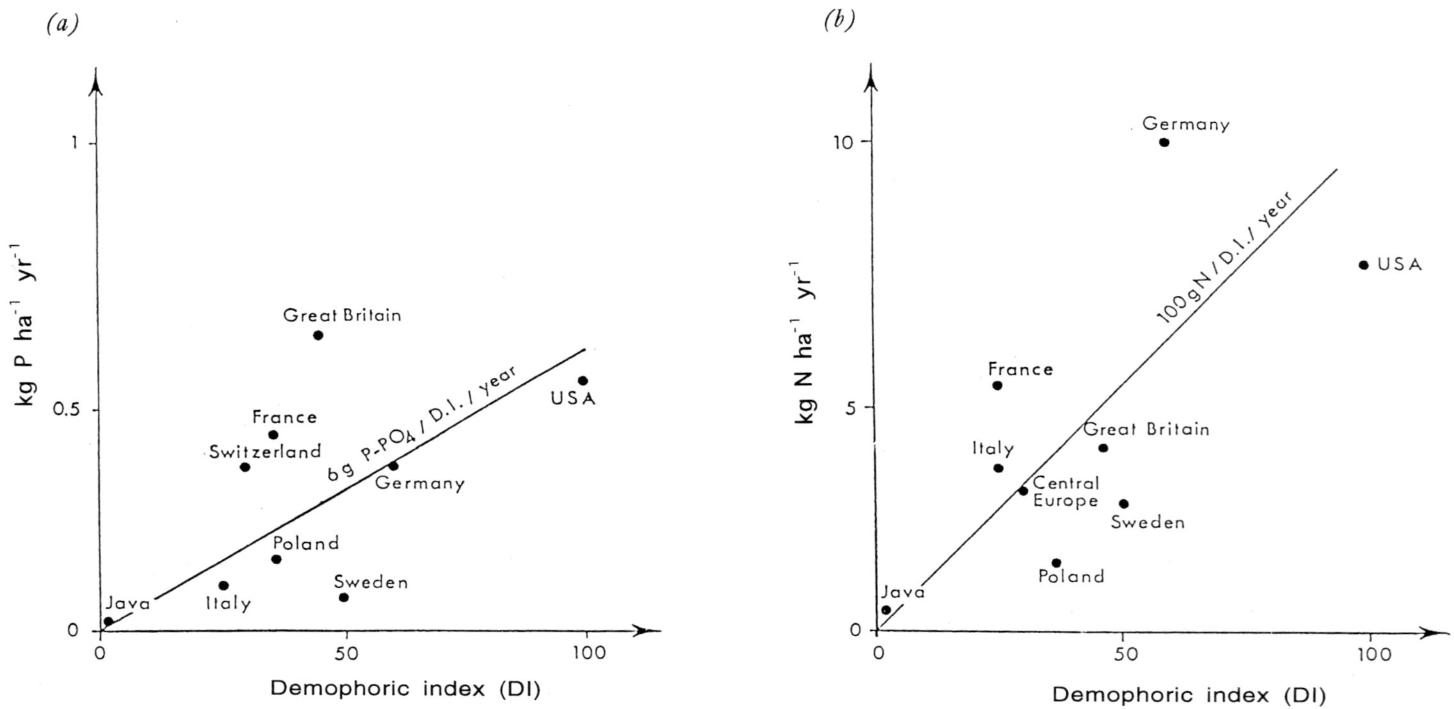

Figure 9.8 *Relationship between the demophoric index (DI) (composite of population and industrial energy consumption relative to subsistence) and excess (a) phosphorus and (b) total dissolved nitrogen loads in rivers. (From Meybeck, 1982.)*

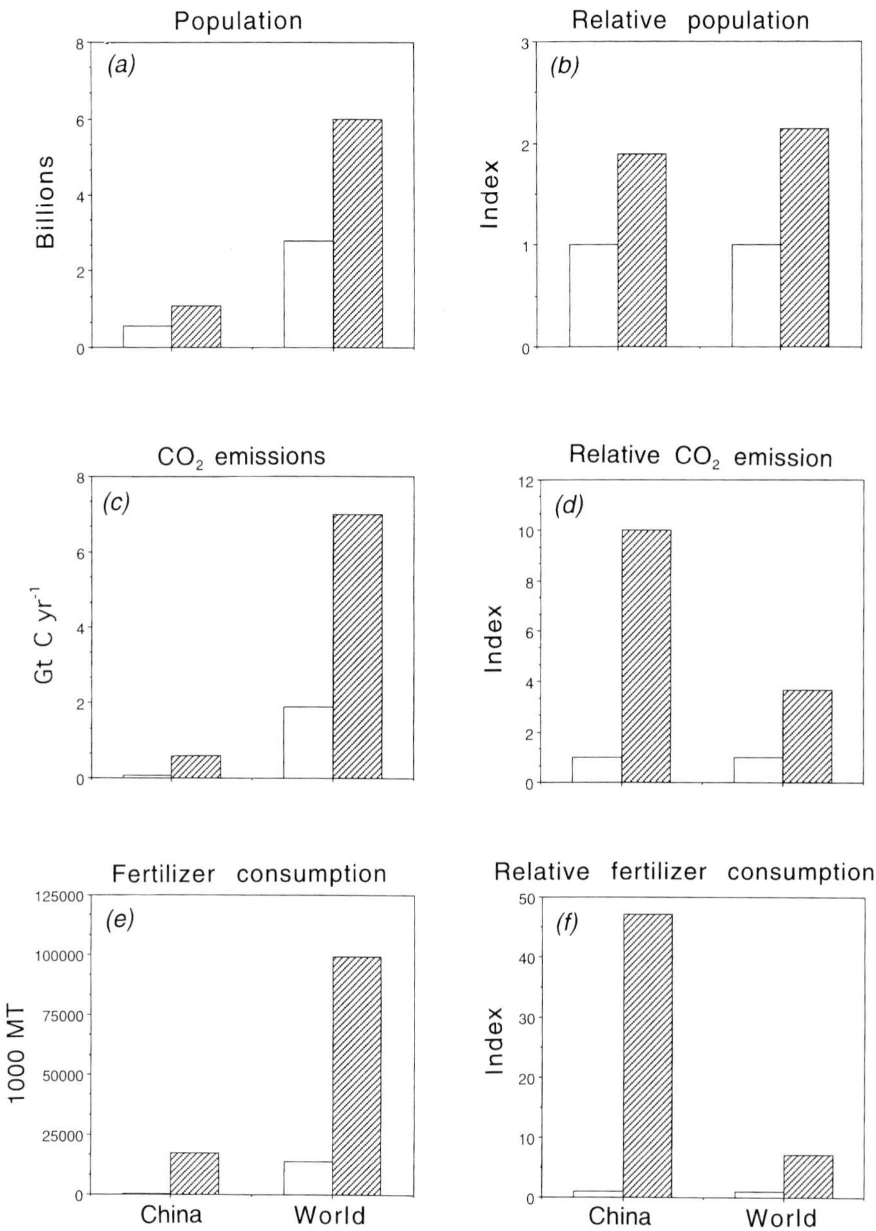

Figure 9.9 *Recent changes in (a, b) population and two key indicators of Chinese and global resource consumption (open column 1955; shaded column, 1990). (c, d) CO_2 emissions are from fossil fuel combustion and (e, f) fertilizer consumption refers to the application of nitrogenous fertilizer. (Sources: State Statistical Bureau of China, 1995; Kang* et al., *1992; He, 1991; FAO, 1987.)*

Table 9.3 *Ambient concentrations and stoichiometric ratios for several key nutrients in pristine, rapidly developing, and industrialized drainage basins*

River	Period	Si (uM)	N (uM)	P (uM)	Si : N	Si : P
Pristine/Near pristine						
Amazon	Before 1972	187	3.2	0.4	58.4	468
	May–June 1976	111–121	7–11	0.3–0.75	12.9	221
Mackenzie	1981–83	143	7.14	0.19	20.0	752
Yukon	1978–85	275	8.35	0.35	32.9	786
Zaire	Nov. 1976	161	7.3	0.72	22.1	224
	May 1978	171	5.9	0.89	29.0	192
Industrialized						
Mississippi	1981–87	108	114	7.7	1.0	14
Po	1981–84	120	147	4.6	0.7	26
Rhine	1976–78	130	310	14	0.4	9
Seine	1976–82	120	372	20	0.3	6
Chinese Rivers						
Yangtze	June 1980	100–105	65	0.5–0.9	1.6	146
Yellow	August 1986	128	64	1.1	2.0	116

Source: Justic *et al.*,1995; compiled from several original sources.

tude (or greater) increase in concentration, an effect commonly identified with severe eutrophication and anoxia in coastal receiving waters (Justic *et al.*, 1995). Such increases are less dramatic, but still significant, in the Yangtze River and Yellow River, particularly for inorganic nitrogen.

As important as the absolute increase is, shifts in the nutrient : nutrient ratios necessary for primary production are important indicators of the state of rivers and their influence on coastal receiving waters. Pristine drainage systems maintain Si : N and Si : P ratios far in excess of metabolic needs of diatomaceous algae (i.e. Si : N : P of 16 : 16 : 1) that are of critical importance to coastal productivity. In drainage basins influenced by human development, the Table shows ratios that are much smaller, which in turn creates a stoichiometric Si limitation on the development of diatoms and a competitive advantage to non-diatomaceous phytoplankton, a flora that is often toxic (Smayda, 1989, 1990). Both the absolute increase and stoichiometric shifts in nutrient status indicate an intermediate position for Chinese rivers in the spectrum from pristine to highly eutrophied drainage systems.

THE ROLE OF MANAGEMENT

We would expect in China a progressive movement toward conditions that now characterize European and US rivers in the wake of expanding industry, urbanization, and continued application of industrial fertilizers. In Europe, mechanized agriculture contributes a significant fraction of all anthropogenic emissions of N and P into surface waters, from 28–82% and 16–38%, respectively, when tabulated on a countrywide or regional basis (Isermann, 1993). Agriculture should therefore play a pivotal role in defining the future status of water quality in China's drainage basins.

We now explore in more detail the nature of losses that might be expected from the sustained use of industrial fertilizer on Chinese cropland, and contrast these with scenarios from the United States, where significant changes in water quality have been linked to industrial agriculture (Gianessi *et al.*, 1981; Briggs & Ficke, 1978). Our focus is on the nitrogen cycle in agroecosystems which we simulated across several sites in China and the US using the DNDC Model of Li *et al.* (1992) (Fig. 9.10). The model is a mechanistic simulation of major pools of organic and inorganic nitrogen, the fluxes of N among these pools and across system boundaries via leaching, ammonia volatilization, and denitrification. DNDC has four submodules representing soil climate, plant growth, denitrification and decomposition. Basic inputs include data on daily air temperature and precipitation, soil parameters, crop type, cropping practices and fertilization. Table 9.4 gives the basic biophysical conditions for each of eight sites. It is important to note that, in keeping with current practices, the yearly fertilizer input is introduced in several small doses in China while, in the US, it is applied once per year. The scenarios for both countries depict a mixture of N delivered as ammonium, urea, anhyrdrous ammonia and nitrate.

Results from these simulations (Fig. 9.11) indicate, for both Chinese and US sites, the expected positive response of crop yields to increasing levels of fertilization. However, the nature and quantity of the loss terms vary significantly between the two agro-ecosystems. Most relevant to the issue of river water quality is the progressive increase in leaching losses predicted for US sites across all ranges of fertilization. In contrast, only at the highest levels of N application do the Chinese systems lose significant quantities of N through leaching, and this relative degree of retention can be traced to the timed application of fertilizers, which maximize uptake into growing crops.

If we consider the widespread utilization of industrial fertilizers in the US to have begun in the 1940s, the progression to a level of 200 kg ha^{-1} or more (typical of many sites throughout the US) (USDA/ERS, 1989) represents a time series of potential pollution that has progressed for 50 years. Results from the DNDC would indicate a relatively high potential for N mobilization that can eventually find its way into inland water systems. Large corresponding increases in water-

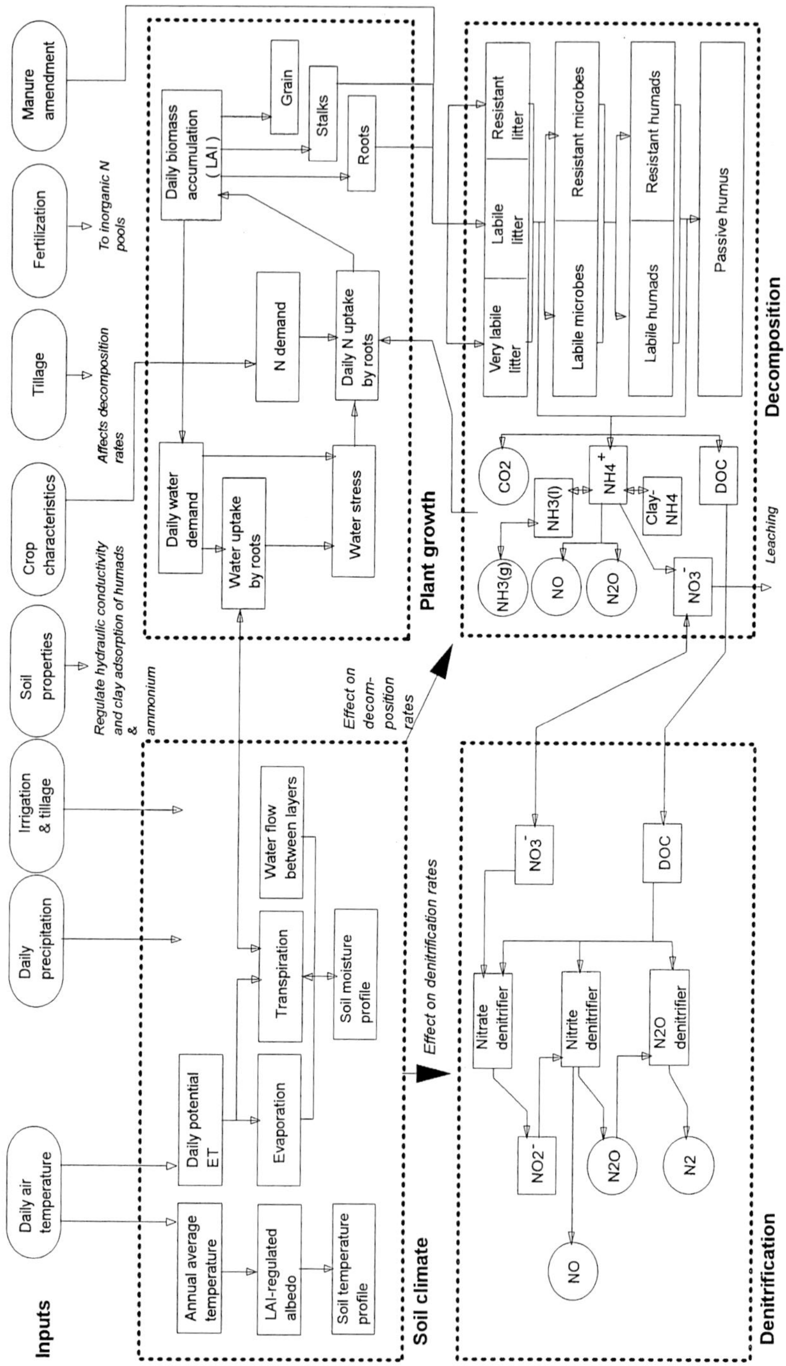

Figure 9.10 *The DNDC model (Li* et al., *1992) used to explore the response of typical Chinese and US agroecosystems to varying levels of industrial fertilizer application.*

Table 9.4 *Basic attributes of the sites tested in the DNDC simulations of fertilization response in Chinese and US agroecosystems*

Site	State/ province	Crop	Soil texture	Temp (°C)	Precip (mm year^{-1})	Soil pH	% soil carbon
China							
Baoding	Hebei	Corn	Silty loam	12.5	520	7.5	0.6
Weifang	Shandong	W. wheat	Silty loam	13.1	670	7.0	1.1
Jian	Jiangxi	1-crop (rice)	Silty clay	17.5	1598	5.5	1.5
Haifen	Guangdong	2-crop (rice)	Clay loam	21.8	1980	5.5	1.0
United States							
Hamilton	Iowa	Corn	Sandy clay loam	9.8	987	6.5	2.3
Dundy	Nebraska	S. wheat	Silty loam	10.5	596	7.0	1.3
Jenkins	Georgia	Oats	Sandy loam	19.5	1043	6.5	0.9
Clay	Florida	Vegetables	Loamy sand	22.9	1116	6.0	0.5

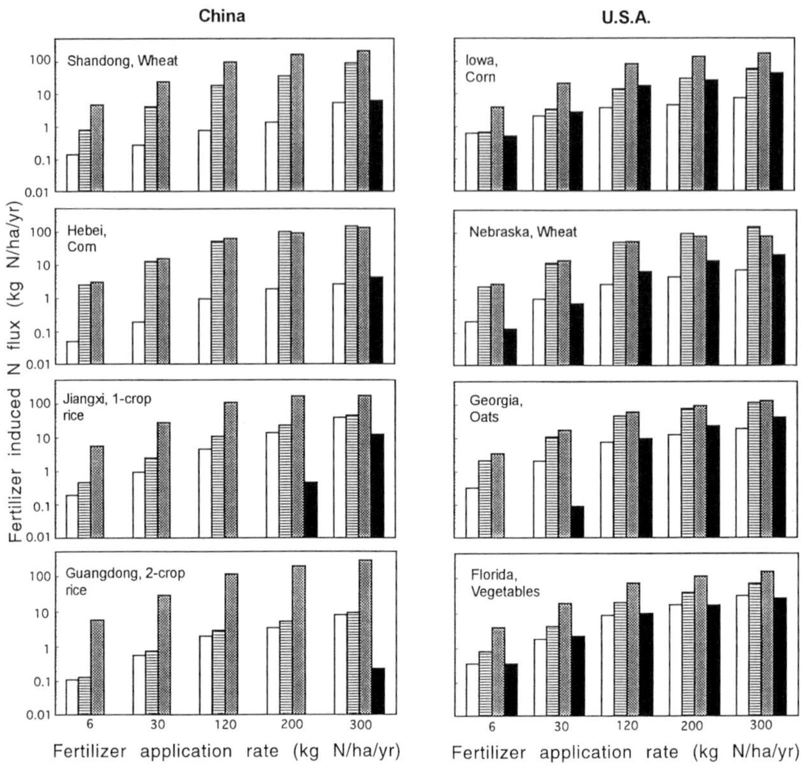

Figure 9.11 *Results from the DNDC model showing outputs of N from Chinese and US cropland as a function of N fertilization. The generally small losses and apparent efficiency of Chinese cropland are related to carefully timed fertilizer application. US farms typically apply a single dose of fertilizer. ($N_2O + NO + N_2$ emission ☐; NH_3 volatilization ▤; N-uptake by crop ▩; NO_3^- available for leaching ■.)*

borne N, in fact, have been observed, as in the case of the Mississippi River (Turner & Rabalais, 1991; Vörösmarty *et al.*, 1986), which drains a vast area of industrial agriculture.

Although the widespread application of nitrogenous fertilizer in China began in earnest only in the late 1960s, it surpassed aggregate US consumption by the early 1980s despite a smaller cropland base (United Nations, 1990). Furthermore, the mean application rate progressed to 188 kg N ha^{-1} $year^{-1}$ in less than 30 years (State Statistical Bureau of China, 1995) and is estimated to rise to above 200 kg N ha^{-1} $year^{-1}$ by the year 2000 (FAO/World Bank, various years). Despite these loadings, the DNDC indicates relatively conservative N budgets for Chinese agro-ecosystems. And, as evidenced by water quality observations in the Yangtze River and Yellow River (Table 9.3), China is in but the early to intermediate stages of water pollution from excessive nutrient loading. Although the question requires significant additional study to identify other factors contributing to the observed water chemistry, management practices in Chinese agriculture may help control land-based N emissions that otherwise would find their way into streams, rivers, and ultimately into the Chinese coastal zone.

An emerging role for IGBP in drainage basin research

Despite our overall judgment that river systems have not received their deserved place in the global change research agenda, attention is beginning to be focused on the issue of large-scale riverine transport, both at the national and international levels. We see several promising, recent developments:

- Formation of the IGBP-PAGES Fluvial Working Group and their ongoing activities in the realm of sediment transport during the period of sedentary agriculture (February 1994 to present).
- Several workshops in which the drainage basin concept at continental and global scales was presented and discussed in the context of Earth System modeling including: the US Interagency Workshop on Earth System Modeling (Washington, DC; May 1994); IGBP Workshop on Modeling River Transports from (Durham, NH; December 1994); SCOPE Estuarine Synthesis Meeting (Orange County, CA; February 1995); GEMS-Water/UNEP/WHO Expert Panel on Land-Based Sources of Pollution (Koblenz, June 1995).
- Development of the GEMS-Water/UNEP/WHO and IGBP-LOICZ Global Registry of River Inputs (GLORI) databases.
- Ongoing planning for the Large-scale Biosphere–Atmosphere Field Experiment in Amazonia (LBA) in which a nested drainage basin approach has been adopted as an integrator of hydrologic and constituent fluxes.
- Several current and future GEWEX projects centered on large drainage systems of the world (e.g. Mississippi River, Mackenzie River, Asian Monsoon region).
- Some key papers quantifying the continental and global-scale impacts of humans on the terrestrial water cycle and material balances (e.g. Dynesius & Nilsson, 1994; Postel *et al.*, 1996; Howarth *et al.*, 1996; L'vovich & White, 1990; Meybeck & Helmer, 1989).

The stage is thus set for significant progress in macro-scale drainage basin analysis, given the improved articulation of the role that rivers and freshwaters play in the Earth System and the growing mobilization of the international research community to study this issue.

It is noteworthy that several of the developments listed above were catalyzed by the IGBP, and it is likely to play an important role in the future. Four Programme Elements within the IGBP have interests in this arena:

- BAHC (Biospheric Aspects of the Hydrological Cycle) Core Project seeks to elucidate reciprocal feedbacks on the water cycle from the land surface, its biospheric components, and the climate system. A major emerging theme is quantification of the linkages between vertical and horizontal fluxes of water that occur across heterogeneous landscapes. Part of this analysis includes estimating the translocation of materials derived from terrestrial ecosystems as they pass through, and are processed by, drainage basins and river systems.
- PAGES (Past Global Changes) Core Project is aimed at securing an improved understanding of the dimensions of contemporary global change by studying the dynamics of past change, both in terms of climatic variability and more direct anthropogenic disturbance. A working group has been assembled within PAGES to identify the impacts of humans on fluvial systems including changes in riverine material transfers since the dawn of sedentary agriculture.
- LOICZ (Land–Ocean Interactions in the Coastal Zone) Core Project is devoted to understanding coastal and near-shore settings in an Earth Systems perspective, and how global change influences these environments. The effects of changes in freshwater-derived constituents provides an important and time-varying boundary condition for modeling studies within the LOICZ Core Project.
- GAIM (Global Analysis, Interpretation, and Modeling) Task Force fosters the integration and assessment of emerging Earth System Models with the aim of developing an improved scientific understanding of the global change issue. GAIM seeks to achieve this objective through intercomparison studies, database development, and technology transfer to the Third World. Recent discussions (April 1996) at the IGBP Congress in Bad Muenstereiffel (GERMANY) articulated the need to consider drainage basin fluxes for closure of global biogeochemical budgets.

Durham workshop recommendations

In December of 1994, a three-day IGBP/UNESCO-sponsored workshop entitled "Modeling the Delivery of Terrestrial Materials to Freshwater and Coastal Ecosystems" was convened in Durham, New Hampshire. The approximately 30 Workshop participants were drawn from several Project Elements within the IGBP including BAHC, PAGES, LOICZ, GAIM, and LUCC. Representatives from UNESCO, IAHS, and GEMS-Water/UNEP/WHO were also in attendance. It also entrained the expertise of the general scientific community active in drainage basin and riverine transport studies. The diversity of Workshop participants reflects the important fact that the scientific issues are wide-ranging and that their successful analysis requires the involvement of both the IGBP and several other international organizations.

The Durham Workshop Report (Vörösmarty *et al.*, 1997*c*) offers a general framework for IGBP activities in river basin research, Earth System modeling, and the broader global change question. Four key contributions could be made by IGBP to the emerging scientific dialogue: (1) provide a global inventory of riverborne constituent flux, (2) identify key controls on these fluxes, (3) identify feedbacks of anthropogenic change in drainage basins on biogeochemical cycling, and (4) quantify the feedbacks of global change on human society. The constituents of interest are water *per se*, sedimentary materials, organic matter, and nutrients. Several time frames were recommended to be considered, including a paleo-component (18 000 years before present to the end of the pre-industrial era), the industrial period, contemporary, and future 100 years. The science would be organized around selected case studies, regional to continental-scale focus areas, and global syntheses. Two general classes of models would be developed – statistical and process-based. The statistical models would be used to generate the flux inventories and to provide guidance in identifying key controlling variables for the process-based models. These process models would, in turn, address the issue of flux controls and feedbacks. A river classification scheme, or "typology", based on river behavior, watershed attributes, and/or recipient coastal ecosystems would be used to extrapolate information from well-monitored to poorly monitored drainage basins. Several supporting data sets would be required: experimental and literature-based data for the case studies, monitoring program and GIS-based biophysical data for developing the typology, and remote sensing information to aid in extrapolation across scales.

An important and emerging aspect of the overall IGBP initiative is coordination of activities across individual Core Projects that will make most efficient use of resources for carrying out field and modeling studies (e.g. IGBP-BIG [BAHC/IGAC/GCTE] coordinated field studies; IGBP-GAIM/DIS/GCTE terrestrial ecosystem model intercomparison studies). To address the drainage basin/anthropogenic change issue effectively, a coordination within IGBP Program Elements and with outside organizations such as UNESCO, IAHS, GEMS-Water/UNEP/WHO is prudent, to foster synergy among Program Elements, to avoid duplication of effort, and to better serve the needs of the broader scientific community. These, and several more detailed scientific and organizational recommendations, are offered in the Durham Workshop Report (IGBP Report #39) and the interested reader is directed to consult this document for further information.

Summary and conclusions

In this chapter we discussed several aspects of anthropogenic change as related to drainage basins and river systems, and promoted these as an emerging Earth System and global habitability issue. We note the traditional lack of mobilization of the scientific community around this issue, in part due to the lack of a unified global signal of such change. We identified three major elements of disturbance: climate change/variability; water engineering works; and land cover change and human settlement. These have individual as well as synergistic effects, and we demonstrated these concepts using several examples from China. China is certainly an actor on the global stage, both affected by, and helping to accelerate, the rates of anthropogenic change.

Chinese river basins, in particular, serve as an excellent example of how humans interact with a changing natural environment and attempt to control several of the forces of such change, often with costly and even disastrous consequences. Climatic variability has conveyed important impacts on Chinese society for several thousand years, and has often led to severe flooding, drought, and famine. In attempts to ameliorate the negative effects associated with such variability, large-scale water engineering works were instituted, dating back several thousand years. Water control projects continue into the present and are likely to be needed well into the future. Substantial uncertainties are evident in projected future changes in regional climate, water balance, and river discharge as projected by recent GCM experiments simulating a $2 \times CO_2$ climate. As a result, rivers in East Asia show large projected increases as well as decreases for several individual drainage basins, providing Chinese planners with troublesome uncertainty, while at the same time pointing to the obvious necessity of improving upon such models. Land use and land cover change are important factors defining the nature of water balances as well as the mobilization and transport of sediments and nutrients. Widespread deforestation favors increased runoff and often severe flooding, as has likely been the case in the Yellow River which has suffered from recurrent, severe flooding for at least the last 1000 years. Deforestation also leads to increased erosion, excess siltation of streambeds and, thereby, increased vulnerability to additional flooding. Enhanced sediment delivery also has a clear economic cost through reservoir siltation and associated losses in storage capacity and hydroelectric potential. Because of point source pollution, accelerated use of industrial fertilizers, and rapid economic expansion, Chinese rivers are progressing toward a polluted state characteristic of more highly developed regions of the globe. Chinese agricultural practices, in particular the timed application of nitrogenous fertilizers, have to some degree delayed the onset of significant net losses of N from cropland to rivers. This points to the importance of management practices carried out within drainage basins as a means to simultaneously improve agricultural efficiency and protect aquatic ecosystems. Nonetheless, several types of Chinese cropping systems show a finite capacity to retain exogenous nutrients and may soon be contributing significant quantities of N to inland waterways.

We also reviewed several institutional issues surrounding the question of global change and drainage basins. We listed numerous recent meetings and workshops devoted to these issues. It is noteworthy that several of these were led by IGBP Program Elements, indicating a potential role for IGBP in the future development of this science. A coordination of individual Program Element activities as well as integration with ongoing work by allied international organizations are prerequisites to a successful IGBP role. In keeping with its overall mandate, the Programme can serve as a catalyst for assembling relevant groups of international scientists, disseminating technical information, developing appropriate databases and models, and interpreting the state-of-the-art science to emerge from these activities.

Acknowledgments

The authors wish to thank B. Fekete, B. Tucker, A. Schloss, and T. Hibschmann for valuable assistance in the preparation of data and graphical output. This work was supported by NASA Mission to Planet Earth Grant #NAGW-2669, the IGBP Secretariat, IGBP-BAHC Core Project, the University of New Hampshire's Institute for the Study of Earth, Oceans, and Space, and the US Department of Energy CHAMMP (DE-FG02-96ER61473).

References

Aber, J. D., Magill, A., McNulty, S. G., Boone, R. D., Nadelhoffer, K., Downs, M. & Hallett, R. (1996). Forest biogeochemistry and primary production altered by nitrogen saturation. *Water, Air, and Soil Pollution*, **85**, 1665 - 70.

Appleton, B. (1984). Li Bing's masterpiece stands test of time. *World Water*, May 1984: 23–5.

Baumgartner, A. & Reichel, E. (1975). *The World Water Balance: Mean Annual Global, Continental and Maritime Precipitation and Runoff.* Amsterdam: Elsevier.

Briggs, J. C. & Ficke, J. F. (1978). *Quality of Rivers of the United States, 1975 Water Year – Based on the National Stream Quality Accounting Network (NASQUAN)*. US Geological Survey Open File Rpt. 78–200. Reston, VA: USGS.

Chang, W. Y. B. (1993). Effects of water management policy on the formation of Lakes Hongtze and Kaoyu, two large Chinese lakes. *Journal of Great Lakes Research*, **19**, 675–80.

Chang, W. Y. B. & B. King (1994). Centennial climate change and their global associations in the Yangtze River. *Climate Research*, **4**, 95–103.

Chau, K. (1995). The Three Gorges Project of China: resettlement prospects and problems. *Ambio*, **24**, 98–102.

Chu, K. (1973). A preliminary study on the climatic fluctuations during the last 5 000 years in China. *Scientia Sinica*, **16**, 226–55.

Cotterell, A. & Morgan, D. (1975). *China's Civilization*. New York: Praeger.

Diaz, H. F. & Fu, C. (1987). Regional precipitation and temperature variability and its relationship to the southern oscillation. In *The Climate of China and Global Climate*, ed. D. Ye, C. Fu, I. Chao & M. Yoshino, pp. 249–55. Berlin: Springer-Verlag.

Dhérent, C. & Petit-Renaud, G. (1994). *Using Archival Resources for Climate History Research*. International Hydrological Programme - IV, SC-94/WS.6. Paris: UNESCO.

Dillon, P. J. & Kirchner, W. B. (1975). The effects of geology and land use on the export of phosphorus from watersheds. *Water Research*, **9**, 135–48.

Douglas, I. (1990). Sediment transfer and siltation. In *The Earth as Transformed by Human Action*, ed. B. L. Turner, W. C. Clark, R. W. Kates, J. F. Richards, J. T. Mathews & W. B. Meyer, pp. 215–34. Cambridge: Cambridge University Press.

Dümenil, L. & Todini, E. (1992). A rainfall-runoff scheme for use in the Hamburg climate model. In *Advances in Theoretical Hydrology, A Tribute to James Dooge*, ed. J. P. O'Kane, *European Geophysical Society Series on Hydrological Sciences*, 1. Amsterdam: Elsevier.

Dynesius, M. & Nilsson, C. (1994). Fragmentation and flow regulation of river systems in the northern third of the world. *Science*, **266**, 753–62.

Famighetti, R. (ed.). (1995). *The World Almanac and Book of Facts 1996*. Mahway, New Jersey: Funk and Wagnalls.

Fang, J-Q. & Xie, Z. (1994). Deforestation in preindustrial China: the Loess Plateau region as an example. *Chemosphere*, **29**, 983–99.

FAO (1987). *Current World Fertilizer Situation and Outlook, 1985/86 – 1991/92*. Rome: UN Food and Agricultural Organization.

FAO/World Bank (various years). FAO Yearbook Tapes (1969–85) and World Bank, Economic Analysis and Projection Dept. (1986–2000).

Feddes, R. (ed.). (1995). *Space and Time Scale Variability and Interdependencies in Hydrological Processes*. Cambridge: UNESCO/ Cambridge University Press.

Galloway, J. N., Levy, H. & Kasibhatia, P. S. (1994). Year 2020: consequences of population growth and development on deposition of oxidized nitrogen. *Ambio*, **23**, 120–3.

Gianessi, L. P., Peskin, H. M. & Young, G. K. (1981). Analysis of national water pollution control policies, 1, A national network model. *Water Resources Research*, **17**, 796–801.

Guo, Q. (1987). The East Asia Monsoon and the southern oscillation, 1871–1980. In *The Climate of China and Global Climate*, ed. D. Ye, C. Fu, I. Chao & M. Yoshino, pp. 249–55. Berlin: Springer-Verlag.

Hahn, C., Warren, S. G., London, J. & Roy, J. L. (1988). *Climatological Data for Clouds over the Globe from Surface Observations*, Oak Ridge, TN: CDIAC, US Department of Energy.

Hammond, Inc. (1992). *Hammond Citation World Atlas*. Maplewood, New Jersey: Hammond.

Hartmann, D. L. (1994). *Global Physical Climatology*. San Diego: Academic Press.

He, K. (ed.). (1991). *China Agricultural Yearbook*. Beijing: Agricultural Publishing House.

Hinga, K. R., Keller, A. A. & Oviatt, C. A. (1991). Atmospheric deposition and nitrogen inputs to coastal waters. *Ambio*, **20**, 256–60.

Hong, Y., Piao, H. & Jiang, H. (1991). Geoenvironmental characters of Yellow (Huanghe) River suspended sediment. *Science in China*, **34B**, 859–70.

Houghton-Mifflin, Inc. (1992). *Webster's II: New Riverside Desk Reference*. Boston: Houghton-Mifflin.

Howarth, R. W., Billen, G., Swaney, D., Townsend, A., Jaworski, N., Lajtha, K., Downing, J. A., Elmgren, R., Caraco, N., Jordan, T., Berendse, F., Freney, J., Kudeyarov, V., Murdoch, P. & Zhao-liang, Z. (1996). Regional nitrogen budgets and riverine N and P fluxes for the drainages to the North Atlantic ocean: natural and human influences. *Biogeochemistry*, **35**(1), 75 - 139.

ICOLD (1988, 1984). *World Register of Dams*. 1984 full edition and 1988 updating, Paris: International Commission on Large Dams.

IPCC (1995). Freshwater systems and hydrology. In *Climate Change 1995 - Impacts, Adaptations and Mitigation of Climate Change: Scientific-Technical Analyses*, ed. R.T. Watson, M.C. Zinyowera & R.H. Moss.

Contribution of Working Group II of the Second Assessment Report of the Intergovernmental Panel on Climate Change, Cambridge: Cambridge University Press.

Isermann, K. (1993). Terrestrial, continental, and global aspects of C, N, P, and S emissions from agricultural ecosystems. In *Interactions of C,N, P, and S Biogeochemical Cycles and Global Change*, ed. R. Wollast, F. T. Mackenzie & L. Chou, pp. 79–121. Berlin: Springer-Verlag.

Justic, D., Rabalais, N. N., Turner, R. E. & Dortch, Q. (1995). Changes in nutrient structure of river-dominated coastal waters: Stoichiometric nutrient balance and its consequences. *Estuarine, Coastal and Shelf Science*, **40(3)**, 339–56.

Kang, D., Zhang, M. & Chen, L. (1992). Changes in carbon and nitrogen contents in the environment of China. In *Global Change and China*, ed. D. Ye & P. Chen. Beijing: Earthquake Publishing House.

Kite, G. W., Dalton, A. & Dion, K. (1994). Simulation of streamflow in a macroscale watershed using general circulation model data. *Water Resources Research*, **30**, 1547–59.

Korzoun, V. I., Sokolov, A. A., Budyko, M. I., Voskresensky, K. P., Kalinin, G. P., Konoplyantsev, A. A., Korotkevich, E. S., & L'vovich, M. I. (eds.). (1978). *Atlas of World Water Balance and Water Resources of the Earth*. Leningrad: USSR Committee for the International Hydrological Decade. (English translation, Paris: UNESCO).

Legates, D. R. & Willmott, C. J. (1988). *Global Air Temperature and Precipitation Data Archive*, Newark, DE: Dept. of Geography, University of Delaware.

Li, C., Frolking, S. & Frolking, T. A. (1992). A model of N_2O evolution from soil driven by rainfall events: 1. Model development and sensitivity. *Journal of Geophysical Research*, **97**, 9759–76.

Likens, G. E., Bormann, F. H., Pierce, R. S., Eaton, J. S. & Johnson, N. M. (1977). *Biogeochemistry of a Forested Ecosystem*. New York: Springer-Verlag.

Liu, C. & Zuo, D. (1987). Environmental issues of the Three Gorges Project, China. *Regulated Rivers: Research and Management*, **1**, 267–73.

L'vovich, M. I. & White, G. F. (1990). Use and transformation of terrestrial water systems. In *The Earth as Transformed by Human Action*, ed. B. L. Turner, W. C. Clark, R. W. Kates, J. F. Richards, J. T. Mathews & W. B. Meyer, pp. 235–52. Cambridge: Cambridge University Press.

Martin, J. M., Zhang, J., Shi, M. C. & Zhou, Q. (1993). Actual flux of the Huanghe (Yellow River) sediment to the western Pacific Ocean. *Netherlands Journal of Sea Research*, **31**, 243–54.

Matthews, P. (ed.). (1992). *The Guinness Book of Records, 1993*. New York: Facts on File.

Mee, L. D. (1992). The Black Sea crisis: a need for concerted international action. *Ambio*, **21**, 278–86.

Melillo, J. M., McGuire, A. D., Kicklighter, D. W., Moore, B., Vörösmarty, C. J. & Schloss, A. L. (1993). Global climate change and terrestrial net primary production. *Nature*, **363**, 234–40.

Meybeck, M. (1982). Carbon, nitrogen, and phosphorus transport by world rivers. *American Journal of Science*, **282**, 401–50.

Meybeck, M. (1993). C, N, P, and S in rivers: From sources to global inputs. In *Interactions of C,N, P, and S Biogeochemical Cycles and Global Change*, ed. R. Wollast, F.T. Mackenzie & L. Chou, pp. 163–93. Berlin: Springer-Verlag.

Meybeck, M. & Helmer, R. (1989). The quality of rivers: from pristine stage to global pollution. *Palaeogeography, Palaeoclimatology, Palaeoecology (Global and Planetary Change Section)*, **75**, 283–309.

Meybeck, M. & Ragu, A. (1996). *River Discharges to the Oceans: An Assessment of Suspended Solids, Major Ions, and Nutrients*. GEMS-Water/UNEP/WHO. Geneva: GEMS-Water.

Miller, J. & Russell, G. L. (1992). The impact of global warming on river runoff. *Journal of Geophysical Research* , **97**, 2757–64.

Milliman, J. D. & Mei-e, R. (1995). River flux to the sea: impact of human intervention on river systems and adjacent coastal areas. In *Climate Change: Impact on Coastal Habitation*, ed. D. Eisma, pp. 57–83. Boca Raton, Florida: Lewis Publishers.

Milliman, J. D., Yun-shan, Q., Mei-e & Saito, Y. (1987). Man's influence on the erosion and transport of sediment by Asian rivers: the Yellow River (Huanghe) example. *Journal of Geology*, **95**, 751–62.

Moore, B., Gildea, M. P., Vörösmarty, C. J., Skole, D. L., Melillo, J, Peterson, B. J., Rastetter, E. & Steudler, P. (1989). Strategy for understanding changes in the metabolic system of the planet. In *Global Ecology*, ed. M. B. Rambler, L. Margulis & D. Sagan, pp. 113–41. New York: Academic Press.

Nixon, S. (1995). Coastal eutrophication: a definition, social causes, and future concerns. *Ophelia*, **41**, 199–219.

Ortner, P. B. & Dagg, M. J. (1995). Nutrient-enhanced coastal ocean productivity explored in the Gulf of Mexico. *EOS*, **76**, 97–109.

Oyebande, L. (1988). Effects of tropical forest on water yield. In *Forests, Climate, and Hydrology: Regional Impacts*, ed. E. R. C. Reynolds & F. B. Thompson, pp. 16–50. Tokyo: United Nations University.

Postel, S. L., Daily, G. C. & Ehrlich, P. R. (1996). Human appropriation of renewable fresh water. *Science*, **271**, 785–88.

Qu, G. (1987). *Environmental Problems in China*. Beijing: Environmental Publishing House.

Qu, G., Wang, Yuqing., Wang, Yanyang, Liu, W., Yu, C. Shen, G., Song, K., Zhang, X., Li, J., Li, X. & Yan, L. (1989). *China's Environment in 2000*. Beijing: Chinese Social Science Publishing House.

Ren, M. E. & Shi, Y. L. (1986). Sediment discharge of the Yellow River (China) and its efect on the sedimentation of the Bohai and the Yellow Sea. *Continental Shelf Research*, **6**, 785–810.

Rosenberg, R., Elmgren, S., Fleischer, S., Jonsson, P., Persson, G. & Dahlin, H. (1990). Marine eutrophication case studies in Sweden. *Ambio*, **19**, 102–8.

Ryding, S-O. & Rast, W. (eds.). (1989). *The Control of Eutrophication of Lakes and Reservoirs*. Man and the Biosphere Series, Volume 1. Paris: UNESCO.

Saito, Y. & Yang, Z. (1995). Historical change of the Huanghe (Yellow River) and its impact on the sediment budget of the East China Sea. In *Global Fluxes of Carbon and Its Related Substances in the Coastal Ocean - Atmosphere System*, ed. S. Tsunogai, K. Iseki, I. Koike & T. Oba. Yokohama: M and J International.

Saito, Y., Matsumoto, E. & Kato, Y. (1995). Cultivation history of the Loess Plateau in China recorded in sediments of the East China Sea. Poster presentation at the IGBP/SAC-IV Symposium, Beijing, November 1995.

Shiklomanov, I. & Krestovsky, O. I. (1988). The influence of forests and forest reclamation practice on streamflow and water balance. In *Forests, Climate, and Hydrology: Regional Impacts*, ed. E. R. C. Reynolds & F. B. Thompson, pp. 78–116. Tokyo: United Nations University.

Sigurdson, J. (1977). Water policies in India and China. *Ambio*, **6**, 70–6.

Sivapalan, M. & Kalma, J. D. (1995). Scale problems in hydrology: contributions to the Robertson Workshop. *Hydrological Processes*, **9**, 243–50.

Smayda, T. J. (1989). Primary production and the global epidemic of phytoplankton blooms in the sea: a linkage? In *Novel Phytoplankton Blooms*, ed. E. M. Cosper, V. M. Bricelj & E. J. Carpenter, pp. 448–83. Coastal and Estuarine Studies #35. New York: Springer-Verlag.

Smayda, T. J. (1990). Novel and nuisance phytoplankton blooms in the sea: evidence for global epidemic. In *Toxic Marine Phytoplankton*, ed. E. Graneli, B. Sundstrom, R. Edler & D. M. Anderson, pp. 29–40. New York: Elsevier Science.

Smil, V. (1992). China's environment in the 1980s: Some critical changes. *Ambio*, **21**, 431–6.

Stanley, D. (1992). *Historical Trends: Water Quality and Fisheries, Albemarle-Pamlico Sounds, with Emphasis on the Pamlico River Estuary*. University of North Carolina Sea Grant College Program Publication UNC-SG-92-04. Greenville NC: East Carolina University, Institute for Coastal and Marine Resources.

State Statistical Bureau of China (1989). *China Statistical Abstract*. New York: Praeger.

State Statistical Bureau of China (1995). *China*

Statistical Yeabook, 1995. Beijing: China Statistical Publishing House.

Swank, W. T. (1988). Stream chemistry responses to disturbance. In *Forest Hydrology and Ecology at Coweeta*, ed. W. T. Swank & D. A. Crossly, pp. 339–57, Ecological Studies 66. New York: Springer-Verlag.

Tobler, W., Deichmann, U., Gottsegen, J. & Maloy, K. (1995). *The Global Demography Project*. Technical Report TR-95-6, National Center for Geographic Information and Analysis. Santa Barbara, California: University of California.

Tregear, T. R. (1980). *China: A Geographical Survey*. New York: John Wiley.

Turner, R. E. & Rabalais, N. (1991). Changes in Mississippi River water quality this century – implications for coastal food webs. *BioScience*, **41**, 140–7.

Turner, R. E. & Rabalais, N. (1994). Evidence for coastal eutrophication near the Mississippi River delta. *Nature*, **368**, 619–21.

United Nations (1990). *UN Statistical Yearbook - 1987*. New York: UNIPUB.

USDA/ERS (1989). *Agricultural Resources: Inputs, Situation, and Outlook*. Washington: USDA Economic Research Service.

Vallentyne, J. R. (1978). Today is yesterday's tomorrow. *International Association of Theoretical and Applied Limnology*, **20**, 1–20.

Vitousek, P. M. (1983). The effects of deforestation on air, soil, and water. In *The Major Biogeochemical Cycles and Their Interactions*, ed. B. Bolin & R. B. Cook, pp. 223–46. SCOPE 21. New York: John Wiley.

Volker, A. & Henry, J. C. (eds.). (1988). *Side Effects of Water Resources Management*. IAHS Publication #172. Wallingford, UK: IAHS Press.

Vörösmarty, C. J. & Moore B. (1991). Modeling basin-scale hydrology in support of physical climate and global biogeochemical studies: An example using the Zambezi River. *Studies in Geophysics*, **12**, 271–311.

Vörösmarty, C. J., Gildea, M. P., Moore, B., Peterson, B. J., Berquist, B. & Melillo, J. M. (1986). A global model of nutrient cycling: II. Aquatic processing, retention, and distribution of nutrients in large drainage basins. In *Watershed Research Perspectives*, ed. D. Correll, pp. 32–56. Washington: Smithsonian Institution Press.

Vörösmarty, C. J., Willmott, C. J., Choudhury, B. J., Schloss, A. L., Stearns, T. K., Robeson, S. K. & Dorman, T. J. (1996). Analyzing the discharge regime of a large tropical river through remote sensing, ground-based climatic data, and modeling. *Water Resources Research*, **32**, 3137–50.

Vörösmarty, C. J., Sharma, K., Fekete, B., Copeland, A. H., Holden, J., Marble, J. & Lough, J. A. (1997*a*). The storage and aging of continental runoff in large reservoir systems of the world. *Ambio*, **26**, 210–19.

Vörösmarty, C. J., Meybeck, M., Fekete, B. & Sharma, K. (1997*b*). The potential impact of neo-Castorization on sediment transport by the global network of rivers. In *Human Impact on Erosion and Sedimentation*, ed. D. Walling & J-C. Probst, IAHS Publ. #245. Wallingford, UK: IAHS Press.

Vörösmarty, C. J., Wasson, R. & Richey, J. E. (eds.) (1997*c*). *Modeling the Transport and Transformation of Terrestrial Materials to Freshwater and Coastal Ecosystems*. Workshop Report and Recommendations for IGBP Inter-Core Project Collaboration. IGBP Report #39. IGBP Secretariat: Stockholm.

Wang, S-W. & Zhao, Z-C. (1981). Droughts and floods in China, 1470–1979. In *Climate and History*, ed. T. M. L. Wigley, M. J. Ingram & G. Farmer, pp. 271–88. Cambridge: Cambridge University Press.

Wang, S., Zhao, Z., Chen, Z. & Tang, Z. (1987). Drought / flood variations for the last 2000 years in China and comparison with global climatic change. In *The Climate of China and Global Climate*, ed. D. Ye, C. Fu, J. Chao & M. Yoshino, pp. 20–9. Berlin: Springer-Verlag. (Beijing: China Ocean Press).

Wollast, R. (1983). Interactions in estuaries and coastal waters. In *The Major Biogeochemical Cycles and Their Interactions*, ed. B. Bolin & R. B. Cook, pp. 385–407. SCOPE 21. Chichester: John Wiley.

Zeng, Q. & Zhou, W. (1989). Soil erosion in the Yellow River basin and its impacts on reservoir sedimentation and the lower Yellow River. In *Sediment and the Environment*, IAHS Publication #184, pp. 123–30. Wallingford, UK: IAHS Press.

Zhang, D. (1984). Synoptic–climatic studies of dust fall in China since historic times. *Scientia Sinica*, **27B**, 825–36.

Zhang, J. & Crowley, T. J. (1989). Historical records in China and reconstruction of past climates. *Journal of Climate*, **2**, 833–49.

Zhao, S. (1994). *Geography of China: Environment, Resources, Population, and Development*. New York: John Wiley.

10 Sediment and nutrient transport to the coastal zone

D. Hu, Y. Saito and S. Kempe

Introduction

Changes in the sediment and nutrient transport from land to coastal seas may have serious impacts upon the coastal system globally. Such effects include reduction or even destruction of living resources, hindrance to marine activities, deterioration of amenities and damage to the coastal ecosystems of mangrove, coral reef, benthic communities. Deforestation and spread of agriculture caused an increase in sediment discharge for the last 2000 years. Fertilization raised the nutrient input in the last 40–50 years. In the areas of southern Asia and Oceania, the sediment discharge has been re-evaluated recently, larger than previously estimated. Some of this increase may be due to the accelecated deforestation. In other areas the sediment discharge decreases due to dam construction for irrigation, flood control and hydroelectric power generation. This decrease of sediment delivery to the coastal zone induces coastal erosion, which is enhanced further by sea-level rise.

In the present chapter progress in the study of sediment/nutrient discharges to the coastal zone is reviewed with special reference to the Yellow River (Huanghe; hereafter, use Yellow River only) as an example for the influence of human activities on sediment discharge and to the Yangtze River (Changjiang; hereafter, use Yangtze River only) on nutrient input.

Global view of sediment and nutrient transport

Under- and over-estimation of sediment discharges from small and large rivers

Sediment discharge of rivers was estimated, based on either erosion rates (Fournier, 1960; Ohmori, 1983) or the sediment loads measured near river mouths (Holeman, 1968; Milliman & Meade, 1983). Milliman and Meade (1983) showed that the total suspended sediment delivered by all rivers to the oceans is about 13.5 $\times 10^9$ t year^{-1} and bedload/flood discharges may account for an additional 1–2 billion tonnes. They also listed the annual discharge of suspended sediment from various drainage basins of the world, and pointed out that about 70% of this total is derived from southern Asia and the larger islands in the Pacific and Indian Oceans. In these papers, however, the possible contribution of smaller rivers was not taken into account, although the authors did recognize that smaller rivers have greater sediment yield (load per unit drainage basin area). Later, Milliman and Syvitski (1992) showed that (1) small rivers (drainage basin less than 10 000 km^2) contribute much more sediment than previously estimated, although these rivers drain only about 20% of the land area, (2) the discharge from larger rivers was overestimated, and (3) rivers discharged about 20 billion tonnes of sediment annually to the ocean before the proliferation of dam construction in the second half of this century. Small rivers, in particular, of the high-standing islands between Australia and Asia discharge about 20% of the global sediment flux to the ocean. Milliman (1995) indicated recently that the rivers draining the island of New Guinea are estimated to discharge about 1.7 billion tonnes of sediment annually to the ocean, which correspond to 8% of the global sediment flux, although the island occupies only 0.9% of the global drainage areas.

Increase of river-sediment discharge

Significant change in sediment discharge in many rivers of the world has been made by human activities. For instance, the very high sediment loads in modern Asian rivers imply considerable influence from human activities, especially from poor soil conserving practices in agriculture.

Milliman and Syvitski (1992) indicated that sediment discharge was less than half of the present level prior to widespread farming and deforestation 2000–2500 years ago. On a global scale, the most massive anthropogenic increases in river sediment loads have resulted from crop farming, especially in the areas where forests were cleared. For example, the eastern parts of Canada and the United States during the 17th, 18th and 19th centuries underwent replacement of native forests by croplands. During the 19th century, prairie grasslands were converted to deeply plowed fields. These conversions caused order-of-magnitude increase in soil erosion and corresponding increases in sediment yields of rivers (Meade, 1969, 1982). Similarly, the intensive and extensive agricultural development of the loess plateau in northern China that began about 200 BC also is believed to have caused an order-of-magnitude increase in sediment discharge of the Yellow River to the coastal zone (Milliman *et al.*, 1987). It must be particularly pointed out that sediment loads in Oceania and south Asia are elevated disproportionately by extensive human influence in this area. Deforested land, which in tropical areas is increasing by 100 000 km^2 annually (Myers, 1988), caused more sediment discharge to the oceans than usual. Saunders and Young (1983) estimated that moderate land use results in a two- to three-fold increase in sediment yield, while intensive land use can result in an order-of-magnitude increase (Milliman, 1992).

Heavy monsoon-related rain can have great erosive power (Walling, 1983). With these facts taken into account, increase of sediment discharge in south Asia and Oceania is a serious matter. Assuming that present Asian and Oceania river loads are five times greater than before humans began deforestation and farming, the world-wide quantity of fluvial sediment reaching the ocean 2500 years ago might be less than 7×10^9 t year^{-1}, and the percentage from Asia and Oceania would account for correspondingly smaller fractions of the world total than they do at present (GESAMP, 1993).

Decrease of river-sediment discharge

On the other hand, due to dam/reservoir construction, the decrease of sediment discharge to the coastal zones is another significant influence of human activities on rivers. More and more rivers in the world are dammed for irrigation, flood control and hydroelectric power generation. Prior to 1900 the rate of dam construction was small. Between 1900 and 1945 dam construction rose obviously, and after 1945 the rate of dam construction continued to rise more sharply, reaching a global average over 900 per year between 1951 and 1982. With the development of industries, there are increasing needs for hydroelectric power, flood control and water resource control. China is a good example of this scenario. Prior to 1950 there were only eight dams in China. As of 1982, the number had increased to 18 600, i.e. 55% of the world total (USA: 5340, 16%; Japan: 2140, 6%). Fig. 10.1 shows the history of world dam construction (Petts, 1984; GESAMP, 1993). The Colorado and Nile rivers are examples for the complete cessation of sediment flux caused by dam construction (GESAMP, 1993). The Rhine River now carries only about 5% of the load it did in the 19th century (GESAMP, 1993), and the Indus River discharges less than 20% of the load it did before construction of barrages in the late 1940s (Milliman *et al.*, 1984; GESAMP, 1993). Sediment discharges by the Mississippi River are now less than 50% of what they were before 1950 (Meade *et al.*, 1990). Small rivers in the Japanese Islands carry also less than 50% of the load they did in 19th century (Asida *et al.*, 1983). Of all river discharges in the world, 13% are presently dammed (Pearce, 1991). The decrease of sediment discharges to the coastal zones by dam construction induces another problem of coastal erosion in many river-mouth areas of the world (Milliman, 1992; GESAMP, 1993). Relative sea level rise also induces coastal erosion. The relative sea level rise is caused mainly from eustatic sea level rise due to global warming and land subsidence accelerated through the removal of groundwater or petroleum from the underlying strata (Milliman, 1992).

Sediment and climate

So far we described the influence of farming, deforestation, and dam/reservoir construction on the river sediment discharge. What is the influence of climate on

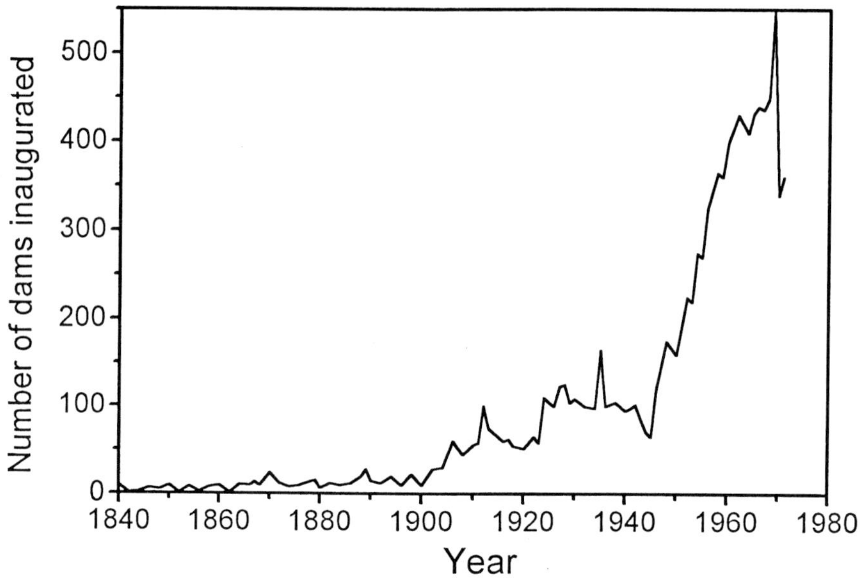

Figure 10.1 *World dam construction. (Redrawn from Petts, 1984.)*

sediment discharge if any? In Fig. 10.2 the sediment discharges of the Yellow River, Yangtze River and Santa Clara River are depicted , compared with the Southern Oscillation Index (SOI), which indicates El Niño events by its pronounced negative values. It seems that most El Niño events (minimal SOIs) correspond to lower sediment discharge of the Yellow River and higher sediment discharge of the Santa Clara River. In contrast, the sediment discharge of the Yangtze River with much smaller interannual variability does not respond to El Niño events. Why is this so? One possibility is due to the size of drainage area. The Yellow River and Santa Clara River are much smaller than the Yangtze River in terms of drainage area and may therefore react more pronounced to climate (rainfall) variation. In addition, they are located in drier regions. So, they are easily affected in terms of water runoff and hence sediment discharge by droughts and floods resulting from El Niño events. During El Niño years, the west coasts of the Pacific Ocean become drier than usual and the east coasts become wetter than usual so that the Yellow River gets less water and hence less sediment discharge and the Santa Clara River gets more water and more sediment. Fig. 10.2 shows a corresponding relationship between the sediment discharge of the Yellow River and SOI with a correlation coefficient of 0.49. For the Yangtze River, its interannual variability of water and sediment is so small because of its larger drainage area and therefore the influence of El Niño is less pronounced. The correlation coefficient between its sediment discharge and SOI is only 0.26. It

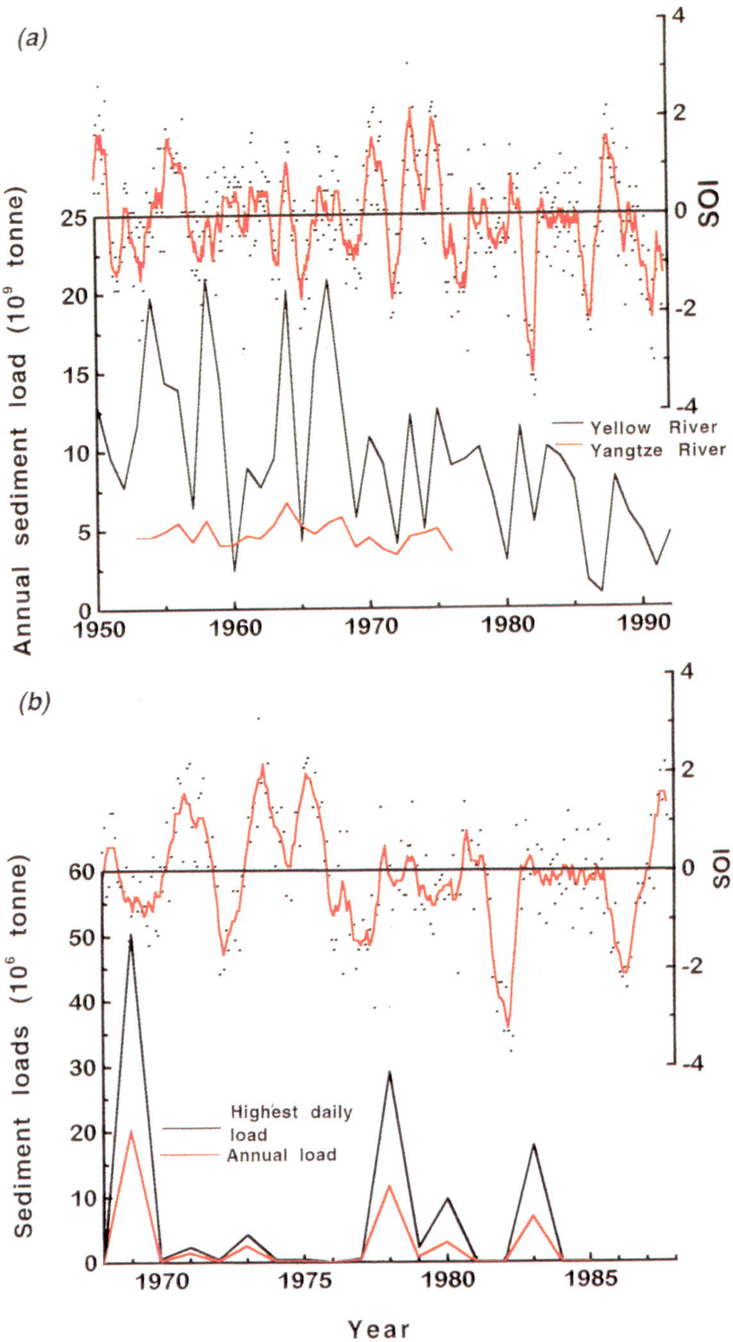

Figure 10.2 *Sediment discharge versus southern oscillation index (SOI) (a) sediment discharges of the Yellow River and Yangtze River and SOI (b) annual sediment loads and highest daily loads of the Santa Clara River (at Montalvo, California USA) and SOI. The sediment data is from Milliman (1991).*

seems that the conclusion "no simple relationship exits between climate and sediment" (Walling & Webb, 1983; GESAMP, 1993) may need to be re-examined.

Increase of nutrient input to the coastal ocean

With increasing economic development, nitrogen is mobilized anthropogenically at a rate which exceeds the natural terrestrial rate through energy production, fertilizer production and crop cultivation. According to Galloway *et al.* (1995), the natural pre-industrial fixation of reactive nitrogen provides 90–130 Tg N year^{-1} (Tg = 10^{12}g) on the continents, of which 35 Tg N year^{-1} was transported to the coastal oceans by rivers. Human activities produce additional fixation of *c.*140 Tg N year^{-1} by energy production (*c.* 20 Tg N year^{-1}), fertilizer production (*c.* 80 Tg N year^{-1}) and cultivation (*c.* 40 Tg N year^{-1}). Among others, the riverine transport to the coastal oceans is *c.* 41 Tg N year^{-1}, anthropogenically produced. Galloway *et al.* predicted that the anthropogenic N fixation rate would be increased by about 60% by the year 2020, mainly due to fertilizer use and fossil fuel combustion. About two-thirds of the increase would occur in Asia, which then will account for more than half of the global anthropogenic N fixation. With increase of anthropogenic N, what would happen to the nutrient (especially N) input to the coastal marine environment and hence to the coastal marine ecosystems?

Van Bennekom and Salomons (1981) calculated and proposed 32 Tg N year^{-1} and 3.7 Tg P year^{-1} for anthropogenic flux to the coastal zone and stated that human influence had increased the river input of nitrogen about five times and that of phosphorus about four times. As reported, the following five rivers are given as example for nutrient increase: (1) in the Mississippi River (1967–1979), NO_3-N increased by 11 times from 1 to 11 mg m^{-3} (Kempe, 1982); (2) in the Rhine River (1963–1978), PO_4-P input increased by three times from 8000 to 25 000 t year^{-1} and NO_3-N by five times: 60–290 μm l^{-1} from 1928 to 1977 (Kempe, 1982); (3) in the Amazon River (1963–1981) NO_3-N increased by five times from 3 to 14.5 μm l^{-1} (van Bennekom & Salomons, 1981); (4) in the Danube River (middle 1950s–1989) NO_3 + NO_2-N flux increased by 22 times from 12.5 to 273 × 10^3 t year^{-1} and PO_4-P flux increased by five times from 5.5 to 29.3 × 10^3 t year^{-1} (Kempe, 1995); (5) in the Yangtze River (1963–1980), NO_3-N increased by four times from 17 to 65 μm l^{-1} (Gu, 1981, 1982; Edmond *et al.*, 1983).

Although different authors provided different values of the nutrient flux from rivers to coastal oceans, for example, 76 Tg N year^{-1} by Galloway *et al.* (1995), 64 Tg N year^{-1} and 24 Tg P year^{-1} by S. V. Smith (personal communication, 1995), 74 Tg N year^{-1} by van Bennekom and Salomons (1981), 190 Tg N year^{-1} (GESAMP, 1987) and 22 Tg P year^{-1} (Howarth *et al.*, 1995), all tell us about, and draw our attention to, the potential problem caused by the additional nutrient input to the coastal zone, which is increased due to human activities. It should be mentioned that, of the 76 Tg N year^{-1} of Galloway *et al.* (1995), 41 Tg N year^{-1} is anthropogenic and 35 Tg N $year^{-1}$ "natural", which means the release of reactive N from continents to coastal oceans via rivers has almost doubled from 35 to 76 Tg N $year^{-1}$, since pre-industrial time, and that of the 76 Tg N $year^{-1}$, 42 Tg N $year^{-1}$ is PON (50% for anthropogenic and natural, respectively) and 34 Tg N $year^{-1}$ is dissolved N (60% for anthropogenic and 40% for natural). Obviously, fertilizer application would be the primary source of the anthropogenic N increase in coastal oceans.

The global fertilizer consumption increased by 2.5 Tg N $year^{-1}$ on average during the period from 1967 to 1989, when a total of 78 Tg N $year^{-1}$ was produced. 50 Tg N $year^{-1}$ is consumed in less-developed regions, of which Asia uses about 46% in 1989 (Galloway *et al.*, 1995). On the basis of the data provided by Galloway *et al.* (1995), the ratio of fertilizer production to energy production of reactive N at present is 2.1 for higher developed regions and 10.3 for Asia. Anyway, Asia has been, and will continue to be, the primary provider for global anthropogenic reactive N by fertilizer application. Of course, in terms of per capita contribution to N fixation and emission, more-developed regions such as North America and Europe have been, and will continue to be, leading.

Impact on coastal marine environment

Changes of sediment and nutrient fluxes, no matter if natural or anthropogenic, to coastal oceans can make a tremendous impact on the coastal marine environment.

The decrease of riverine sediment caused by dam/reservoir construction and irrigation is the most serious concern in terms of impact of sediment change on coastal marine environment. Big dams in the world have increased from about 5000 to 38 000 since 1950, of which 85% were built in the last 35 years. Rivers are originally essential to the earth ecosystem and to the global hydrological cycle. Owing to anthropogenic impacts, the rivers as natural system have been changed. Some of the changes caused by the decreasing sediment input are the erosion of the coastline, deterioration of marine ecosystem and even collapse of living resources. After the Aswan Dam was completed (1964), the catch of sardine fish was reduced by 95% in the eastern Mediterranean and the delta shrank rapidly (Abdel-Aal, 1985; Wahby & Bishara, 1981). After the Colorado River dams were completed, decreases of sediment and nutrients caused rapid reduction of shrimp catch in the California Bay and many species depending on the ecosystems in the Colorado downstream reaches and in the California Bay were imperiled. After completion of the Kotri Barrage on the Indus River in 1956, fish-catch decreased by a factor of three (Milliman *et al.*, 1984). As a result of the decrease in Yellow River sediment discharge and of the artificial control of the sandspit direction, severe coastline erosion north of the river mouth and rapid progradation in the south took place, and the marine ecosystem and the environment in the Bay of Bohai deteriorated, leading to a reduction of even the living resources. The catch of shrimps in the Bay of Bohai has been reduced by 85% since 1960s when it was 19.8×10^6 kg year^{-1}, and the percentage of high-quality fish in total catch had dropped from 50–65% in the 1950s to only 4.8% in 1993.

Owing to continuous boost of fertilizer production in the last 30 years, eutrophication caused red tides in coastal oceans near rivers, such as the Mississippi, Yangtze, Rhines, Po, and so on (Nixon, 1995).

Asian case with special reference to the Yellow River and Yangtze River

Among the first five rivers on the globe with the largest annual suspended sediment load, four are situated in Asia. These are the Ganges/Brahmaputra, Yellow, Yangtze and Irrawaddy (Milliman, 1991). In addition, the Ganges/Brahamputra and Yangtze occupy the fourth and fifth place in the world in terms of average water discharge. Fig. 10.3 shows that about 47% of the world total of fluvial sediment is derived from Asia. In Asia, there are about 3.4 billion people, which accounts for *c*. 60% of the world population. Asian economy is growing with one of the highest growth rates in the world and is expected to rapidly develop further in the next century. Right now, dam/reservoir construction and fertilizer production are still growing fast (see above).

Therefore, Asia is the proper region for studying the anthropogenic influence on sediment and nutrient transport to the coastal zone.

The Yellow River

The Yellow River is well known for its course changes, 26 major ones happened over the past 4000 years (EGWCHYR, 1982; Saito & Yang, 1995). Not all these changes can be attributed to natural causes. Some result from human activities, including even wars. The change of the Yellow River course is a unique phenomenon in the world, especially for the switch of the mouth from one coastal sea to another. From 2278 BC to AD 1128, the river emptied into the Bay of Bohai and, from 1128 to 1546, it bifurcated into the Yellow Sea (main stream) and the Bay of Bohai. The Yellow River flowed only to the Yellow Sea from 1546 to the year of 1855, when it switched to the Bay of Bohai again up to the present. Fig. 10.4 provides a rough picture of the historical change of the sediment discharge to the Yellow Sea and East China Sea (Saito *et al.*, 1994).

Northern China, in addition, suffered from flooding disasters, which occurred over 1500 times within the last 4000 years. These floods are attributed mainly to heavy rainfall in association with the change of discharged sediment, resulting from human activities such as cultivation and so forth. In terms of impacts of human activities on the Yellow River, the change of sediment discharge is distinguished, including both increase and decrease (EGWCHYR, 1982).

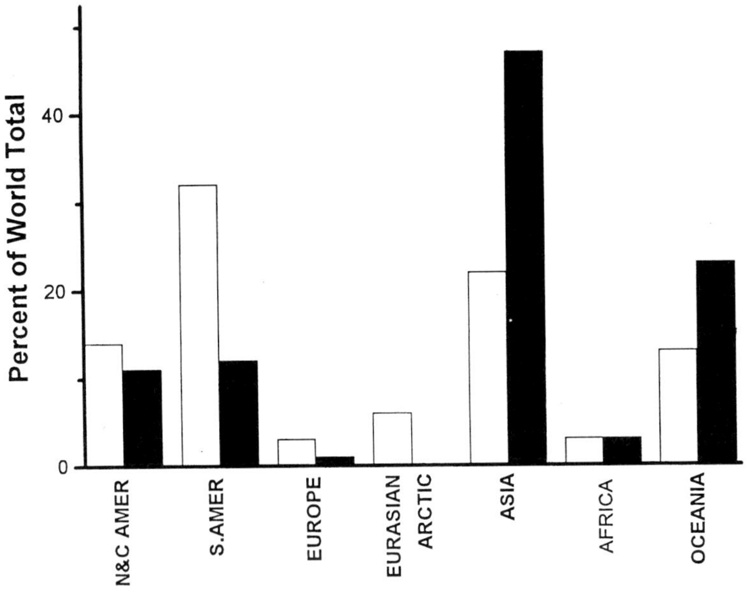

Figure 10.3 *Land sources for fluvial sediment (solid columns) and water (empty columns) discharged into the ocean. (Courtesy, Milliman, 1991.) Sediment discharge from large rivers of continents is over-estimated. Some of the sediment transported by large rivers is deposited on the flood plain. Sediment discharge from Oceania and small islands of the western Pacific is under-estimated, which was re-estimated to be 9 × 10^9 tonnes/year by Milliman and Syvitski (1992) and accounts for 40 – 50% of the world total.*

Increase of sediment discharge by cultivation

The sediment discharge of the Yellow River changed remarkably in the last 5000 years. About 2500 years ago the river was called the Dahe, which means "large river". The river was not muddy at that time. Its sediment discharge should account for only one-tenth of its present level, based on Holocene sediment budgets (Milliman *et al.*, 1987) and delta growth (Saito & Yang, 1994). During the Han Dynasty (206 BC–AD 220) it was called the Zhuohe, which means "turbid river" and during the Tang Dynasty (AD 618–AD 907) the present name – the Yellow River appeared. This change in the river's name illustrates the impact of human activity on the sediment discharge. In Chinese history, large-scale migrations occurred in two different periods: during the Han Dynasty, 700 thousand people, during the Tang Dynasty 60–70 million moved to the loess plateau and about 600–700 years ago pastoral areas and forests were turned to agricultural areas. The green area in the loess plateau was decreased from 53% to 3% (Wang & Aubrey, 1987). The reason of this decrease is thought to be cultivation of loess plateau (EGWCHYR, 1982; Milliman *et al.*, 1987). The rate of land formation can be used as an estimate of the increase of sediment discharge of the Yellow River. For instance, from 1122 BC to AD 1168 the present Yellow River delta area, includ-

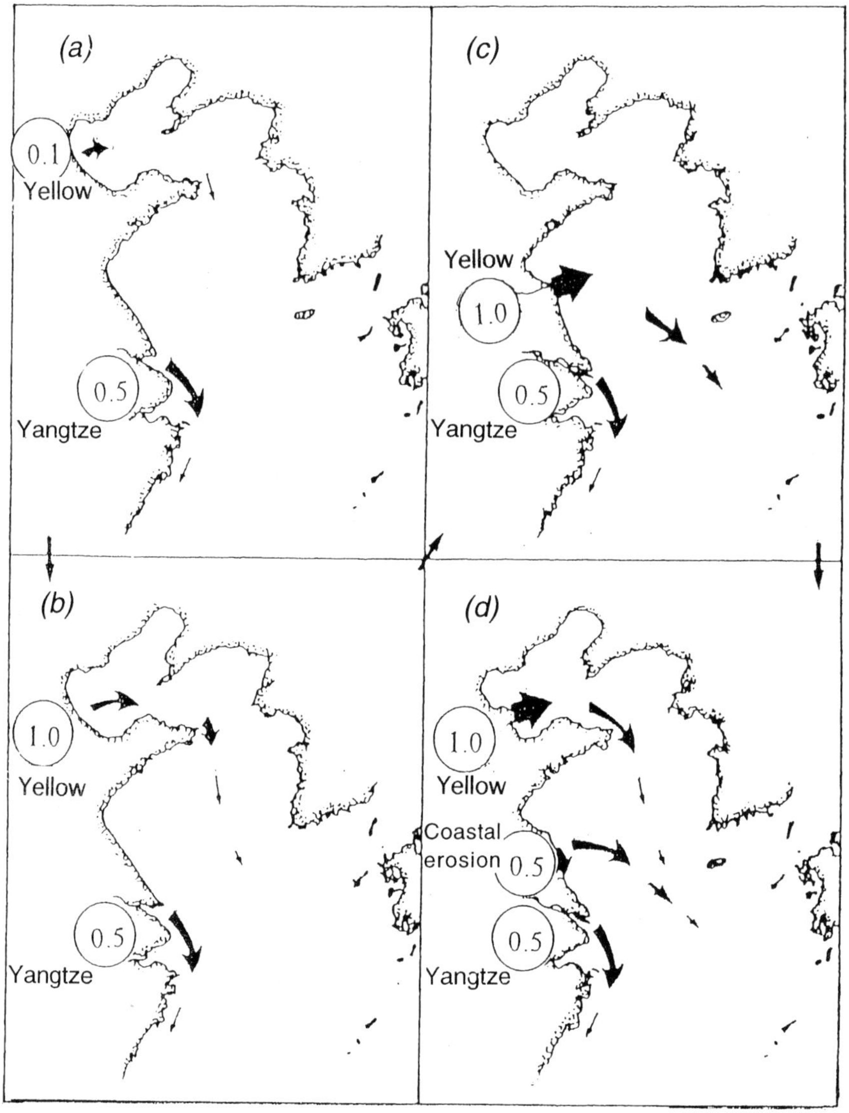

Figure 10.4 *Historical change of the sediment supply (× 10⁹ tonnes/yr) to the East China Sea: (a) before* c. *200* BC*; (b)* c. *200* BC*–*c. *1400; (c)* c. *1400–1855; and (d) 1855–present. (After Saito* et al., *1994.)*

ing Tianjin, grew by *c.* 2.4 km^2 $year^{-1}$. It enlarged by 1.2 km^2 $year^{-1}$ at the north Jiangsu coast when the river emptied into the Bay of Bohai. However, during the periods AD 1168–AD 1855 and 1938–1947, when the Yellow River switched to the Yellow Sea, the average land formation rate near the north Jiangsu coast increased to 14.5 km^2 $year^{-1}$. After 1855, when the river shifted its course to the Bay of Bohai again, and until the middle 1980s, the average land formation rate increased to 20.2

km^2 $year^{-1}$. All these figures illustrate the abrupt increase of sediment discharge from the Yellow River about 1200 years ago due to the cultivation of the loess plateau.

During the 700 year period when the Yellow River flowed to the Yellow Sea, the old Yellow River delta formed to a area of 27×10^3 km^2 with the north Jiangsu coastal plain of 10×10^3 km^2.

Decrease of sediment discharge by dam construction

The Yellow River sediment discharge has decreased rapidly in the last 40–50 years. Fig. 10.5 shows the fluctuations of water and sediment discharges through the middle reaches of the Yellow River from 1740 to 1989 (Ye, 1994). A sharp decline from about 1.5 to 0.8×10^9 t $year^{-1}$ of sediment discharge had occurred since the middle 1960s right after the Sanmenxia Dam was completed. Within 30 years the sediment discharge dropped by *c*. 45%. After the course shift of the lower reaches of the Yellow River from Jiangsu to Shangdong in 1855, sediment discharge was $1.1–1.6 \times 10^9$ t $year^{-1}$ and was 1.27×10^9 t $year^{-1}$ on average between 1919–1959 before construction of dams. In more detail, the sediment and water discharges for the 43 years from 1950 to 1992 are shown on Fig. 10.6. The average annual sediment discharge of the Yellow River at Lijin station, about 100 km upstream from the river mouth is 0.95×10^9 t $year^{-1}$ with an average water discharge of 38.0×10^9 m^3 $year^{-1}$ for the whole period (1950–1992). From 1950 to 1969 the mean sediment discharge was 1.2×10^9 t $year^{-1}$ with a water discharge of 49.2×10^9 m^3 $year^{-1}$ and decreased to a mean of 0.7×10^9 t $year^{-1}$ with a water discharge of 28.3×10^9 m^3 $year^{-1}$ during the period (1970–1992). The average annual sediment and water discharges dropped by *c*. 40% and *c*. 43%, respectively. The decadal average annual sediment discharge is 1.32 for 1950s, 1.02 for 1960s, 0.90 for 1970s and 0.65 billions for 1980s. The Sanmenxia and other main dams were completed between 1960 to 1968. Sediment volume deposited behind major dams has reached about 7.5 billion m^3, which corresponds to 9 billion tonnes with a dry bulk density of 1.2 g cm^{-3}. This value is equivalent to a decrease of sediment discharge of $0.3–0.4 \times 10^9$ t $year^{-1}$, which is larger than the total sediment discharge of 0.23×10^9 t $year^{-1}$ from all of Europe (Milliman & Meade, 1983).

Measures for soil preservation were taken particularly in the 1980s. By 1989 about 10 000 km^2 of the loess plateau had been treated. Sediment yield from the loess plateau was reduced to a half. Moreover, the amount of sediment transported to irrigation areas of 3.5×10^4 km^2 accounted to 0.09 billion tonnes for 1950s, 0.07 for 1960s, 0.18 for 1970s and 0.16 for 1980s (Qian *et al.*, 1993). In short, dam construction, green recovery of the loess plateau and irrigation are the major reasons for the recent (especially since the 1970s) decrease of sediment discharge of the Yellow River.

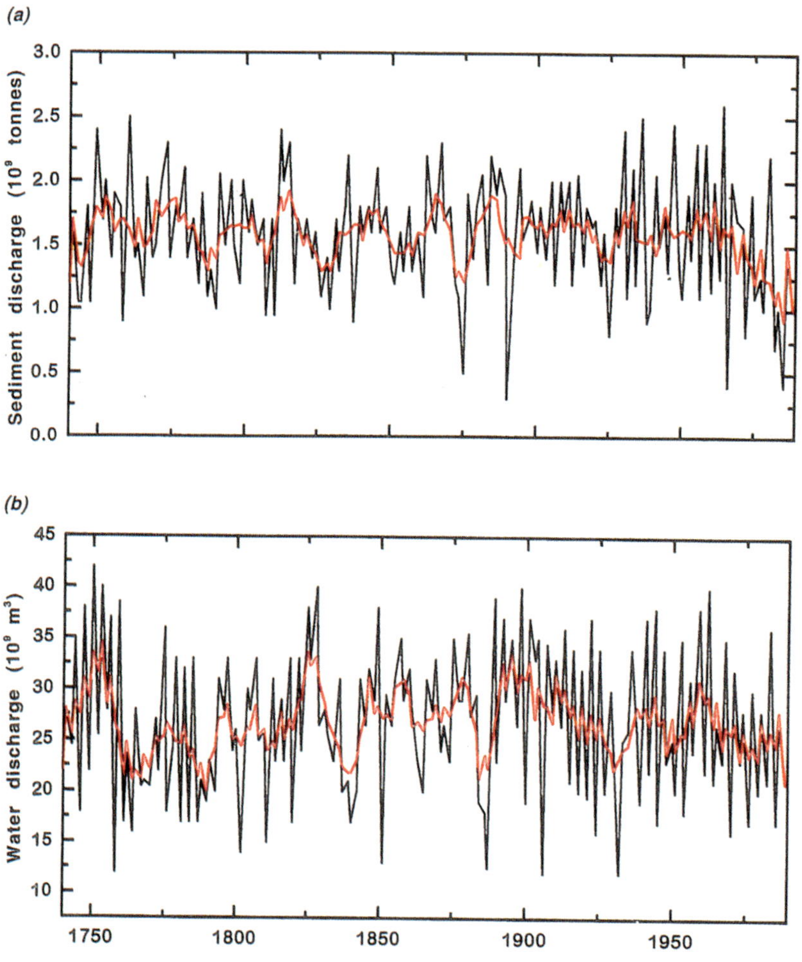

Figure 10.5 *Variations of (a) sediment and (b) water discharges along the middle reaches of the Yellow River (1740 – 1989). (Courtesy, Ye, 1994.) The middle reaches include Huayuankou, Lanzhou and Shanxian. (Black line is annual mean; red line is five-year running mean.)*

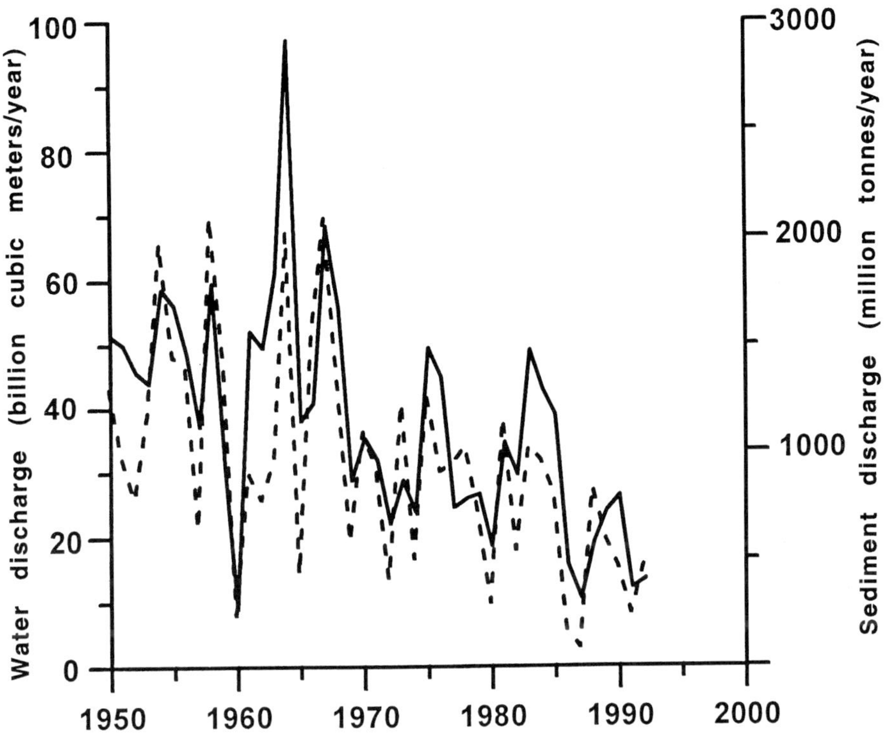

Figure 10.6 *Variation of water (solid line) and sediment (broken line) discharges of the Yellow River at Lijin Station (1950–1992).*

Up to now, 3380 reservoirs and about 30 000 diversion works of various scales have been constructed along the Yellow River basin and diverted water during the 1980s is of the order of $27.4 \times 10^9 m^3$ year^{-1} on average (Qian *et al.*, 1993), which is 48.9% of the water discharge of the Yellow River at Huayuankou in the middle reaches. The average water diversion is 7.1×10^9 m^3 year^{-1}, and average sediment is 0.15×10^9 t year^{-1}. Together with the water stored in reservoirs, so much water was withdrawn that in recent years the Yellow River often dried up in spring, as example, in 1995 about 650 km upstream from the river mouth, where the river dried up for 130 days from March to July.

Coastal erosion and progradation

Since 1855, when the Yellow River switched its mouth from the Yellow Sea to the Bay of Bohai north of Shandong Peninsula, the coastline of north Jiangsu has endured serious erosion with an average rate of about 63 m year^{-1}. Since the river changed its course in 1976 to flow east instead of north, coastal erosion has become serious in the northern part of the Yellow River delta, where the river flowed before 1976. The erosion rate reached a maximum of 1 km year^{-1} at the beginning of the course change. An annual maximum land loss occurred in 1986 with an area of 66 km^2 $year^{-1}$. Coastal erosion takes place not only along the shoreline but also in the shallow sea with depths less than 20 m. Fig. 10.7 shows the erosion and progradation (accumulation) near the Yellow River mouth from 1976 to 1988 (Fan & Guo, 1992). Within 12 years, severe erosion both to the coastline and the bottom near shore in the north and rapid progradation (accumulation) to the southeast took place. The erosion depth exceeded 6 m in place and the maximum accumulation accounted to larger than 14 m. In terms of land formation by the river, it was 2.4 km^2 $year^{-1}$ in the modern Yellow River delta area and 1.2 km^2 $year^{-1}$ in north Jiangsu from 1122 BC to AD 1168; 14.5 km^2 $year^{-1}$ from 1168 to 1855 in the old Yellow River delta, and 20.2 km^2 $year^{-1}$ from 1855–1948 in the modern Yellow River delta. With a present sandspit extension rate of 3 km $year^{-1}$, the remaining 50 km between the sandspit and the opposite coast of the Laizhou Bay will diminish quickly and the Laizhou Bay is expected to become a lagoon early next century. Once this happens, the coastal marine environment and ecosystem will change dramatically.

Summary of sediment discharge of the Yellow River

Prior to 2000–2500 years ago, the sediment discharge of the Yellow River was 0.1–0.2 × 10^9 t $year^{-1}$ and the river was clear. By cultivation of the loess plateau, sediment discharge increased up to 1.1–1.6 × 10^9 t $year^{-1}$ about 1200 years ago. However, by dam construction, soil reservation and irrigation, annual sediment discharge decreased to 0.90 × 10^9 t $year^{-1}$ in the 1970s and 0.65 × 10^9 t $year^{-1}$ in the 1980s and the Yellow River often dried up in the spring, such as in early 1995.

In order to create future scenarios on sediment discharge, we must reconsider the influence of dams. Over half of all dams constructed in the world between 1951 and 1982 were in China (van der Laeden *et al.*, 1990). The Xiao Langdi Dam is reaching completion (69.4 × 10^4 km^2 in area and 12.6 billion m^3 in volume) and will lead to a further reduction in Yellow River sediment transport. The Sanmenxia Dam on the Yellow River was originally completed in 1960 with no provision for bypassing sediment. After only 4 years of operation, more than 60% of the water-storage capacity was filled by sediment. With a reconstruction project in 1965–1973, new sluicing outlets were installed. The influences of construction and reconstruction of the Sanmenxia Dam have been recognized in particular in the 700-km long lower reaches of the Yellow River between the dam site and the river mouth: net deposition of 0.36 billion tonnes of sediment per year in the decade before completion; net scour of 0.58 × 10^9 t $year^{-1}$ during the 4-year period when the reservoir impounded floods; net deposition of 0.44 billion tonnes per year during the 9-year period when the reservoir was operated to retard floods and to sluice sediment (Zhao *et al.*, 1989; GESAMP, 1993). Based on the experience at Sanmenxia, similar sluicing works were incorporated in the Gezhouba Dam on the

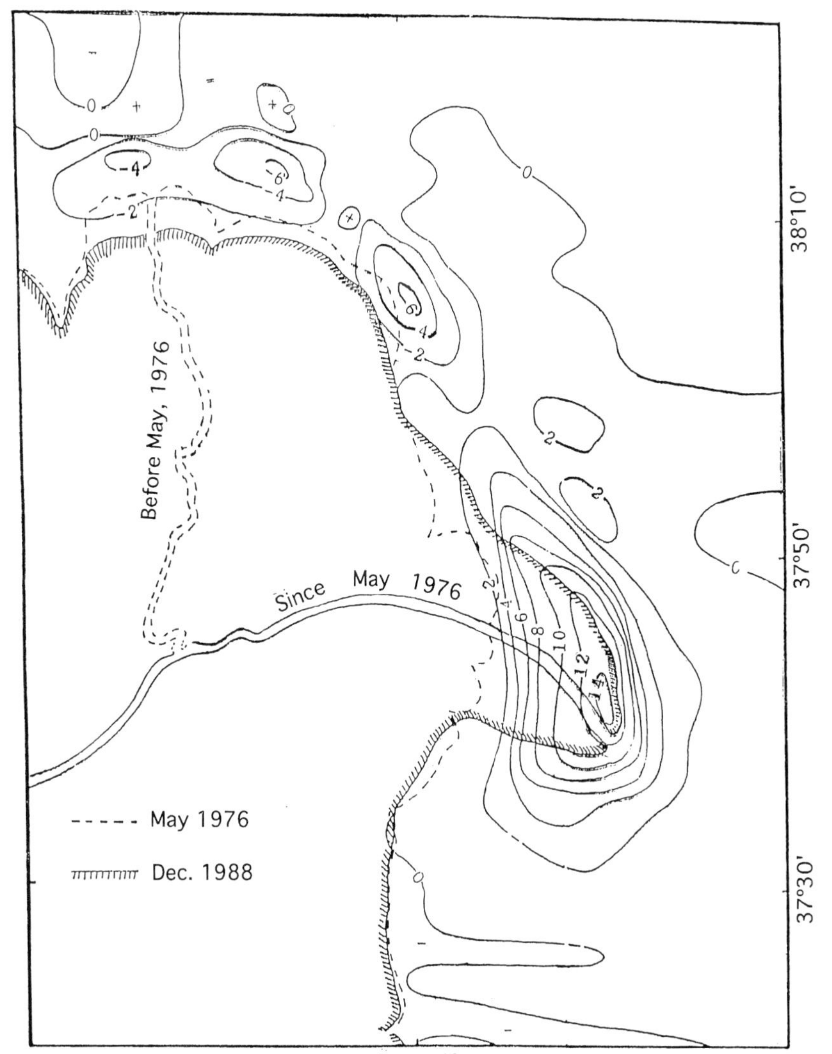

Figure 10.7 *Accumulation and erosion near the Yellow River mouth (from 1976 to 1988). (Courtesy, Fan & Guo, 1992.)*

Yangtze River, and similar structures are being included in the Three Gorges Project on the same river, and in the Xiao Langdi Dam Project on the Yellow River. Therefore, the effects of these large dams on sediment discharge to the coastal seas are expected to be slight (Qian *et al.*, 1987; Tang & Lin, 1987). However, sediment continues to accumulate at dams and the decrease of sediment discharge from the Yellow River and other large rivers still continues.

The existing and future reservoirs will be filled up by sediment, eventually. Once that happens, the sediment discharges may increased again at some point in the future (GESAMP, 1993). As dams increase, sediment discharged to the coastal zone decreases and the coastal environment and marine ecosystems are badly influenced. This is the present situation. For the Yellow River, dams are being built, needed for economy development. It is, on the other hand, also possible that the Yellow River will follow the Nile, Colorado, and Indus. The water could be consumed in the upper and middle reaches and hardly any water could be left for the lower reaches. There would be no water, no sediment and no nutrients discharged to the coastal zone. Then what would happen to the coastal marine environment and ecosystem?

The Yangtze River

The Yangtze River discharges almost the same amount of water as the Ganges/Brahmaputra Rivers, but has a much greater basin and particularly a much larger drainage area and many more paddy fields. It therefore is an excellent example for the impact of human activities on nutrient input to the coastal zone in Asia.

The concentration of inorganic nitrogen as NO_3-N increased by about four times during the period of 1963–1980 from 17 to 65 μm l^{-1} (Gu *et al.*, 1981, 1982; Edmond *et al.*, 1983), supported by Shen *et al.* (1987). Z. S. Yang (1996, personal communication) reported decadal changes of inorganic nitrogen and phosphorus near 122°30' E east of the Yangtze River estuary. They were 15, 24 and 45 μm N l^{-1} for 1960s, 1980s and 1990–1995 and 0.85, 0.80, 0.82 and 0.73 μm P l^{-1} for 1960s, 1970s, 1980s and 1990–1995, respectively, indicating that inorganic nitrogen increased by a factor of three east of the Yangtze River estuary from the 1960s to the early 1990s, while inorganic phosphorus stayed more or less the same. Compared with other rivers in the world, Wang *et al.* reported (1990) that the average (1958–1984) dissolved nitrogen concentration of the Yangtze River is larger than the Amazon's and the world rivers' average by five times, while the phosphorus concentration of the Yangtze River is lower than the world average. The reason for the increase in nitrogen is controversial. Gu *et al.* (1981, 1982) pointed out that the increase of NO_3-N is attributed to rapid increase of fertilizer production along the basin of the Yangtze River. However, Edmond *et al.* (1983) deemed that this high value of 65 μm l^{-1} measured in 1980 resulted from the biological fixation of nitrogen by a water fern which is associated with a symbiotic cyanobacterium.

They think that eutrophicatiion by fertilizer is negligible. Recently, Shen (1996) claimed that precipitation is responsible for the high concentration of inorganic nitrogen (see below for details).

To identify what is the real cause for the increase of NO_3-N, first of all, let us check the impact of human activities on nitrogen fixation, which primarily includes fertilizer production and energy production. Obviously, at present, fertilizer production accounts for *c.* 78% of the two (Galloway *et al.*, 1995). In terms of fertilizer production, Asia provides 36 Tg N $year^{-1}$, which occupies about 46% of the world total of fertilizer. So far, China produces about 20 Tg N $year^{-1}$. The Yangtze River drainage basin contains large area of paddy fields. Therefore, fertilizer application should be an important source of increase of nutrients. With growth of the population, agriculture needs to be further developed in China, for which fertilizer production has to be enhanced. This tendency can be easily seen on Fig. 10.8. From 1960 to 1993 the fertilizer production was increased by about 100 times. The average increase rate between 1975 and 1993 is 0.9 Tg N $year^{-1}$ and has become about 1.6 Tg N $year^{-1}$ for the last 5 years. China now, after the USA is the second largest fertilizer producer in the world. According to statistics, 50% and 40% of the nitrogenous fertilizer applied in China were lost from paddy fields and dry fields, respectively (S. Zhang, 1995, personal communication). It can be immagined that large amount of fertilizer could be drained to the Yangtze River and then to the East China Sea. In Fig. 10.9 the flux of dissolved inorganic nitrogen at Datong, which is about 300 km upstream from the Yangtze River

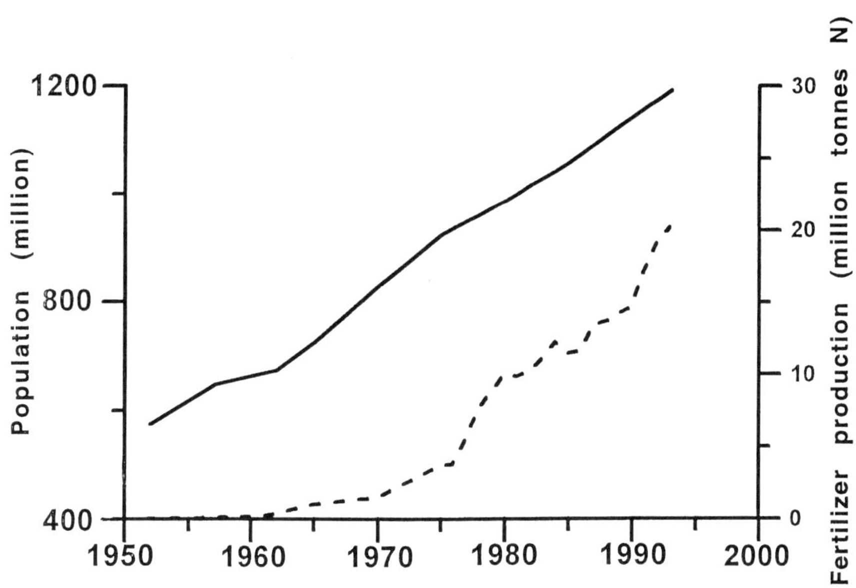

Figure 10.8 *Growth of population (solid line) and fertilizer production (N) (broken line) in China.*

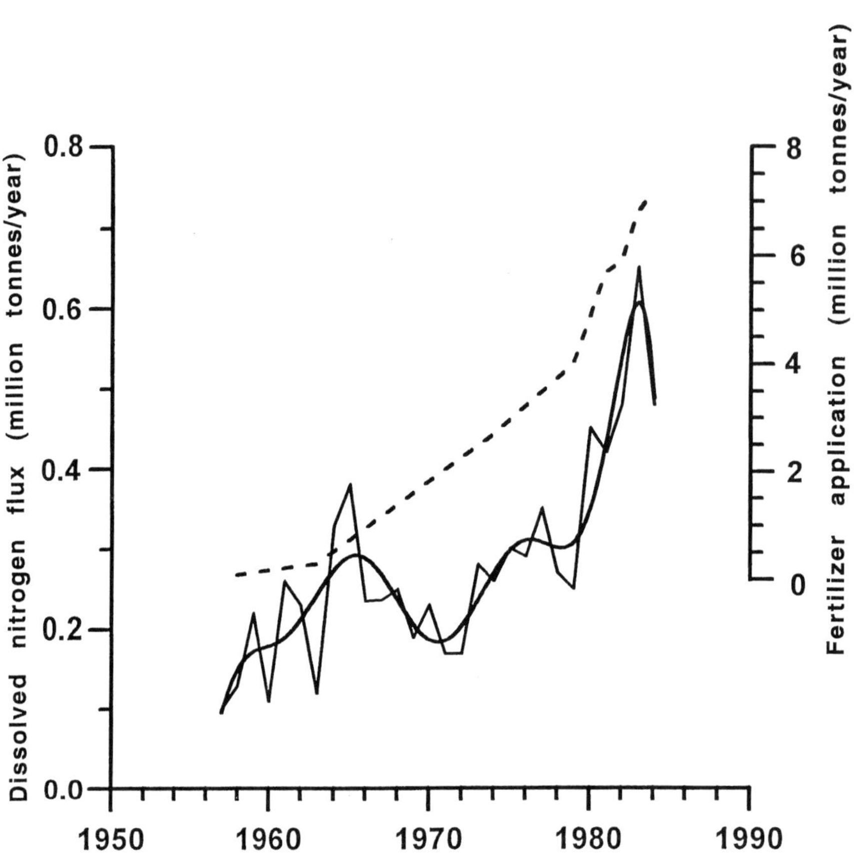

Figure 10.9 *Dissolved nitrogen flux (solid line) at Datong and fertilizer application (broken line) in Yangtze basin. (Courtesy, Wang* et al., *1990.)*

mouth, and the fertilizer application along the Yangtze River basin from 1958 to 1984 are compared (correlation coefficient of 0.88, Wang *et al.*, 1990). From the middle 1970s to the middle 1980s, both the fertilizer application along the Yangtze River basin and dissolved nitrogen flux at Datong were doubled. The annual average nitrogen flux of 1958–1984 is about 55×10^4 t N $year^{-1}$ (Wang *et al.*, 1990), which is smaller than 64×10^4 t N $year^{-1}$ given by Shen (1993) for 1985–1986.

Actually, the main sources of inorganic nitrogen for the Yangtze River are fertilizer application, drainage, soil erosion and precipitation. On the basis of the data from ten cruises during a full year from 1985 to 1986 conducted in the mouth of the Yangtze River, Shen (1996) stated that the average concentration of total inorganic nitrogen (TIN) is about 90.3 μm l^{-1}, of which NO_3-N accounts for about 65% and NH_4-N for about 35%, and then the annual flux of TIN is about 88.8×10^4 t N $year^{-1}$. He calculated and integrated all the components of fluxes between upstream and downstream reaches and came to the conclusion that precipitation is the main source of inorganic nitrogen of the Yangtze River with the following numbers. The Yangtze River receives inorganic nitrogen of about 56×10^4 t N year^{-1} from precipitation, which accounts for about 63% of the total, and 21.3×10^4 t N year^{-1} from fertilizer loss (24%). The question is therefore "where does the inorganic nitrogen in precipitation come from?". It may be due either to horizontal advection in the air, or the evaporated NH_3 from the local lands where fertilizer is applied, or both. However, most likely NH_3 evaporated locally is the main source, as the Yangtze basin is more heavily fertilized than the surrounding areas (in this respect, horizontal advection does not make sense) and NH_3 is dominant in fertilizer application in China (Galloway *et al.*, 1996). According to Zhang (personal communication, 1995) 50% N-fertilizer applied to paddy fields was lost as NH_3 to the atmosphere and redeposited with precipitation and drained to the river. So, it would be fair to deem that precipitation input of nitrogen is attributed to fertilizer application, also. Therefore, according to Shen (1996), the contribution of fertilizer to the inorganic nitrogen flux of the Yangtze River to the East China Sea is about 78×10^4 t N year^{-1}, which is about 86% of the total for the period of 1985–1986.

In what follows, a simple estimation can be made in order to evaluate the relative contribution of fertilizer application and non-fertilizer component to the nitrogen flux of the Yangtze River to the coastal ocean, based on the data available from the early 1960s and the early 1980s.

Before 1960, the fertilizer production is negligible (less than 2×10^5 t N year^{-1}). In the 1980s it grew to about 1.2×10^7 t N year^{-1} on average. In contrast, the other developments would not be that fast. So, we can suppose non-fertilized-affected N flux was more or less the same from 1963 to the mid-1980s, denoted by X. The average fertilizer-affected N flux in the first half of the 1980s is denoted by Y. Now, we have the measured TIN flux of 27.4×10^4 t N year^{-1} for 1963 (Gu *et al.*, 1981) and average of 88.8×10^4 t N year^{-1} for the period of 1980–1985 (Edmond *et al.*, 1983; Huang *et al.*, 1983; Wang *et al.*, 1983; Shen *et al.*,1987; Shen 1991, 1993,1996). Also, the fertilizer production numbers are available, which are 68.8×10^4 t N year^{-1} (1963) and 1072×10^4 t N year^{-1} (average for 1980–1985). Again, suppose the fertilizer-affected N flux is proportional to fertilizer production, then we have the following equations:

$$X + Y = 88.8 \quad \text{and} \quad \frac{27.4 - X}{68.8} = \frac{Y}{1072} \quad \text{for X and Y.}$$

The solution is easily obtained as follows:

$$X \cong 23.2 \times 10^4 \text{ t N year}^{-1} \quad \text{and} \quad Y \cong 65.6 \times 10^4 \text{ t N year}^{-1}$$

This means that the average non-fertilizer-affected inorganic nitrogen flux (by soil erosion, drainage, and so on) of the Yangtze River to the East China Sea for the period of 1963–1985 is about 23.2×10^4 t N year^{-1}, and the average annual fertilizer-produced N flux for the period of 1980–1985 is about 65.6×10^4 t N year^{-1} accounting for about 74% of the TIN flux. Then, the fertilizer-produced N flux for 1963 is easily derived to be about 4×10^4 t N year^{-1}. It must be mentioned that the estimate of fertilizer-produced N flux is less by *c*. 17% than that estimated by Shen (1996). The difference might result from the representativeness of the data used or from the actual increase of the N flux from the early 1980s to 1986.

In addition to the impact of fertilizer input on the TIN flux from the Yangtze River to the East China Sea, dam construction, especially the Three Gorges Project, is another important factor to be considered. Although some pre-studies have been done (LGEETGP, 1987; Liu, 1992) on the impacts of the Three Gorges Project on ecology and environment along the Yangtze River basin, and of the adjacent coastal zone, the question is still suspended about what could really happen to the environment and ecology in the short- or long-term, or to what extent the impact could be. Without a monitoring program along the river and in the estuarine and adjacent waters, we probably could not find out.

Fate of fine sediment and nutrients from the rivers

Yangtze River nutrients will participate in the complex biogeochemical cycling in the coastal sea, and will eventually be exported to the deep ocean. van Bennekom and Salomons (1981) pointed out that the output of nutrients from coastal zones to the ocean may be quite different from the input by rivers into the coastal zone. With present knowledge, it is very difficult to separate recent and long-term processes in coastal zones. So, nutrient cycle or dynamics on the shelf really needs to be intensively and extensively studied.

Where does the fine sediment go after being discharged from rivers? Hu (1984) suggested that wherever there is upwelling on the continental shelf, there must be a mud (fine sediment) patch under it on the seafloor with the Yellow Sea and East China Sea as examples (Fig. 10.10). That means upwelling is a trap for most fine sediment on the shelf. The question is how much of the fine sediment is transported to the ocean and how. Hu (1995) proposed a conceptual model for fine sediment transport from the shelf to the ocean by wind-driven vertical circulation, which is favorable for in winter but blocks the fine sediment transport from the East China Sea to the Okinawa Trough in summer.

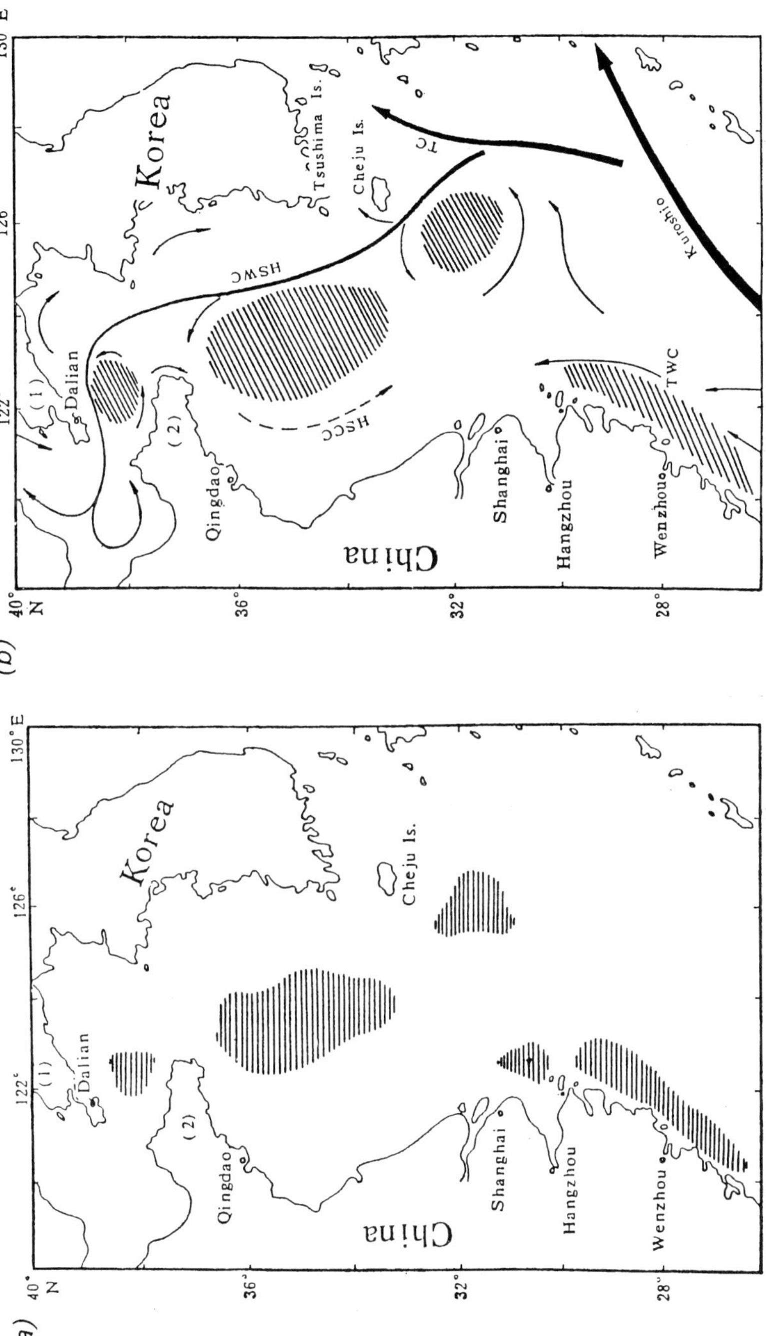

Figure 10.10 *(a) Mud patches and (b) scheme of current systems with upwelling areas (shaded) in the Yellow Sea and East China Sea. (After Hu, 1984.)*

Summary and recommendations

Global inputs of sediment and nutrient from the river to the coastal zone are *c.* 20 $\times 10^9$ t year^{-1} (Milliman & Syvitski, 1992) and *c.* 75 Tg N year^{-1} (van Bennekom & Salomans, 1981; Galloway *et al.*, 1995), respectively, with long historical changes, natural and anthropogengic. The anthropogenic influence has become imperiling, especially for the last 40–50 years, during which time nutrient input has increased about four times substantially by fertilizer application, and sediment discharge has decreased mainly by dam/reservoir construction. Because of that, coastal marine ecosystem and environment have deteriorated, that is, many species have disappeared, eutrophication and red tide occur more often and coastal erosion has become severe.

At present, about 13% of the global total of fluvial sediment is caught in reservoirs, and the percentage will rise with the increasing dam construction, while deforestation and cultivation will increase sediment discharge, as shown in Fig. 10.11. Which way will the sediment discharge go, up or down? Most probably the sediment discharge in the near future at least will still decrease due to the continuation of dam construction on big rivers such as the Yangtze and the Yellow Rivers. However, it is difficult to predict what will happen about 40–50 years from now. The sediment discharge to the coastal zone may increase again after the reservoirs constructed since the beginning of this century have filled up with sediment at some time in the next century.

The increase of nutrient input to the coastal zone will undoubtedly continue, owing to the intensive and extensive fertilizer application under way, coupled with

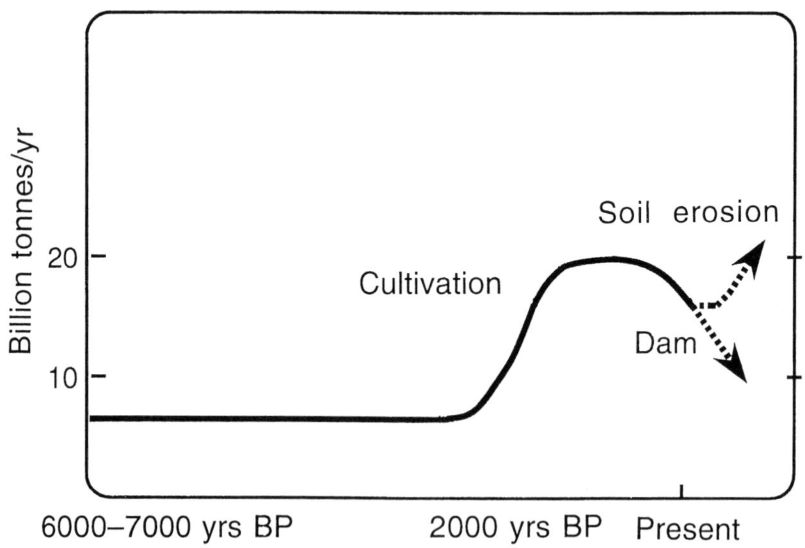

Figure 10.11 *Scheme of world sediment discharge (past, present and future).*

population growth. The input materials, nutrient and sediment, will be involved in the physical, chemical, biological and sedimentological processes of the coastal ocean (seas) and be exchanged with the open ocean. However, so far, little is known about the processes on the shelf which control the fate of those materials. To track these changes, to understand the processes and to find measures to deal with what would happen to society, a long-term global programme is needed to survey, monitor and understand the changes of these material fluxes. Actually, there have been several local programs such as MFLECS* (1992–1995), MASFLEX* (1992–1997), MAFLECS* (1993–1999), POFLECS* (1996–1999), and so on. It would be appropriate to integrate such local programs into a global one under the auspices of LOICZ with special reference to its focus 1 and focus 3 (IGBP, 1994).

Acknowledgments

We are especially grateful to Professor F. Li for his generous offer of time for valuable discussion and material. We thank Professors Z. L. Shen, S. Zhang, H. K. Gu, P. C. Li for discussion. We also thank Professors H. Lin, H. S. Yang, Y. Q. Chen and S. V. Smith for material provided. Last but not the least, we would express our thanks to Ms H. Bai, Mr X. Bai, Ms Z. Jin and Mr S. Lian for typing the draft and redrawing the figures. The preparation of this paper is supported by the National Natural Science Foundation of China (NSFC 49636210).

**Note*: MFLECS: Margin Flux in the East China Sea (January 1992-December 1995, National Natural Science Foundation of China's (NSFC) Program); MASFLEX: Marginal Sea Flux Experiment in the West Pacific (April 1992–March 1997, Science and Technology Agency of Japan's (STA) Program); MAFLECS: Material Flux in the East China Sea (August 1993–August 1999, China–Japan Joint Program); POFLECS: Key Processes of Ocean Flux in the East China Sea (January 1996–December 1999, National Natural Science Foundation of China's (NSFC) Program).

References

Abdel-Aal, F. M. (1985). Erosion of the Nile Delta coast. In *Proceedings of Coastal Zone '85*, pp. 1601–11. New York: American Society of Civil Engineers.

Asida, K., Takahashi, T. & Michigami, M. (eds.) (1983). *Sediment Disasters and Treatment*, 260 pp. Tokyo: Morikita Publishers.

Editorial Group of the Water Conservancy History of the Yellow River (EGWCHYR) (ed.) (1982). *Summary of the Water Conservancy History of the Yellow River*. China Water and Electricity Press, Beijing. [in Chinese].

Edmond, J. M., Spivak, A., Grant, B. C., Hu, M. X. & Chen, Z. X. (1983). Chemical dynamics of the Changjiang Estuary. In *Proceedings of International Symposium on Sedimentation on the Continental Shelf with Special Reference to the East China Sea*, pp. 229–39. Beijing: China Ocean Press.

Fan, Z. M. & Guo, Y. S. (ed.) (1992). *Atlas of Remote Sensing Dynamical Analysis of the Coastal Zone Near the Yellow River Delta*, pp.

106–7. Beijing: China Ocean Press [in Chinese].

Fournier (1960). *Climate Erosion*, 201 pp. Paris: Press University de Frocnce.

Galloway, J. G., Schlesinger, W. H., Levy, H., Michaels, A. & Schnoor, J. L. (1995). Nitrogen fixation: anthropogenic enhancement–environmental response. *Global Biogeochemical Cycles*, **9(2)**, 235–52.

Galloway, J. N., Zhao, S., Thomson, V. E. & Chang, L. H. (1996). The mobilization of nitrogen in the United States of America and the People's Republic of China. *Atmospheric Environment*, **30**, 1551-61.

GESAMP (1987). Land/sea boundary flux of contaminants: contributions from rivers. *Report Studies GESAMP no. 32* 172 pp., *Group Experts Scientific Aspects Marine Pollution*. Switzerland: United Nations Geneva.

GESAMP (1993). Anthropogenic influences on sediment discharge to the coastal zone and environmental consequences. *GESAMP Reports and Studies*, No. 52, 67 pp.

Gu, H. K., Xiong, X. X., Liu, M. X. & Li, Y. (1981). Nitrogenous geochemistry near the Yangtze mouth. I. Nitrates near the Yangtze estuary. *Journal of Shandong College of Oceanography*, **11(4)**, 37–46 [in Chinese with English abstract].

Gu, H. K., Xiong, X. X., Ma, X. N., Li, Y., Wei, Q. R., Liu, M. X., Shen, W. T., Ren, G. F., Diao, H. X., Li, G. J., Jiang, C. X., Wu, Y. Y., Chen, Z. & Chang, L. Y. (1982). The nitrogen cycle near the estuary of the Yangtze River. *Collected Oceanic Works*, **5(1)**, 48–66.

Holeman, J. N. (1968). The sediment yield of major rivers of the world. *Water Resources Research*, **4**, 737–47.

Howarth, R. W., Jensen, H. S., Marino, R. & Postma, H. (1995). Transport to and processing of P in near-shore and oceanic waters. In *Phosphorus in the Global Environment*, ed. H. Tiessen, pp. 324–58. Chichester, UK: John Wiley.

Hu, D. (1984). Upwelling and Sedimentation I. The role of upwelling in sedimentation in the Yellow Sea and East China Sea. *Chinese Journal of Oceanology and Limnology*, **2(1)**, 12–9.

Hu, D. (1995). The role of vertical circulation in sediment dynamics. In *China Contribution to Global Change Studies*, ed. D. Ye, H. Lin, pp. 168–70. Beijing, China: Science Press.

Huang, S. G., Yang, J. D., Ji, W. D., Yang, X. L. & Chen, G. X. (1983) Silicon, nitrogen and phosphorus in the Changjiang River mouth water. In *Proceedings of International Symposium on Sedimentation on the Continental Shelf with Special Reference to the East China Sea*, pp. 220 - 8. Beijing: China Ocean Press.

IGBP (1994). Implementation plan of land–ocean interactions in the coastal zone. *Global Change Report*, **33**, 215 pp.

Kempe, S. (1982). Long-term records of the CO_2 pressure fluctuation in fresh water. In *Transport of Carbon and Minerals on Major World Rivers*, ed. E. T. Degens, Vol.1, pp. 91–331. SCOPE/UNEP. Mitteilungen aus dem Geologisch-Palacontologischen Institut der Universitaet Hamburg, 52.

Kempe, S. (1995). Coastal seas: a net source or sink of atmospheric carbon dioxide? *LOICZ Reports & Studies*, **1**, 27 pp.

Leading Group of the Project on Ecology and Environment of the Three Gorges Project of the Chinese Academy of Sciences. (LGEETGP) (1987). *Collection of papers on impacts of the Three Gorges Project on the ecology and environment and the countermeasure*, 1126 pp. Beijing: Science Press.

Liu, R. Y. (1992). Impacts of the Three Gorges Project on the ecology and environment of the Changjiang River estuary and adjacent waters. *Studia Marina Sinica*, **33**, 359 pp. [in Chinese].

Meade, R. H. (1969). Errors in using modern stream-load data to estimate natural rates of denudation. *Geological Society of American Bulletin*, **8**, 1265–74.

Meade, R. H. (1982). Sources, sinks and storage of river sediment in the Atlantic drainage of the United States. *Journal of Geology*, **90**, 235–52.

Meade, R. H., Yuzyk, T. R. & Day, T. J. (1990). Movement and storage of sediment in rivers of the United States and Canada. In *Surface Water Hydrology: The Geology of North America*, ed. M. G. Wolman & H. C. Riggs, V.

O-1, pp. 255–80. Boulder, CO: Geological Society of America.

Milliman, J. D. (1991). The flux and fate of fluvial sediment and water in coastal seas. In *Ocean Margin Processes in Global Change*, ed. R. F. C. Mantoura, J-M. Martin & R. Wollast, pp. 69–89. Dahlem Conference PC9, Wiley.

Milliman, J. D. (1992). Management of the coastal zone: impact of onshore activities on the coastal environment. In *Use and Misuse of the Seafloor*, ed. K. J. Hsu & J. Thiede, pp. 213-27. New York: John Wiley.

Milliman, J. D. (1995). Sediment discharge to the ocean from small mountainous rivers: the New Guinea example. *Geo-marine Letters*, **15**, 127–33.

Milliman, J. D. & Meade, R. H. (1983). World-wide delivery of river sediment to the oceans. *Journal of Geology*, **91**, 1–21.

Milliman, J. D. & Syvitski, J. P. M. (1992). Geomorphologic/tectonic control of sediment discharge to the ocean. The importance of small mountainous rivers. *Journal of Geology*, **100**, 525–44.

Milliman, J. D., Quraishee, G. S. & Beg, M. A. A. (1984). Sediment discharge from the Indus River to the ocean: past, present and future. In *Marine Geology and Oceanography of Arabian Sea and Coastal Pakistan*, ed. B. U. Hag & J. D. Milliman, pp. 65–70. New York, Cincinnati, Stroudsbury, Toronto, London, Melbourne: Van Nostrand Reinhold Company Scientific and Academic Editions.

Milliman, J. D., Qin, Y. S., Ren, M. E. & Saito, Y. (1987). Man's influence on the erosion and transport of sediment by Asian rivers: the Yellow River (Huanghe) example. *Journal of Geology*, **95**, 751–62.

Myers, N. (1988). Tropical deforestation and climate change. *Environmental Conservation*, **15**, 293–8.

Nixon, S. W. (1995). Coastal marine eutrophication: a definition, social causes, and future concerns. *Opelia*, **41**, 199–219.

Ohmori, H. (1983). Erosion rates and their relation to vegetation from the view point of world-wide distribution. *Bulletin Department of Geography*, University of Tokyo, **15**, 77–91.

Pearce, F. (1991). A dammed fine mess. *New Scientist*, May 4, 36–9.

Petts, G. E. (ed.) (1984). *Impounded Rivers*, 326 pp. New York: John Wiley.

Qian, N., Zhang, R. & Chen, Z. C. (1987). On some sedimentation problems of the Three Gorges Project. *International Journal of Sediment Research* (Beijing), **1**, 5–38.

Qian, Y., Ye, Q. & Zhou, W. (eds.) (1993). *Variations of Water Discharge and Sediment Discharge and Evolution of River Bed of the Main Channel of the Huanghe.* Publishing House of China Construction Material Industry, 230 pp. [in Chinese].

Saito, Y. & Yang, Z. (1994). The Huanghe River: Its water discharge, sediment discharge, and sediment budget. *Journal of Sediment Society Japan*, **40**, 7–17. [in Japanese with English abstract]. TAGS, **7**, 3–10.

Saito, Y. & Yang, Z. (1995). Historical change of the Huanghe (Yellow River) and its impact on the sediment budget of the East China Sea. In *Global Fluxes of Carbon and Its Related Substances in the Coastal Seas–Ocean–Atmosphere System*, ed. S. Tsunogai, K. Iseki, I. Koike & T. Oba, pp. 7–12. Yokohama: M & J International.

Saito, Y., Ikehara, K., Katayama, H., Matsumoto, E. & Yang, Z. (1994). Course shift and sediment-discharge changes of the Huanghe recorded in sediments of the East China Sea. *Chishitsu News*, **476**, 8–16. [in Japanese].

Saunders, I. & Young, A. (1983). Rates of surface processes on slopes, slope retreat and denudation. *Earth Surface Processes and Landforms*, **8**, 473–501.

Shen, Z. L. (1991). A study on the effects of the Three Gorges Project on the distributions and changes of the nutrients in the Changjiang River estuary. *Oceanologia et Limnologia Sinica*, **22(6)**, 540–6 [in Chinese with English abstract].

Shen, Z. L. (1993). A study of the relationship of the nutrients near the Changjing River estuary with flow of the Changjiang River water. *Chinese Journal of Oceanology and Limnology*, **11(3)**, 260–7.

Shen, Z. L. (1996). A preliminary discussion on the control of inorganic nitrogen in the Changjiang estuary. *Marine Sciences*, **1**, 61–62 [in Chinese with English abstract].

Shen, Z. L., Liu, X. J. & Lu, J. P. (1987). Distribution of inorganic nitrogen and phosphate in the water of downstream of the Changjiang River and their removal processes in the estuary. *Studia Marina Sinica*, **28**, 69–77 [in Chinese with English abstract].

Tang, R. C. & Lin, W. Q. (1987). A study on sedimentation problems of the Gezhouba Project. *International Journal of Sediment Research* (Beijing), **1**, 69–101.

van Bennekom, A. J. & Salomons, W. (1981). Pathways of nutrients and organic matter from land to ocean through rivers. In *River Inputs to Ocean Systems*, UN, UNEP, IOC and SCOR. pp. 33–51.

van der Laeden, F., Troise, F. L. & Todd, D. K. (eds.) (1990). *The Water Encyclopedia*, 808 pp. Chelsea, MI: Lewis Publishers.

Wahby, S. D. & Bishara, N. F. (1981). The effect of the river Nile on Mediterreanean water, before and after the construction of the high dam at Aswan. In *Proceedings of a Review Workshop on River Inputs to Ocean Systems*, pp. 311–18. New York: United Nations.

Walling, D. E. (1983). The sediment delivery problem. *Journal of Hydrology*, **65**, 209–37.

Walling, D. E. & Webb, B. W. (1983). Patterns of sediment yield. In *Background to Pataeohydrology*, ed. K. J. Gregory, pp. 69–100. Chichester: John Wiley.

Wang, Y. & Aubrey, D. G. (1987). The characteristics of the China coastline. *Continental Shelf Research*, **7**, 329–49.

Wang, M. Y., Zhao, G. J. & Zhang, S. (1990). A study of the mass transport rate of carbon, nitrogen, phosphorus and sulphur of the Changjiang River. In *Study on environmental backgound of chemical elements in waters*, ed. S. Zhang, L. C. Zhang, L .Q. Tian & B. W. Zeng, pp. 121–31. Survey and Drawing Press, Beijing, 1990 [in Chinese with English abstract].

Wang, Z. F., Yao, L. K. & Yuan, X. Z. (1983). Distributions and changes of nutrients (N, P and Si) in the Yangtze estuary. *Oceanlogia et Limnologia Sinica*, **14(4)**, 324–32 [in Chinese with English abstract].

Ye, Q. C. (ed.) (1994). *Researches on Environmental Changes of the Yellow River Basin and Laws of Water and Sediment Transportation*. 40 pp. [in Chinese].

Zhao, Y. A., Pan, X. D., Fan, Z. Y. & Han, S. F. (1989). Sedimentation in the lower reaches of the Yellow River and its basic laws, In *Yellow River: Silt and Floods*, ed. L. M. Brush, pp. 477–516. Wolman, Dordrecht: Kluwer Academic Pubilshers.

11 Carbon and nutrients in the ocean

C-T. A. Chen and S. Tsunogai

Introduction

Of the estimated anthropogenic input of carbon dioxide (CO_2), less than 60% is now present in the atmosphere. The ocean is believed to be taking up much of the remainder, at a rate of about 2 Gt C year^{-1}, where 1 Gt C is 10^{15} g of carbon. Ocean waters contain a very large inventory of carbon, about 39 000 Gt C. This is more than 50 times more than the inventory of carbon in the atmosphere, and about 20 times more than the amount of carbon tied up in the terrestrial biota and soils. As long as one is considering processes ranging over hundreds of years, the carbon reservoirs of the sedimentary cycle can be neglected. As a result, the marine carbonate system, being the largest carbon pool in the atmosphere, biosphere, and hydrosphere, is of primary importance for the partition of atmospheric excess carbon dioxide produced by human activities. A relatively small change in the oceanic carbon cycle can have large atmospheric consequences.

Moreover, the ocean has a more complicated carbon cycle than the atmosphere, involving many inorganic and organic forms. Various models of the atmosphere and ocean give different estimates of how much CO_2 is exchanged and where. Many of these models calibrate with the first world-wide data set collected by GEOSECS (Geochemical Ocean Sections Study) between 1972 and 1978. Recognizing that we must improve the observational and conceptual basis of our estimates and predictions, a new world-wide survey has been undertaken under the auspices of the Joint Global Ocean Flux Study (JGOFS) with close collaboration with WOCE (World Ocean Circulation Experiment). JGOFS has been designed to increase our understanding of the ocean carbon cycle, its sensitivity to change, and the regulation of the atmosphere–ocean CO_2 balance. More formally, it has two goals (SCOR, 1990):

1. To determine and understand, on a global scale, the processes controlling the time-varying fluxes of carbon and associated biogenic elements in the ocean, and to evaluate the related exchanges with the atmosphere, seafloor, and continental boundaries;

2. To develop a capability to predict on a global scale the response of oceanic biogeochemical processes to anthropogenic perturbations, in particular those related to climate change.

Clearly, in order to understand the carbon cycle, the "associated biogenic elements" must also be studied. The most important biogenic elements associated are the major nutrients, nitrogen, phosphorus and silicon (SCOR, 1992).

Carbon cycle

One of the most important systems in the oceans is the CO_2–carbonate system, a buffering system that helps to maintain the pH of seawater to within a narrow range. Over 95% of the oceanic carbon is in the form of inorganic, dissolved carbon; the remainder is composed of various forms of organic carbon (living organic matter, particulate and dissolved organic carbon).

CO_2 (gas) hydrates with water rather rapidly (in milliseconds) to form carbonic acid (H_2CO_3). Once CO_2 is hydrated, it is involved in a series of even more rapid proton transfer steps (submicroseconds) to form bicarbonate (HCO_3^-) and carbonate (CO_3^{2-}) ions:

$$CO_2\,(\text{gas}) + H_2O \Leftrightarrow H_2CO_3 \quad (1)$$
$$H_2CO_3 \Leftrightarrow H^+ + HCO_3^- \quad (2)$$
$$HCO_3^- \Leftrightarrow H^+ + CO_3^{2-} \quad (3)$$

The sum of the three aqueous carbonate species is defined as total CO_2 (TCO_2). A frequently used property is titration alkalinity (TA), the amount of acid needed to neutralize all bicarbonate and carbonate ions, plus a few minor species.

Because of the large buffer factor (or *Revelle* factor) of seawater, at equilibrium, a 10% change in the partial CO_2 pressure (pCO_2) results in merely 1% change in TCO_2. At the current pH of seawater (about 8), the excess CO_2 added to the oceans reacts with the carbonate ion to form the bicarbonate ion without the use of H^+:

$$CO_2 + CO_3^{2-} + H_2O \Leftrightarrow 2HCO_3^- \quad (4)$$

The concentration of the carbonate ion is limited by the solubility of calcium carbonate as follows:

$$Ca^{2+} + CO_3^{2-} \Leftrightarrow CaCO_3 \quad (5)$$

In this reaction the solubility of $CaCO_3$ increases with pressure but decreases with temperature. The surface layer is usually supersaturated with respect to

$CaCO_3$, but except for isolated regions such as the Bahama Banks, spontaneous precipitation does not occur because organic material and magnesium in seawater retard $CaCO_3$ formation. Many marine organisms, however, utilize calcium and carbonate ions to form $CaCO_3$ skeletons and shells. This process occurs mainly in the surface layer. As the organisms die, their shells and skeletons sink to a level where the temperature becomes low enough or the pressure becomes high enough; thus the water becomes undersaturated with respect to $CaCO_3$, at which point they start to dissolve. Accumulation of undissolved $CaCO_3$ in the sediments over millions of years makes the sediments a major sink of carbon over the geological time (Chen & Drake, 1986). Since calcium and carbon are always in abundant supply, they are not considered as nutrients. However, in the "silica ocean" such as the Southern Ocean and the western North Pacific where diatoms prosper, siliceous shells are formed and deposited instead (Tsunogai *et al.*, 1986; Noriki & Tsunogai, 1986). Since silicon can be in short supply, it is considered as a nutrient.

Inorganic carbon, nitrogen and phosphorus are also used by marine organisms to grow soft tissue. Soft tissue decomposes quickly in the water column. As a result, the marine organisms act as a "biological pump", thus removing CO_2 and nutrients from the surface ocean and transferring these elements into the deep ocean and ocean bottom.

The role of the oceans in the global carbon cycle is two-fold: first, it represents a passive reservoir which absorbs excess atmospheric CO_2. Second, changes in the physical state of the ocean (temperature, circulation) and the marine biota may affect the rate of air–sea exchange, and thus future atmospheric CO_2.

The flow of carbon dioxide between atmosphere and ocean is dominated by ocean upwelling and downwelling. High latitude water cools and dissolves more CO_2 before it sinks; water upwelling at the Equator warms and releases CO_2. The rate of carbon entering the oceans at high latitudes and leaving at low by this route is about 40 ($\pm$ 10) Gt C year^{-1} (Moore & Bolin, 1986). The gas flux across the sea surface is a product of the partial CO_2 pressure difference (ΔpCO_2) between the air and surface waters and a surface transfer coefficient that increases with wind speed. Thus exchanges of carbon between the atmosphere and the oceans (net air–sea fluxes) can potentially be deduced from measurements of ΔpCO_2. It is clearly shown in Fig. 11.1 that tropical waters are generally supersaturated in pCO_2 as a result of upwelling and warming. Consequently, the tropical oceans are a source of CO_2 to the atmosphere. On the other hand, at around 40° S in the Southern Hemisphere and in the high latitude regions in the Northern Hemisphere, especially in the northern North Atlantic, the pCO_2 is generally undersaturated, depending on the productivity and the nutrients remaining in the water. As a result, these regions become a sink for CO_2.

Many carbonate species and nutrients, especially when normalized to a constant salinity to remove the effects of evaporation and precipitation, are known to

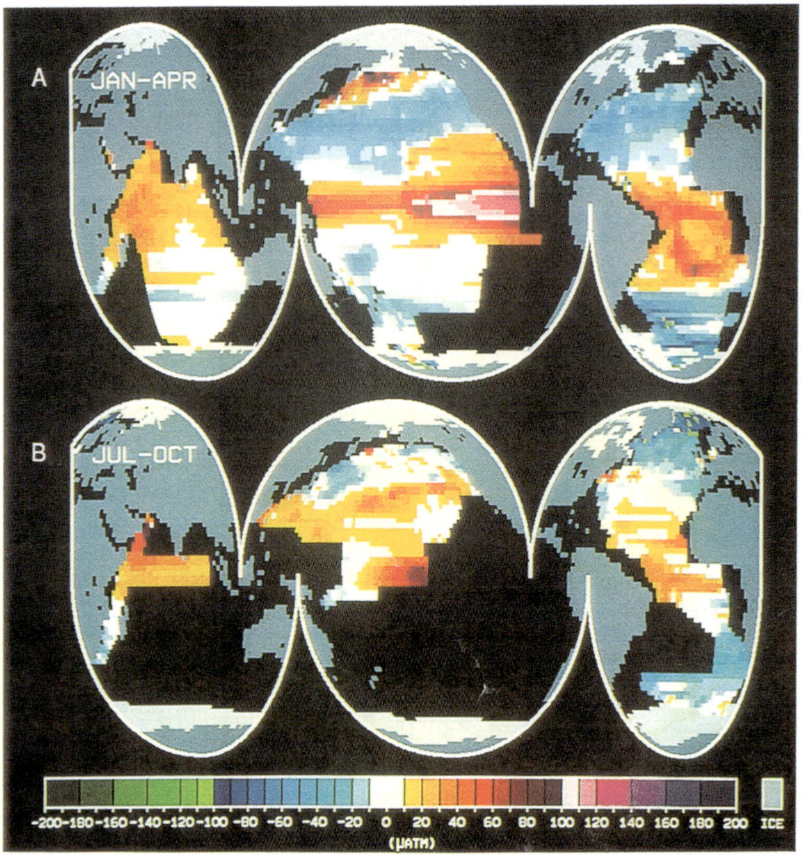

Figure 11.1 *The distribution of ΔpCO_2 in the surface ocean: implications for climate and global change on seasonal-to-century time scales. Blue and green show zones where CO_2 is undersaturated; yellow, red and purple show zones where CO_2 is supersaturated. Black areas are data gaps. (Courtesy, T. Takahashi.)*

correlate linearly with temperature, although temperal and spatial variations of these trends exist (e.g. nitrate: Chen *et al.*, 1982*a*: pH, phosphate and silicate: Chen, 1984; calcium: Chen *et al.*, 1982*b*; alkalinity and total CO_2: Chen & Millero, 1979). The normalized nitrate ($NNO_3 = NO_3 \times 35/S$) values for surface waters are also found to correlate linearly with surface temperature between 2 °C and 17 °C (Fig. 11.2*a*). There is essentially no nitrate above 17 °C.

The pH, normalized alkalinity ($NTA = TA \times 35/S$), and normalized total CO_2 ($NTCO_2 = TCO_2 \times 35/S$) also correlate linearly with temperature (Fig. 11.2*b–d*). There may be a slight change in slope at 13 °C for NTA and $NTCO_2$ near the Subtropical Front near 40° S. The NTA slope also changes slightly at 4 °C near the Antarctic Front.

It is not clear why such linear correlations exist. Chen (1992), based on the INDIVAT 1 data, assumed that the biological activity reduces the NNO_3 concentration from 28 μmol kg^{-1} in the Circumpolar Current near 50° S to almost zero in the subtropical region near 35° S at a rate linearly correlated with the warming of the water from 2 °C to 17 °C. Given the Redfield C/N ratio of 106/16, this consumption of nitrate must produce 185.5 μmol kg^{-1} of organic carbon. At the same time, the $NTCO_2$ must be reduced by the same amount. In addition, the warming of seawater drives out dissolved CO_2 and reduces the $NTCO_2$ by 8 μmol kg^{-1}, assuming that the surface seawater remains at the same degree of saturation with respect to CO_2.

Production of organic carbon as soft tissue is associated with the production of inorganic carbon as hard tissue and shells at roughly a four-to-one ratio. Thus, the production of 185.5 μmol kg^{-1} in organic carbon should result in a further reduction of 46.4 μmol kg^{-1} in $NTCO_2$ and 63 μmol kg^{-1} in NTA after taking into consideration the effect of nitrate and phosphate on alkalinity. Consequently, a total reduction in CO_2 of 240 μmol kg^{-1} should be expected and the observed reduction was 215 μmol kg^{-1}. Chen (1992) also observed a NTA reduction of 70 μmol kg^{-1} vs. the expected 63 μmol kg^{-1}. These correlations, in a way, quantify the relationships between the carbon cycle and the nutrient cycle and indicate that biological activities contribute to most of the reduction in nitrate, alkalinity and total CO_2.

For waters north of 35° S (waters warmer than 17 °C), nitrate concentration is so low that other sources of nitrogen, such as ammonia, or nitrogen fixation may be important in biological consumption. Thus, the Redfield Ratio is no longer applicable. Furthermore, the effect of equatorial upwelling becomes important, and simple linear relations cease to exist.

The decrease in NNO_3 of 28 μmol kg^{-1} between 4 °C and 17 °C for INDIGO 1 corresponds to the decrease of 239 μmol kg^{-1} in $NTCO_2$ and 63 μmol kg^{-1} in NTA. Chen (1992) observed a similar decrease in $NTCO_2$ and NTA of 200 μmol kg^{-1}, respectively. These results strengthen the suggestion that biological activities contribute to most of the reduction in nitrate, alkalinity and total CO_2 before the Antarctic waters reach the tropical zone.

We have mentioned the transfer of CO_2 into the deep waters by the "biological pump". Sinking particles, downward mixing of dissolved organic matter, and the vertical zooplankton migration all result in the transfer of biogenic material from the surface to the deep sea. The dissolution of $CaCO_3$ increases alkalinity, calcium, and total CO_2. The decomposition of organic matter, be it from calcareous or siliceous organisms, also increases total CO_2 and pCO_2, but it decreases pH and alkalinity (Brewer *et al.*, 1975). Overall, the dissolution of $CaCO_3$ contributes approximately 15–30% of the total CO_2 input into the deep waters (Tsunogai *et al.*, 1973; Chen *et al.*, 1982*b*; Chen, 1990; Tsunogai & Noriki, 1991). As a result,

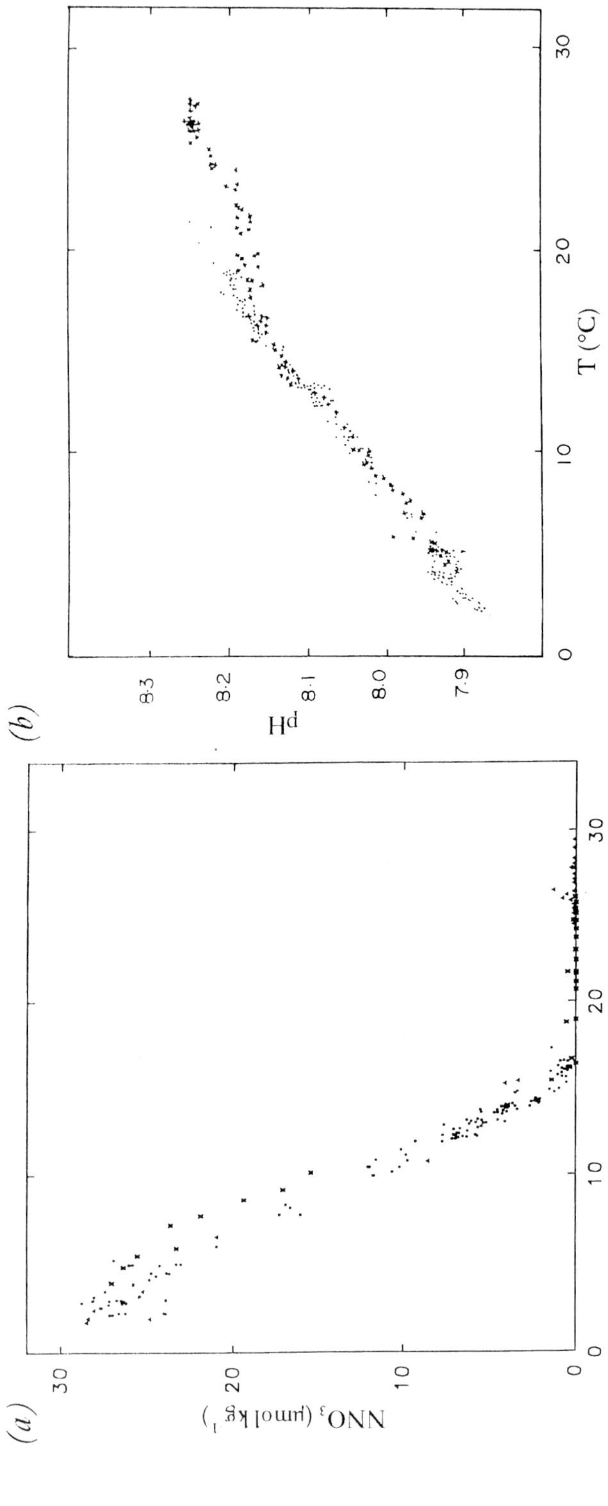
(a)
NNO3 (μmol kg-1)
30
20
10
0
T (°C)
0
10
20
30
(b)
pH
8.3
8.2
8.1
8.0
7.9
T (°C)
0
10
20
30

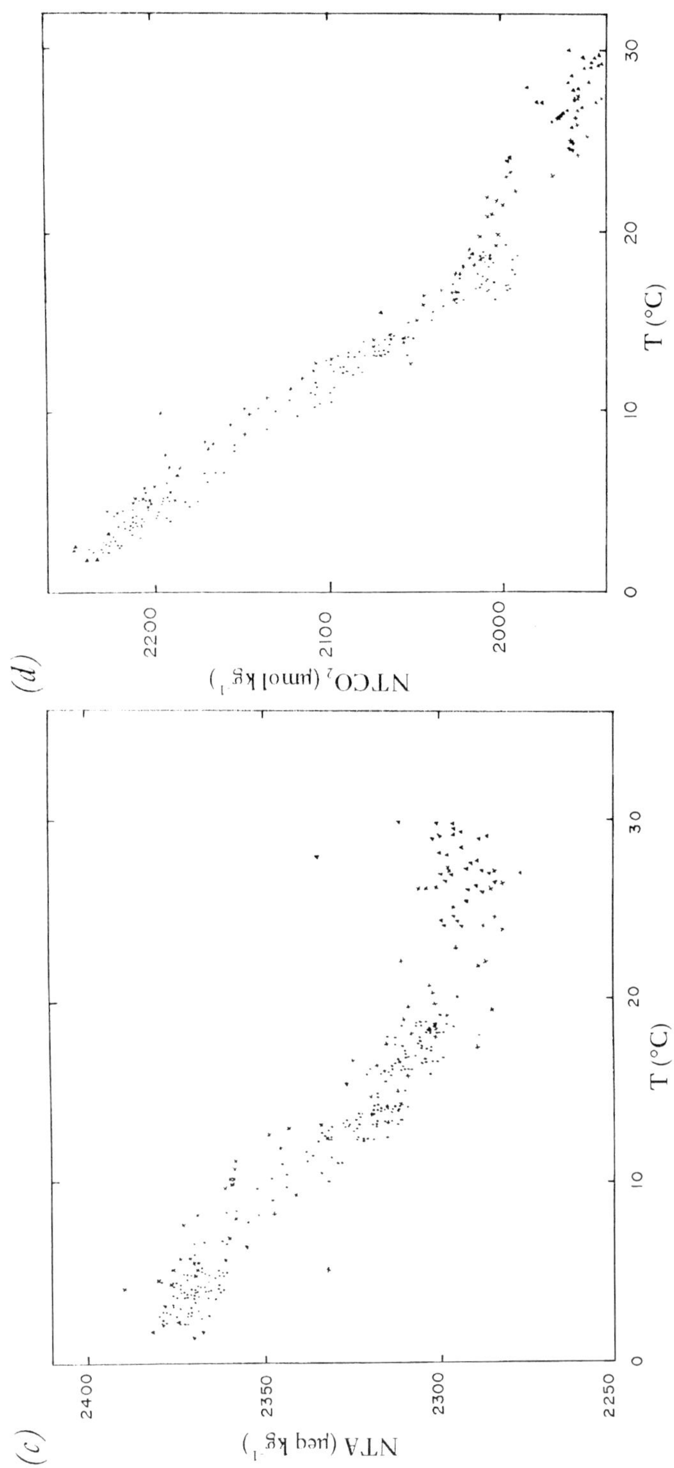

Figure 11.2 *The correlations of surface (a) NNO_3, (b) pH, (c) NTA and (d) $NTCO_2$ with respect to water temperature in the south Indian Ocean, based on the INDIVAT 1 (July, 1984), INDIGO 1 (February–March, 1985) and GEOSECS (December, 1977–April, 1978) data.*

deep waters all have higher TA and TCO_2 values than the surface (Fig. 11.3*a,b*). The Pacific deep and bottom waters have higher concentrations than those in the Atlantic because of their older age (hence more $CaCO_3$ and organic carbon dissociation). The northward flowing bottom water in the North Pacific Ocean upwells to form the deep water and flows southward. Consequently, the Pacific deep water, older than the bottom water, has the highest TA and TCO_2 concentrations.

Nutrient cycle

A nutrient element is one which is sometimes in short supply, yet is functionally involved in the processes of living organisms. Traditionally, in oceanography, the term has been applied almost exclusively to nitrogen, phosphorus and silicon, although it is known that minor elements, such as iron, also play an important role (Martin, 1990). The primary processes influencing the nutrient concentrations in the sea are the geophysical and geochemical processes which control the addition to seawater of these elements and those responsible for their dispersion and removal. Rock weathering and the decay of organic material, together with waste discharges, are the major sources of most forms of nutrients to the sea, to which they are usually carried by terrestrial drainage (Spencer, 1975; Mackenzie *et al.*, this volume). For coastal oceans, the human inputs are beginning to be significant compared to weathering (see Hu *et al.*, this volume; Vörösmarty *et al.*, this volume).

The primary biological removal of inorganic nitrogen, phosphorus and silicon from seawater is by phytoplankton. On the shelves, the littoral and benthic algae also remove these elements from seawater, but the total amounts involved are relatively small. The primary and other consumers of the marine food web regenerate these elements to the soluble form. Bacteria, protozoa and autolysis are involved in these regeneration processes.

Assuming C : N : P ratios of 106 : 16 : 1, remineralization can be represented by the equation (Redfield *et al.*, 1963):

$$(CH_2O)_{106}\,(NH_3)_{16}H_3PO_4 + 138O_2 \Leftrightarrow 106CO_2 + 122H_2O + 16HNO_3 + H_3PO_4 \qquad (6)$$

The replenishment of the supply of the nutrients in the surface water depends upon the action of physical processes. The rates of accumulation of inorganic nitrogen, phosphorus and silicon in deep water are controlled by the productivity of the surface layers, by the rates of *in situ* regeneration, by the vertical mixing and eddy diffusivity, by lateral adjective movements of the deep water, and by diffusion out of the sediments (Spencer, 1975; Gordon *et al.*, 1996).

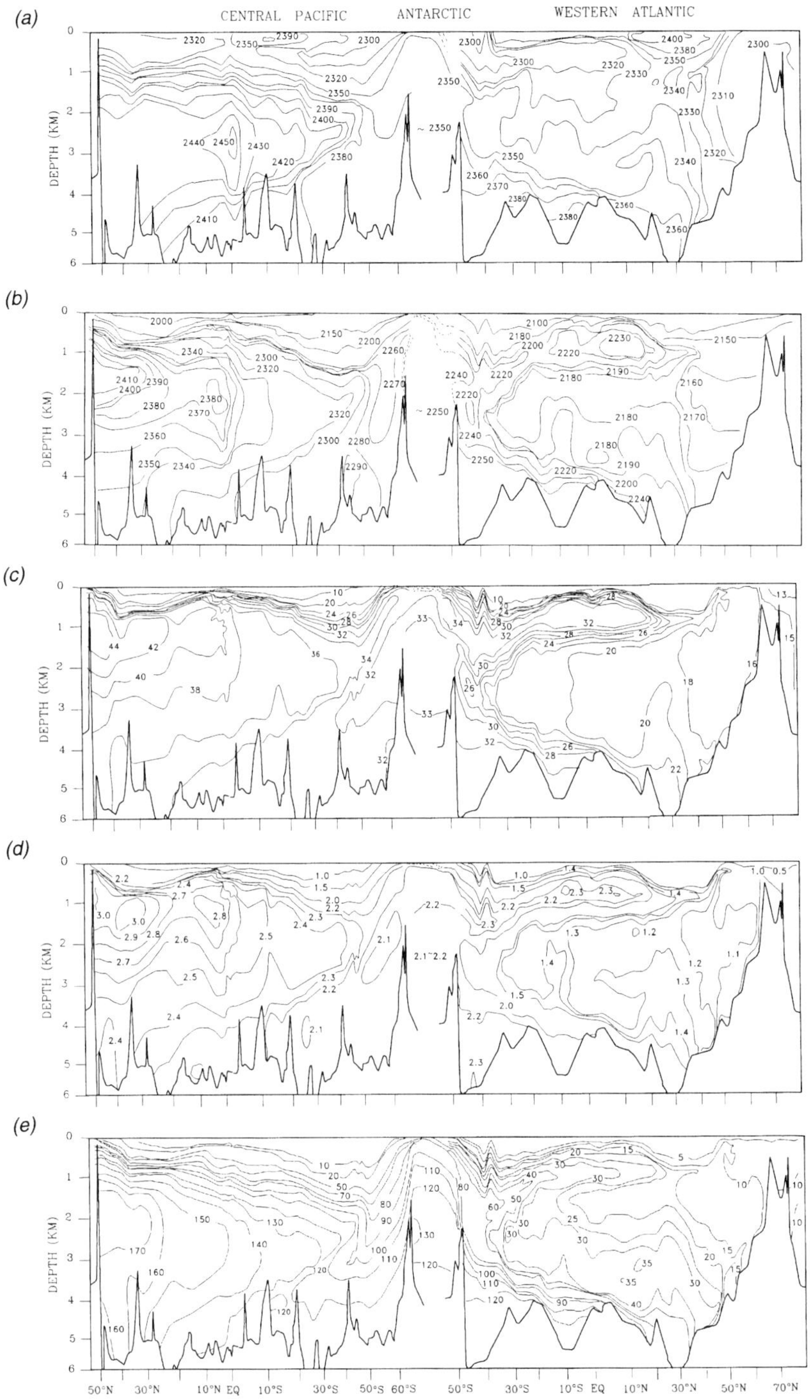

Figure 11.3 *The N–S cross-section of (a) TA, (b) TCO_2, (c) NO_3, (d) PO_4 and (e) SiO_2 in the Atlantic and Pacific Oceans (based mainly on the GEOSECS and JGOFS data).*

Nitrogen

Nitrogen is found in the sea primarily in the following forms: NH_4^+, N_2, N_2O, NO_2^-, NO_3^- and peptides. By far the most abundant species of nitrogen in the sea is nitrogen gas (N_2) , but it is essentially unreactive and is not consumed except for nitrogen fixation (Simpson, 1977; Tsunogai *et al.*, 1985; Tsunogai, 1991). N_2 is found throughout most oceanic and coastal waters at near saturation values (Kester, 1975). The next most abundant species, and a biologically active one, is the nitrate ion (NO_3^-). The other bioactive inorganic ions, nitrite (NO_2^-) and ammonium (NH_4^+) are less abundant overall, but are of local significance. There are significant stocks of organic nitrogen in the surface ocean but the biological availability of this nitrogen is not clear. Organic nitrogen is usually considered to be found in the amino form NH_2^- (Spencer, 1975; Sharp, 1983; Galloway *et al.*, 1995).

Nitrate, as a limiting plant nutrient, is almost exhausted in surface waters in the majority of the world's oceans; exceptions are at high and low latitudes and in coastal and estuarine waters. Beneath the photic zone, nitrate increases, usually reaching a maximum near the oxygen minimum layer, and then usually decreases slightly below when the Antarctic Bottom Water (AABW) is encountered.

A nitrate cross-section is shown in Fig. 11.3*c*, starting in the North Atlantic, running south down the western basin to the Antarctic, and then from south to north through the center of the Pacific Ocean based on the GEOSECS and JGOFS data. The North Atlantic Deep Water (NADW) can be seen between about 1.5- and 4-km depths in the Western Atlantic with a profound maximum (isopleths of 16–30 μmol kg^{-1}). The Antarctic waters show little variation with depth (32–34 μmol kg^{-1}), and are the obvious source of AABW, which is found at the bottom of both basins even beyond the Equator. In the Atlantic, the Antarctic Intermediate Water (AAIW) overlies the NADW as a distinct northward intrusion going as far as 40° N at about 1-km depth. The AAIW is less distinct in the Pacific, but it is obvious that the source of subsurface water is largely Antarctic. Distinct equatorial upwelling can also be seen in both oceans, but it is more dramatic in the Pacific (Sharp, 1983).

The South Pacific also has a deeper, broader nitrate maximum, whereas the North Pacific has a shallower maximum. The concentration of nitrate in the maximum is two-fold greater in the North Pacific (*c*. 45 μmol kg^{-1}) than in the North Atlantic (*c*. 22 μmol kg^{-1}). The Antarctic profile shows much higher surface values than those of lower latitudes and a very shallow nitrate maximum; Antarctic surface values are usually around 25 μmol kg^{-1}, and a fairly consistent deep-water value of 32–34 μmol kg^{-1} is characteristic of these waters.

Nitrate values in coastal waters usually decrease going seaward in surface waters and increase along the same axis in deep waters. Upwelling regions have major inputs of new nitrogen from the subsurface open ocean and the surface nitrate concentrations are higher than regions without upwelling. Upwelling regions often have strong oxygen depletion and periodically show high subsurface ammonium, N_2O and nitrite values (McCarthy & Carpenter, 1983). An example based on the upwelling near the East China Sea continental shelf break is given in a later section.

The primary release of inorganic nitrogen during decomposition is as ammonium which is oxidized microbially into nitrate (nitrification), if not utilized by phytoplankton. Nitrate, however, undergoes denitrification and turns into nitrite, ammonium, N_2O and N_2. Denitrification, and the subsquent loss of N_2, is the main sink for oceanic fixed nitrogen but whether or not this process is at a steady state is currently a matter of concern (Codispoti, 1995; Galloway *et al.*, 1995; Gordon *et al.*, 1996). The reaction is as follows:

$$(CH_2O)_{106}\,(NH_3)_{16}\,H_3PO_4 + 94.4\,HNO_3 \Leftrightarrow$$
$$106\,CO_2 + 177.2\,H_2O + H_3PO_4 + 55.2\,N_2 \qquad (7)$$

Phosphorus

Inorganic phosphorus occurs in seawater as orthophosphate. In seawater at pH 8.0 about 1% of the orthophosphate is present as $H_2PO_4^-$, 87% as HPO_4^{2-} and 12% as PO_4^{3-} (Spencer, 1975). The maximum concentrations of dissolved inorganic phosphate in the surface waters are in the range of 0.5 to 1.5 μmol kg^{-1} found in the northern North Pacific Ocean, in the Antarctic Ocean and in areas of upwelling. In general, the lower the temperature, the higher the concentration (Chen, 1984). Lower concentrations occur in the surface water of the tropics where maximum concentrations of 0.1 to 0.2 μmol kg^{-1} are common. The concentration of phosphate in coastal waters is often higher than in the water further offshore due to land runoff, sewage outflow, and coastal upwelling.

In common with the other nutrients, the concentration of inorganic phosphate increases with depth. The vertical distribution of phosphate in the oceans generally shows a gradual increase from the surface to a maximum value at between 500 and 2000 m. The phosphate maximum layer is most marked in the Atlantic Ocean based on the GEOSECS and JGOFS data (Fig. 11.3*d*). The layers below this usually have slightly decreasing concentrations with increasing depth. The highest concentrations in deep water occur in the North Pacific (over 3.0 μmol kg^{-1}).

The principal route of removal of inorganic phosphate from the surface ocean is assimilation by the phytoplankton. Subsequently, the inorganic phosphate is released as a result of autolysis, and is rapidly regenerated by microbial action from natural plankton and detritus. Usually, the N/P ratio is 16 during photosynthesis or remineralization (eqn. 6) but when denitrification occurs N_2O or N_2 may escape from the water column leaving phosphate behind (eqn. 7; Gordon *et al.*, 1996).

Silicon

Silicon (Si) is present in seawater in solution and as solid material in suspension. Si plays a key role in the marine ecosystem and is essential for the growth of the extracellular structures of diatoms, silicoflagellates and radiolarians, thus removing Si from the surface ocean and transporting it downward. The higher the Si/N ratio in the surface water, the larger the downward flux of the particulate organic carbon. So, the silicon is also removed more effectively than nitrogen by these organisms. The Si/N ratios in these planktons are higher than the ratio in seawater (Tsunogai, 1979; Tsunogai & Watanabe, 1983).

The particulate silicon consists of the silicious planktons, both living and dead, and of inorganic silicon in various mineralogical forms (Spencer, 1975). The ultimate fate of much of this particulate silicon in suspension is to enter the sediments where most dissolution occurs. The concentration of dissolved silicon in the form of silicic acid, $Si(OH)_4$, in seawater varies from as low as 0.5 μmol kg^{-1} for surface water in the tropical region to as high as 50 μmol kg^{-1} at the surface of the Antarctic region. The concentration of silicic acid in deep ocean waters reaches 125 μmol kg^{-1} in the Atlantic and higher values of up to 175 μmol kg^{-1} in the Pacific Ocean (Fig. 11.3*e*) based on the GEOSECS and JGOFS data. The concentration reaches 240 μmol kg^{-1} in the deep Bering Sea (Tsunogai *et al.*, 1979). In common with many other elements which are incorporated into living organisms, the concentration of dissolved silicic acid increases with depth. The circulation and exchange of the deep water differs from ocean to ocean, and this clearly affects the accumulation and dispersion of the silicic acid.

Human perturbations

Carbon

The oceanic uptake of CO_2 proceeds by (1) transfer of the CO_2 gas through the air–sea interface, (2) chemical interactions with the oceanic dissolved inorganic carbon, and (3) transport into the thermocline and deep waters by means of water mass transport and mixing processes. Once the CO_2 enters the oceans, the biological pump also works to remove CO_2 from the surface oceans and transports it downward. As a result, the pCO_2 of the surface oceans is reduced, thus favoring the uptake of more CO_2.

Although there are large geographical and seasonal variations of the surface ocean pCO_2, especially in the high latitude regions, averaged annually, the mid- and low-latitude surface waters are close to a constant degree of saturation with respect to the atmosphere. Therefore, processes (2) and (3) mentioned above are the main factors limiting the capacity of the ocean to serve as a reservoir on decadal and centennial time scales. The potential CO_2 uptake, however, occurs mainly in those parts of the deep ocean that are mixed with surface waters on decadal time scales or less. Therefore, high latitude regions need to be studied more extensively, especially in the winter time.

While the annual productivity of plants in the oceans is as great as that on land, their instantaneous effect on the atmosphere is greatly damped owing to the slowness of gas exchange between air and sea (Peng & Broecker, 1985). Thus the marine biota, if in steady state, are believed to play a minor role, if any at all, in the uptake of excess anthropogenic CO_2. The marine biota, however, play a crucial role in maintaining the steady-state level of atmospheric CO_2. About three-quarters of the vertical gradient in dissolved inorganic carbon is generated by the "biological pump". In the open ocean, however, this process is believed to be limited by the availability of nutrients, light, or by phytoplankton population control via grazing, and not by the abundant carbon (Falkowski & Wilson, 1992). Therefore, a direct effect of increased dissolved inorganic carbon (less than 2.5% since pre-industrial times) on carbon fixation and export is thought to be unlikely (Schimel *et al.*, 1995) although Riebesell *et al.* (1993) suggested that, under particular conditions, the rate of photosynthesis might be limited by the availability of CO_2.

The ocean's inventory of carbon is estimated to be increasing at a rate of about 2 Gt C $year^{-1}$. In principle, the net uptake of CO_2 by the oceans can be calculated by global integration of the net flux across the air–sea interface based on the pCO_2 data. However, the gas exchange coefficient is a relatively poorly known quantity (Wanninkhof, 1992; Watson, 1993). For instance, Takahashi *et al.* (1995) calculated the CO_2 flux in the North Atlantic. Results based on the most widely used coefficients of Tans *et al.* (1990) and Liss and Merlivat (1986) differ by more than a factor of two. Furthermore, large geographical and seasonal variations in pCO_2 exist in the atmosphere, and particularly in the surface oceans. These are induced both by physical processes (storms, upwelling, mixing, sea surface temperature fluctuations) and by the activity of the biota (Wong & Chan, 1991; Keeling, 1993; Inoue *et al.*, 1995; Takahashi *et al.*,1983, 1993, 1995; Boden *et al.*,1994). Representative estimates of the seasonally and regionally averaged net air–sea carbon transfer thus necessitates pCO_2 and wind measurements with high spatial and temporal resolution. Correction for the skin effect is also necessary (Garcon *et al.*, 1992; Robertson & Watson, 1992).

Estimates of the regional net air–sea carbon fluxes have been obtained, albeit with considerable error margins. However, the global oceanic carbon balance is more difficult to deduce by this method, as it represents a small residual computed from summing up the relatively large emissions from super-saturated and uptake in under-saturated regions. Effects of sporadic storm and interannual changes due to El Niños on the ocean biology need also to be studied (Wong *et al.*, 1993).

Characterization of the air–sea disequilibrium of the $^{13}C/^{12}C$ ratio of inorganic carbon with respect to that of the atmosphere has been used to estimate the net air–sea flux of CO_2 (Tans *et al.*, 1993; Okuda *et al.*, 1995). Tans *et al.* argue that the signal-to-noise ratio for the air–sea isotopic disequilibrium is potentially much better than that for the air–sea pCO_2 disequilibrium. However, the existing ^{13}C database is very small.

The possibility of using net meridional CO_2 fluxes within ocean basins to infer global sources and sinks of CO_2 has been explored by Brewer *et al.* (1989), Tans *et al.* (1990), Broecker and Peng (1992) and Goyet and Brewer (1993). Goyet and Brewer gave a net southward transport of 0.42 Gt C $year^{-1}$ across 25° N in the Atlantic. More CO_2 data coming out of the JGOFS and WOCE expeditions should make this method more widely used.

Beside the approaches mentioned above, inventory-based estimates have been used. For instance, in certain locations, a time series of CO_2 measurements is becoming available (Wong *et al.*, 1984). By the end of WOCE, it will be possible to examine carbon inventories over a 20–25-year interval at locations which were previously sampled during the GEOSECS expeditions. Preliminary data indicate that a significant increase of TCO_2 can be detected within the upper 1000 to 1200 m in the Brazil Basin over a period of 20 years (Wallace, 1995). The JGOFS also runs a network of time series stations at which regular measurements of key carbon related properties and processes are made.

Other inventory approaches were suggested: those using the $^{13}C/^{12}C$ isotope anomaly caused by the burning of fossil-fuel (Quay *et al.*, 1992), those based on the distribution of bomb-derived ^{14}C or ^{3}H, or models validated through comparison with the distribution of transient tracers such as bomb-derived ^{14}C or ^{3}H (Bolin, 1983; Broecker *et al.*, 1985; Bolin *et al.*, 1983; Sarmiento *et al.*, 1992; Orr, 1993; Siegenthaler & Sarmiento, 1993; Stocker *et al.*, 1994).

More recently, other transient tracers such as chlorofluorocarbons (CFCs) and CCl_4 have also been used to predict excess CO_2 penetration (Gammon *et al.*, 1982; Cline *et al.*, 1985; Krysell & Wallace, 1988; Joos *et al.*, 1991*a*, *b*; Wallace *et al.*, 1992, 1994; Watanabe *et al.*, 1994; Tokieda *et al.*, 1996). These data have high signal-to-noise ratios and thus the results are precise. In addition, the CFC results agree with those based on CO_2 directly (Cline *et al.*, 1985; Wallace,1995).

The oceanic contribution to the global carbon budget can also be assessed by direct observations of changes in the oceanic carbon content. This approach also suffers from the problem of determining a very small signal against large spatial and temporal background variability. Model estimates of the spacially and temporally variable rate of change induced by the anthropogenic perturbation are of the order of 1 part in 2000 per year on the average. Therefore, on a 10-year time scale, variations in dissolved inorganic carbon would have to be measured with an accuracy of better than 1% in order to determine the carbon balance in a particular oceanic region. While such accuracy can be obtained, a substantial sampling effort is required in some critical regions, together with some other tracers such as

CFCs (Tsunogai *et al.*, 1993*b*; Keeling, 1993; Schimel *et al.*, 1995). Repeated observational surveys, especially those in the GEOSECS regions occupied 20 years ago, would reveal regional carbon inventory changes, and thus provide a cross-check for other approaches.

Recently, there are indications that the excess CO_2 signal can be detected by using carbonate data directly (Brewer, 1978; Chen & Millero, 1978, 1979; Chen & Pytkowicz, 1979; Chen, 1982*a*, *b*; Anderson *et al.*, 1991; Tsunogai *et al.*, 1993*a*). The method involves a back-calculation of the CO_2 concentration of a parcel of seawater to its initial concentration at the sea surface by correcting for changes due to the decomposition of organic material (Redfield *et al.*, 1963) and dissolution of carbonate particulates based on the TA, TCO_2 and oxygen data. These back calculated CO_2 concentrations of waters with various ages are then compared with each other and with the contemporary surface CO_2 concentrations to obtain the oceanic CO_2 increase (Chen, 1993*b*; Tsunogai *et al.*, 1993*a*). Uncertainties in the Redfield Ratios and in the estimation of the initial TA and TCO_2 values give this approach a low signal to noise ratio of only about 8. Nevertheless, the approach provides model-independent estimates of the spatial variability of the excess CO_2 distribution which can be used to validate model predictions (Wallace,1995).

Chen (1993*b*) defined the lower boundary of the excess CO_2 penetration as the depth at which the excess CO_2 concentration is less than 5 $\mu mol\ kg^{-1}$, approximately 12% of the magnitude of the signal. Chen (1993*b*) reported that the anthropogenic CO_2 has penetrated below the thermocline everywhere in the Atlantic, deeper on the western side.

The deepest penetration of excess CO_2 in the Atlantic occurs in the Northwest region where the excess CO_2 has reached the seafloor. Other than the northwest Atlantic Ocean, the deepest excess CO_2 penetration occurs near 40° S, especially in the region off South America where the Falkland Current joints the Brazil Current Extension.

Winter surface sea ice coverage and intensive upwelling of old Circumpolar Deep Water (CDW) in the Weddell Sea prevent excess CO_2 from moving freely across the air–sea boundary or being advected downward. As a result, little excess CO_2 is found below 200 m in the Weddell Gyre (Poisson & Chen, 1987).

An isogram map of the lower boundary of the excess CO_2 penetration in the Pacific Ocean is shown in Fig. 11.4. The major difference between the Pacific and the Atlantic Ocean is that excess CO_2 does not penetrate much deeper than the permanent thermocline in the Pacific Ocean except in the Sea of Japan (Chen *et al.*, 1995*a*) because there is no bottom water formation in the North Pacific. In the Sea of Japan, the entire water column contains excess CO_2 due to rapid turnover of the water.

The shallowest penetration outside of the Southern Ocean occurs in the eastern equatorial region where the excess CO_2 only penetrates to 400 m, or shallower. The

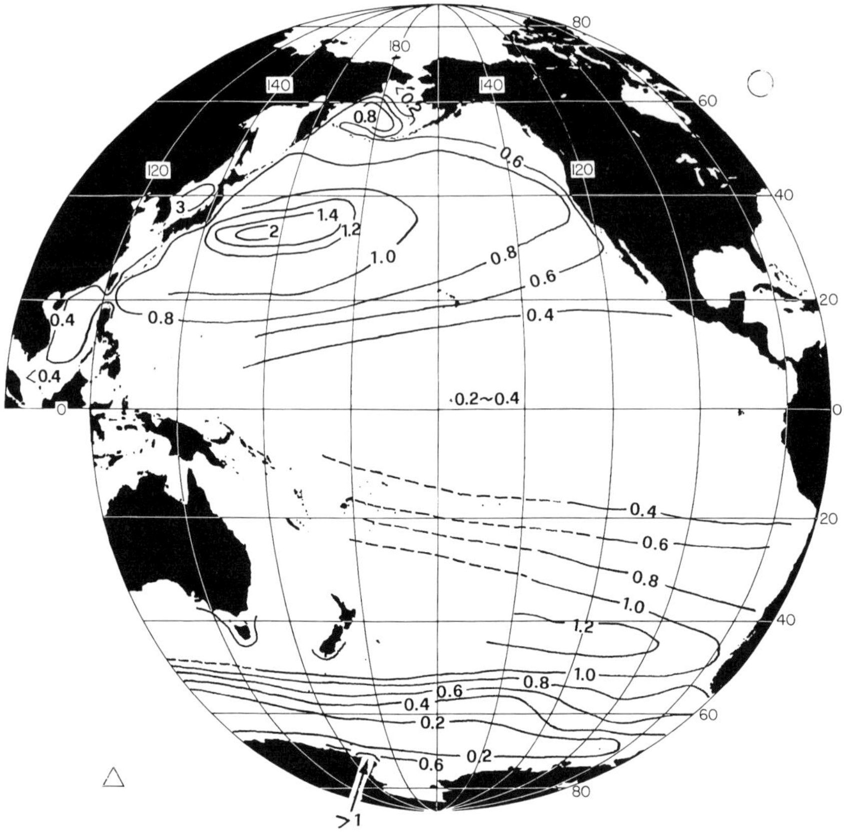

Figure 11.4 *Isogram of the lower boundary of excess CO_2 penetration in the Pacific Ocean (depth contours in km, modified from Chen, 1993*a*).*

general trend indicates a deeper penetration (800 m) in the western Pacific. Overall, the excess CO_2 penetrates to a shallower depth in the equatorial Pacific than in the Atlantic, perhaps reflecting the higher equatorial upwelling rate in the Pacifc and less influence of the newly formed water advected from the north. Higher upwelling rates also results in higher supersaturation of CO_2 in the surface water in the equatorial Pacific (Fig. 11.1). The ΔpCO_2, however, is decreasing with time because of the increasing CO_2 concentration in the atmosphere. As a result, less CO_2 is expected to be released from the equatorial region in the future.

Similar to what was found in the Weddell Sea, intensive upwelling prevents excess CO_2 from reaching more than 200 m deep in the region around 65° S. The excess CO_2 penetrates deeper further south, and reaches more than 1000 m off Cape Adare at the northwest corner of the Ross Sea (Chen, 1987).

The deepest excess CO_2 penetration in the South Pacific occurs at around 45° S near where the Subantarctic Mode Water is located, but the depth of penetration is slightly shallower than that found in the South Atlantic Ocean. Instensive vertical mixing in the Drake Passage also seems to result in deeper excess CO_2 penetration compared to the situation found elsewhere at the same latitude. The deepest penetration in the North Pacific occurs in the confluence of the Kuroshio and Oyashio currents where the North Pacific Intermediate Water (NPIW) originates (Talley *et al.*, 1995; Tsunogai *et al.*, 1995). This is the region off Japan near the area of circulation of the North Pacific variety of the Subtropical Mode Water (Chen, 1987). Here the ΔpCO_2 is near -60 μatm (Inoue *et al.*, 1995) and the excess CO_2 may have penetrated to a depth of 2000 m (Fig. 11.4).

Talley *et al.* (1995) reported that the newly formed NPIW has a potential density up to 27.4 σ_θ, corresponding to water as deep as 1200 m. Watanabe *et al.* (1994) estimated the 27.4 σ_θ water at 40 years old, based on CFC data at 175° E. Tokieda *et al.* (1996) estimated the 27.2 σ_θ water at 33 years old or younger in the northwest Pacific. This relatively young water contains much excess CO_2. Indeed, Tsunogai *et al.* (1993*a*) reported that the NPIW is an important excess CO_2 sink, large enough to account for the missing CO_2. Tsunogai *et al.* (1995) further showed that between the 1970s and 1980s (for a time span of 14 ± 4 years) the bomb ^{14}C has increased rapidly in the NPIW to as deep as 1000 m. In general, deeper penetration occurs on the western side of the Pacific (Chen, 1987) but, in the South China Sea, intensive upwelling prevents excess CO_2 from penetrating more than 500 m deep.

The contours of the excess CO_2 penetration in the Indian Ocean are quite similar to those in the Pacific Ocean. No deep penetration can be found in the northern part except in the Red Sea because there is no bottom water formation in the North Indian Ocean. The excess CO_2 penetrates the entire Red Sea because of intensive winter vertical mixing in the Gulf of Elat (Krumgalz *et al.*, 1990). The major difference between the Indian and the Pacific Ocean is that excess CO_2 penetrates much deeper in the equatorial Indian Ocean, perhaps because of weaker upwelling as compared to the Pacific Ocean (Chen & Chen, 1989; Chen & Wu, 1991).

Nutrients

Anthropogenic nutrients are carried to the oceans mainly via rivers (Table 11.1). Some of the nitrogen oxides released to the atmosphere due to fossil fuel burning and the nitrate subsequently formed find their way into the ocean (Laska *et al.*, 1980). Release of combined N in sewage discharges amounts to about 10 g N per person per day (Garside *et al.*, 1976). If we assume, as an upper bound, that the sewage from the entire world population of 5.5×10^9 persons is discharged to the oceans, the input rate would be about 2×10^{13} g N per year. A recent IGBP report (1994) estimated the total anthropogenic inorganic nitrogen and phosphorus inputs to the coastal zone as 2.1×10^{13} g N and 1.6×10^{12} g P year^{-1}. Galloway *et*

Table 11.1 *Estimations of anthropogenic fluxes of dissolved nutrients in rivers (in 10^{12} g year^{-1}; modified from GESAMP, 1987)*

Authors	N	P
Stumm (1973)	–	0.6
Lerman *et al.* (1975)	–	1.8
Soderlund and Svensson (1976)	18	–
Delwiche and Likens (1977)	35	–
van Bennekom and Salomons (1981)		
(domestic)	11	1.7
(agricultural*)	8	0.35
(industrial)	13	1.7
Total	32	3.75
Meybeck (1982)	7	1.0
Wollast (1983)	21	1.7
Galloway *et al.* (1995)	41	–

*Particulate N and P fluxes due to land erosion estimated as 28×10^{12} g N year^{-1} and 10^{12} g P year^{-1} by these authors.

al. (1995), however, estimated that 5.9×10^{13} g N year^{-1} of the nitrogen fixed by human activities are injected into oceans, of which 4.1×10^{13} g N year^{-1} are from river flow and the rest are from the atmospheric deposition.

The flux of fixed N to the oceans has not been altered by human activity to an extent that the concentration of combined nitrogen in the oceans has significantly changed (Laws, 1983; Denman *et al.*, 1995; Galloway *et al.*, 1995). On a local scale, however, there is clear evidence of changes in the concentrations of fixed N in coastal systems due to human activities (Laws, 1983; Kempe, 1995). For instance, the NO_3 concentration near the Yangtze River (Changjiang) estuary increased four-fold between 1963 and 1983, because of the 2.25-fold increase in the use of nitrogen-containing fertilizers in China. The N/P ratio in the estuarine water also doubled in 20 years, but the productivity has not shown much variation (Gu, 1991; Zhang & Gu, 1994). Bolin (1977) and Walsh *et al.* (1981) hypothesized that acceleration of the organic carbon storage in the coastal sediments has occurred partially because the increased primary productivity resulting from large increases in anthropogenic nitrate inputs from rivers. This hypothesis needs to be tested further because the productivity increase over the continental marginal seas is probably still small when compared with the total primary productivity. The excess organic carbon production, due to the nutrient input, is estimated at only 3% of the gross primary productivity. Furthermore, much of the unused shelf primary productivity is not available for export, but is rather remineralized on the shelf (Kempe & Pegler, 1991; Biscaye *et al.*, 1994; Hong *et al.*, 1995; Tsunogai *et al.*, 1997).

Joint JGOFS/LOICZ task team on continental margin studies

Biogeochemical processes occur principally in the top 200 meters of the sea, often associated with the continental margins. The continental margins, with waters shallower than 200 m occupy 8% of the ocean surface but less than 0.5% of the ocean volume, also play a major role in oceanic biogeochemical cycling. Around 14% of global ocean production, and up to 50% of global ocean denitrification, occurs in the coastal ocean. The burial sites of 80% of organic carbon derived both from oceanic processes as well as from terrestrial sources, and in excess of 50% of present day global carbonate deposition also occur in the coastal ocean (IGBP, 1994). A potential feedback may result from anthropogenic eutrophication of continental shelf areas due to increased nutrient availability, which may also lead to reduced oxygen concentration as a result of increased microbial respiration. Denitrification and methanogenesis are thus promoted, hence the release of N_2, N_2O and CH_4. The eutrophication accelerates the cycle of organic synthesis and regeneration of nutrients other than silicate but calcareous shells and skeletons are preserved. These processes, however, cannot yet be accurately quantified (Christensen, 1994; Codispoti, 1995; Galloway *et al.*, 1995; Denman *et al.*, 1995).

The flux of carbon from the terrestrial biosphere to the oceans takes place via river transport. Global river discharge of carbon in organic and inorganic forms may be approximately 1–1.4 Gt C $year^{-1}$ (Schlesinger & Melack, 1981; Degens *et al.*, 1991; Meybeck, 1993; Tsunogai *et al.*, 1997). A substantial fraction of this transport (up to 0.8 Gt C $year^{-1}$), however, reflects the natural geochemical cycling of carbon and thus does not affect the global budget of the anthropogenic CO_2 perturbation (Sarmiento & Sundquist, 1992). Furthermore, the anthropogenically induced river carbon fluxes reflect, to a large extent, increased soil erosion and not a removal of excess atmospheric CO_2 (Schimel *et al.*, 1995). As mentioned in the preceding section, the discharge of organic matter and nutrients from coastal communities does not have a significant impact on the open world ocean, but can have important effects on the coastal ocean (Weyl, 1981).

The discharge of excess nutrients by rivers might have significantly stimulated carbon fixation (up to 0.5–1 Gt C $year^{-1}$). At present, however, it is not known how much of this excess organic carbon is simply deoxidized, and how much is permanently sequestered by export to the deep ocean, or in sediments on the shelves and shallow seas. Because of the limited surface area, a burial rate significantly exceeding 0.5 Gt C $year^{-1}$ is not very likely, as it would require all coastal seas to be undersaturated in pCO_2 by more than 50 μatm on annual average, in order to supply the carbon from the atmosphere. Such undersaturations have been documented, e.g. in the North Sea and the East China Sea (Kempe & Pegler, 1991; Tsunogai *et al.*, 1997), but it is not certain how these measurements are representative of all coastal oceans. For instance, Holligan and Reiners (1992) estimated a net flux of CO_2 (approximately 0.4 Gt C year^{-1}) to the atmosphere from the coastal ocean. Based on the above considerations, the role of the coastal ocean cannot be accurately assessed at present (Schimel *et al.*, 1995). In order to address various aspects of the oceanic cycles of carbon and associated elements, and to understand their horizontal and vertical fluxes in the marginal seas, the JGOFS/LOICZ Task Team on Continental Margin Studies was established (Chen *et al.*, 1994).

Case studies on Asian marginal seas

East Asia has some of the largest marginal seas in the world, including the Bering Sea (2.27×10^6 km^2), the Sea of Okhotsk (1.53×10^6 km^2), the Sea of Japan (1.0×10^6 km^2), the East China Sea (0.77×10^6 km^2 plus the Bohai Gulf at 0.077×10^6 km^2 and the Yellow Sea at 0.38×10^6 km^2), and the South China Sea (3.5×10^6 km^2). These seas form the linkage between the largest continent and the largest ocean, and receive much land runoff, ventilate the deep oceans, and exchange much material with the open oceans. The role of these marginal seas in the context of carbon and nutrient cycles are briefly discussed below, with more emphasis on the CO_2 sink.

The Bering Sea and the Sea of Okhotsk

The Bering Sea and the Sea of Okhotsk are divided into a wide shelf and a deep basin. The shelf water is probably supersaturated with excess CO_2 but the deep basin waters are not. Chen (1993*c*) reported that the excess CO_2 seems to penetrate to approximately 1000 m in the Bering Sea, deeper in the eastern and southern regions, with slightly shallower penetration off the Kamchatka Peninsula. Tritium shows similar penetration, but the ^{14}C concentration does not level off until a slightly deeper depth. There is no published work on the excess CO_2 input in the Sea of Okhotsk, but preliminary work suggests that the excess CO_2 penetrates to no more than 1000 m in the deep basin (C. T. A. Chen & A. S. Bychkov, unpublished observations, 1996).

The Bering Sea and the Sea of Okhotsk contained, respectively, about 0.21 (± 0.05) and 0.18 (± 0.08) Gt excess carbon around 1980. These values are small compared with 149 Gt C generated by fossil fuel (Stuiver *et al.*, 1984). The carbonate deposits on the vast Bering Sea and the Sea of Okhotsk shelves, however, could provide a sink for excess CO_2 in the near future. The shelf waters are currently about 210 and 150% saturated with respect to calcite and aragonite, respectively. Because of the shallow depth of the shelves (less than 200 m) relative to the saturation horizons of calcite (400 m) and aragonite (350 m), the upward migration of these saturation horizons due to the excess CO_2 input (Feely & Chen, 1982) probably has not changed the dissolution rates of carbonates.

An increase of approximately 200 μatm in pCO_2, however, would make the shelf water undersaturated with respect to aragonite and high magnesium calcite, which would then dissolve and neutralize the excess CO_2. A doubling of the current CO_2 level in the atmosphere by the latter part of the next century would cause the calcites on the shelves to dissolve, providing another large sink for CO_2 (Chen, 1993*c*). Possible removal of anthropogenic CO_2 by excess primary productivity, enhanced burial of organic carbon sediments, and offshore transport of regenerated organic matter need also be tied into the carbon cycle studies (Walsh *et al.*, 1990; Tsunogai *et al.*, 1997).

Strong winds coupled with intense vertical mixing and inter-leaving in the winter enhance the oceanic penetration of excess CO_2 in the Bering Sea, and in the Sea of Okhotsk. Although winter carbonate data are available in the eastern Bering Sea, no winter carbonate data are available in the western Bering Sea near the Kamchatka Peninsula and in the Sea of Okhotsk, where vertical penetration is enhanced and where subsurface water is believed to be formed (Reid, 1965, 1973; Talley, 1991).

Such winter data are needed in order to obtain complete information on the excess CO_2 penetration in the western Bering Sea and in the Sea of Okhotsk. Although these seas have limited capacity to store the excess CO_2, they act as conveyer belts which transport excess CO_2 to the North Pacific Ocean. Because this subsurface water spreads into the northwest Pacific Ocean, the excess CO_2 penetration in the northwest Pacific cannot be studied adequately without knowledge of the source water chemistry (Chen, 1993*c*).

The Sea of Japan

The Sea of Japan is a deep basin with a narrow shelf except in the far north. It is characterized by the unique homogeneity of subsurface waters. Numerous results of hydrological measurements show an anomalously high concentration of dissolved oxygen, suggesting rapid turnover of the subsurface waters and a potential sink for excess CO_2 (Chen *et al.*, 1995*a*).

The calculated vertical profile of excess CO_2 indicates that the entire Sea of Japan has been penetrated by excess CO_2, totaling 0.31 (± 0.05) Gt C in 1992 (Chen *et al.*, 1995*a*). The carbonate data have been used to calculate the ratio of $CaCO_3$ dissolution and the organic carbon decomposition (Chen *et al.*, 1982*b*). The results vary from 0.05 at 300 m to about 0.17 below 2000 m. These values are lower than the ratios of 0.14 and 0.36 found in the South and North Pacific, respectively (Chen *et al.*, 1982*b*; Chen, 1990), or 0.54 in the Bering Sea (Chen, 1993*a*).

The Sea of Japan is rather deep, with an average depth of over 1500 m, yet the straits connecting it to the outside are all narrow and shallower than 150 m. As a result, the deep waters have no direct exchange with the Pacific, the East China Sea and the Sea of Okhotsk. Since the deep waters of the Sea of Japan are all formed inside the closed basin, they provide a unique opportunity to estimate the Redfield Ratios based on the mass-balance method (Chen *et al.*, 1996). The results are in good agreement with the traditional values represented by eqn. (6).

The East China Sea

The East China Sea (ECS), including the Bohai Gulf and the Yellow Sea, has a wide shelf where the water is probably also saturated with excess CO_2. Between the shelf break and the Ryukyu Island is a deep trench. The hydrography is such that there is upwelling toward the west, bringing older deep water upward (Liu *et al.*, 1992; Chen *et al.*, 1995*b*). As a result, the excess CO_2 penetrates only to approximately 600 m. The entire ECS contained 0.07 (± 0.02) Gt C excess carbon in 1992.

Since the waters on the ECS shelf are highly supersaturated with respect to calcite and aragonite, sediments on the ECS shelf are not expected to neutralize excess CO_2 in the coming century. The high produtivity, however, makes a large portion of the ECS inner and mid-shelf waters undersaturated with respect to CO_2 near the surface (Fig. 11.5, the fugacity is very close to the partial pressure). As a result, the flux of CO_2 into the ECS is large (40 g C m^{-2} $year^{-1}$; Tsunogai *et al.*, 1997). If the rate were applicable to the whole continental shelf in the world it would amount to 1 Gt C $year^{-1}$. The outer shelf and the Kuroshio region (east of 126° E) is depleted in PO_4 (Fig. 11.6) and the surface water pCO_2 is near saturation.

The South China Sea

The South China Sea (SCS) also has a large shelf in the south and the southwest. The basin is very deep but since there is no deep or intermediate water formation, excess CO_2 cannot go very deep. In addition, there is strong upwelling in the SCS. Thus, even the subsurface water that contained some excess CO_2 would be pushed upward.

Chen and Huang (1995) have calculated the excess CO_2 based on data collected in the northeastern SCS. The deepest penetration is no more than 600 m, shallower than in the west Philippine Sea. The entire SCS probably contained 0.43 (± 0.1) Gt C anthropogenic carbon in 1994.

The saturation horizon of calcite is deeper than 2000 m in the SCS; thus little enhanced dissolution is expected due to the excess CO_2 penetration (Chen &

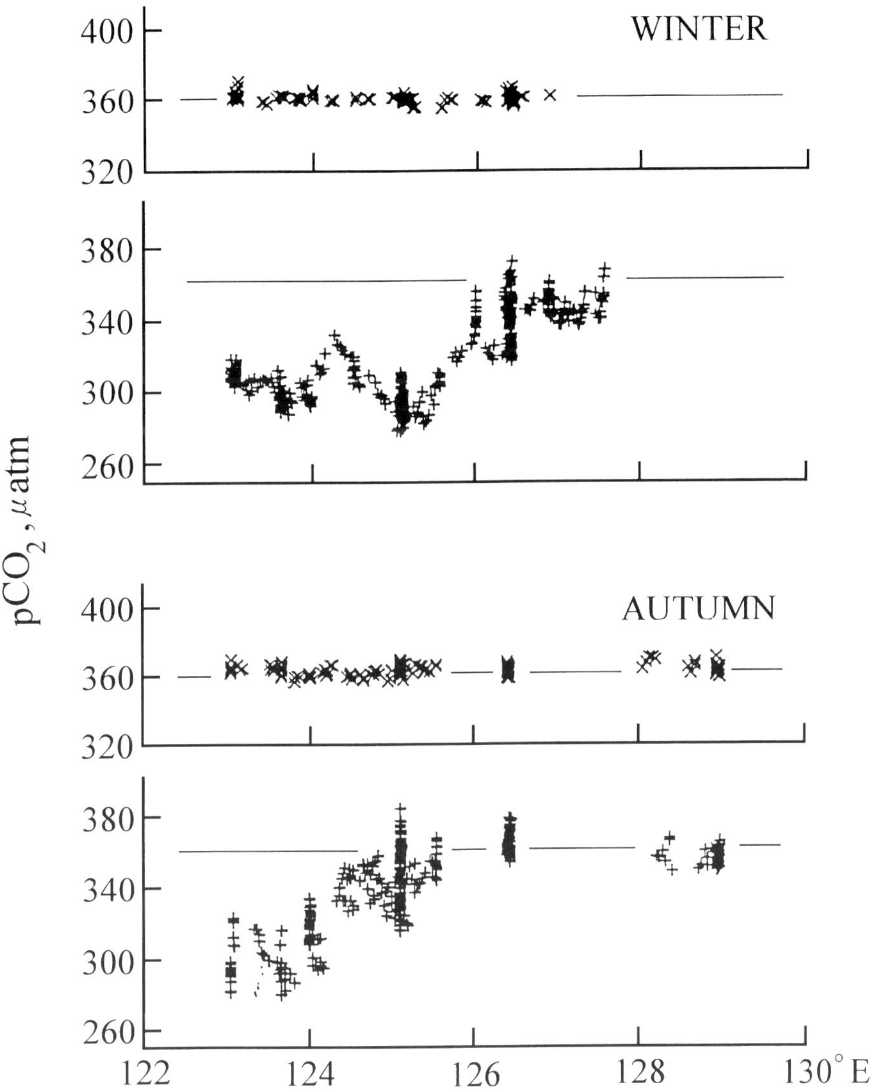

Figure 11.5 *The observed CO_2 fugacity in the surface water and the atmosphere at a cross-section perpendicular to the ECS shelf extending from the mouth of the Yangtze River to Ryukyu Island. (Taken from Tsunogai* et al., *1997.)*

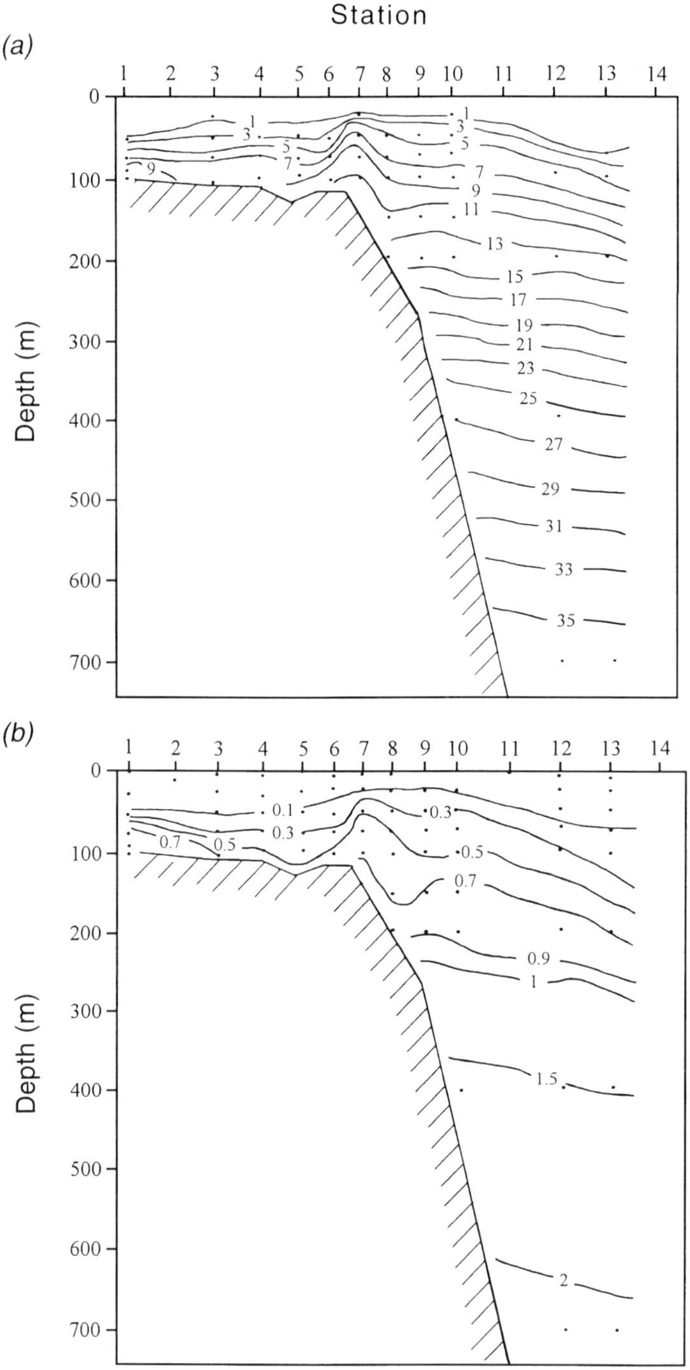

Figure 11.6 *(a) NO_3 and (b) PO_4 concentrations (μ mol kg^{-1}) in a cross-section perpendicular to the East China Sea Shelf. (Taken from Chen* et al., *1995b.)*

Huang, 1995). The saturation horizon of aragonite is 600 m, about the same as the excess CO_2 penetration depth but deeper than the saturation horizon found in the Bering Sea (350 m). The upward migration of the saturation horizon (Feely & Chen, 1982) would not affect the calcareous deposits on the SCS shelf to the degree that it affects the Bering Sea deposits.

Since two of the largest rivers in the world, the Yangtze River and the Yellow River (Huanghe), empty into the East China Sea, it has frequently been assumed that the high productivity on the shelf is a result of the riverine nutrient input. However, such an input is not nearly enough to support the high productivity. Further, the Kuroshio is known to be low in nutrient concentration in the surface water (Fig. 11.6). As a result, if only near-surface Kuroshio water moved onto the shelf, the water would not contribute much to the productivity. There must be other sources of nutrients.

Nutrients from the South China Sea contribute to the high productivity on the East China Sea shelf

The Taiwan Strait connects the East and South China Seas. The SCS is rather deep but because of its shallow depth, averaging only 50 m with no deep channels, the Strait does not permit exchange of waters between these two seas except for low-nutrient near-surface waters. In summer, the water flows northward through the Strait but, in winter, there is a southward flowing current along the coast of Fujian Province on the western side of the Strait (Guan, 1994).

On the other hand, the Kuroshio feeds the SCS year-round. The nutrient-depleted near-surface Kuroshio water exchanges freely with the SCS all year. However, the nutrient-rich intermediate water, arbitrarily taken as the layer between 350 and 1350 m encompassing the S_{min} core, seems only to flow out of the SCS and forms the western portion of the Kuroshio (Chen & Huang, 1996). The core of the outflowing SCS S_{min} layer sits at about 600 m, roughly 200 m shallower than the S_{min} core of the eastern part of the Kuroshio originating from the west Philippine Sea. As a result, the now northward flowing SCS intermediate water is able to cross the ridge northeast of Taiwan to enter the Okinawa Trough, which is adjacent to the ECS continental shelf.

Ito *et al.* (1994) and Chen *et al.* (1995*b*) have shown that the Kuroshio Intermediate Water (KIW) contributes up to 30% in the upwelling and cross-shelf mixing at the East China Sea shelf break. The upwelled KIW actually originates from the SCS and is high in nutrients (Fig. 11.6), thus contributing a great deal of nutrients to the shelf as compared with the much smaller riverine fluxes (Table 11.2). Furthermore, PO_4 flux from the Yangtze River could be identified only in the estuary. The NO_3 flux extends further but could still be identified only in the Yangtze River Plume. This suggests that the productivity in much of the East China Sea is influenced by the upwelled subsurface Kuroshio water (Chen, 1996).

Table 11.2 *Nutrient inputs to the East China Sea shelf (Chen, 1996)*

	NO_3 (10^9 mol year^{-1})	PO_4 (10^9 mol year^{-1})	SiO_2 (10^9 mol year^{-1})
Riverine	29	0.23	87
Kuroshio surface water	1.6	0.31	12.6
Kuroshio tropical water	42	3.1	41.7
Kuroshio intermediate water	161	10.9	385

Any potential change in the upwelling rate would thus have a much larger effect than the change in the nutrient input from rivers.

Fig. 11.7 shows the NO_3 and PO_4 vs. pH (25° C) for the East China Sea. The light-limited KIW and KTW, although rich in nutrients, do not contribute to primary productivity until reaching the surface. The higher pH values for Kuroshio Surface Water and surface shelf water are a result of primary productivity, which is evidently limited by the availability of PO_4 even when there is still abundant NO_3 near the Yangtze River Plume. Consequently the phosphorus-containing Kosa (yellow sand from the loess in China, containing 0.3% P_2O_5) fallout would enhance the productivity. According to Tsunogai (1991), the Kosa flux amounts to 50 g m^{-2} for 3 months in late spring and leads to the new production of 2.40 g C m^{-2} year^{-1} in the Kuroshio region. Similarly, anthropogenic iron input may increase primary productivity in the high nutrient-low chlorophyll regions (Martin, 1990) or in waters where molecular nitrogen-fixing blue-green algae (cyanobacteria) may bloom (Tsunogai, 1979; Tsunogai & Watanabe, 1983), but the global effect is probably small. Denman *et al.* (1995) concluded that intentional iron fertilization to promote phytoplankton growth, and hence the drawdown of atmospheric CO_2, may not be very effective.

Summary

The schematic diagrams of flow patterns of the marginal seas discussed above are given in Fig. 11.8. The Bering Sea and the Sea of Okhotsk do not have very deep water formation but may contribute to the formation of NPIW (Reid, 1965, 1973; Talley, 1991), thus pumping excess CO_2 to the North Pacific. The Sea of Japan has deep and bottom water formation, but does not export excess CO_2 because of the shallow sills. Sediments in the shelves of these three seas may neutralize excess CO_2 in the latter part of the next century when the shelf waters become under-saturated with respect to calcite and aragonite.

The East China Sea absorbs a large amount of CO_2 because of its high productivity, hence low pCO_2. Yet the East and South China Seas are not important reservoirs for excess CO_2 because of their small size and upwelling. The sediments

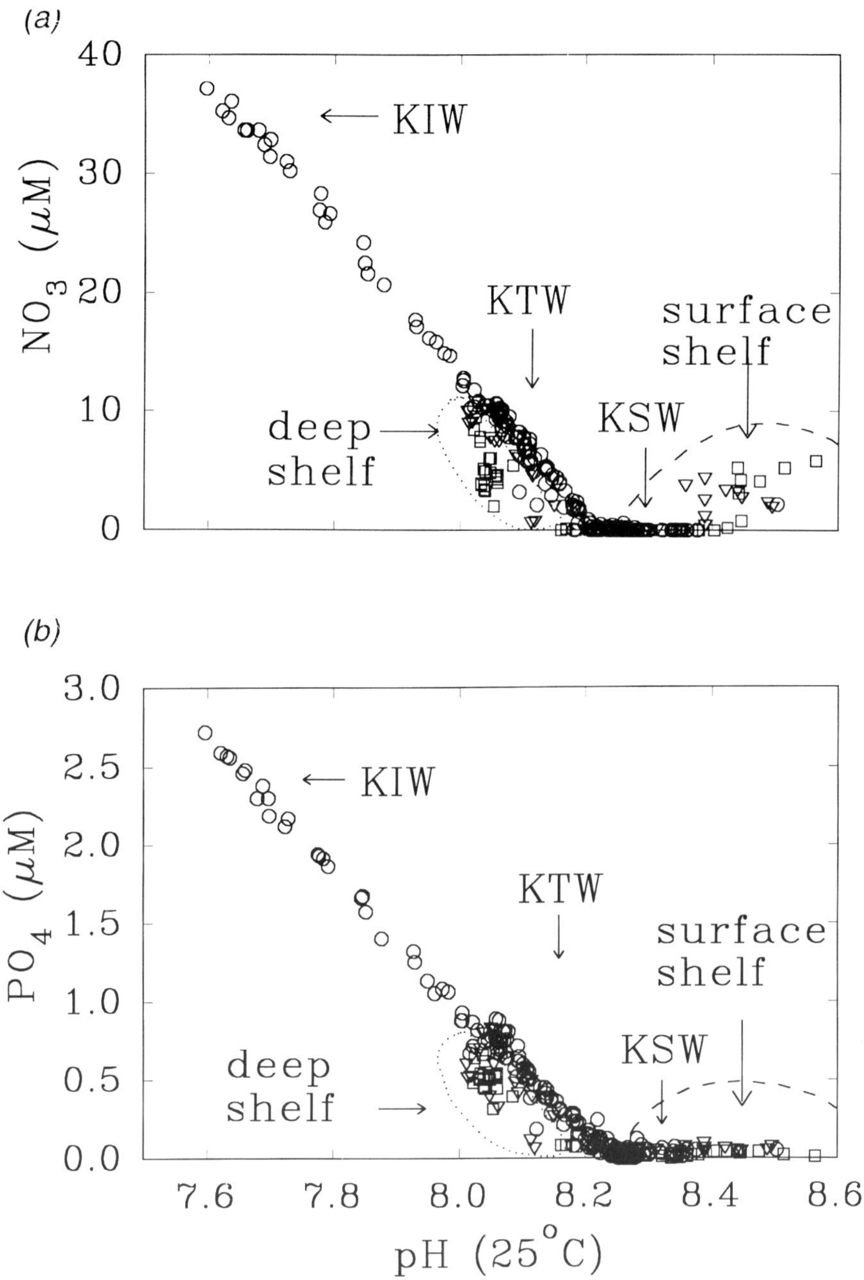

Figure 11.7 *(a) NO_3 and (b) PO_4 concentrations vs. pH (25° C) for the East China Sea (KIW is the Kuroshio Intermediate Water; KTW is the Kuroshio Tropical Water; KSW is the Kuroshio Surface Water; data taken from Chen* et al., *1992). The surface shelf waters with elevated NO_3 concentrations are all near the Yangtze River plume.*

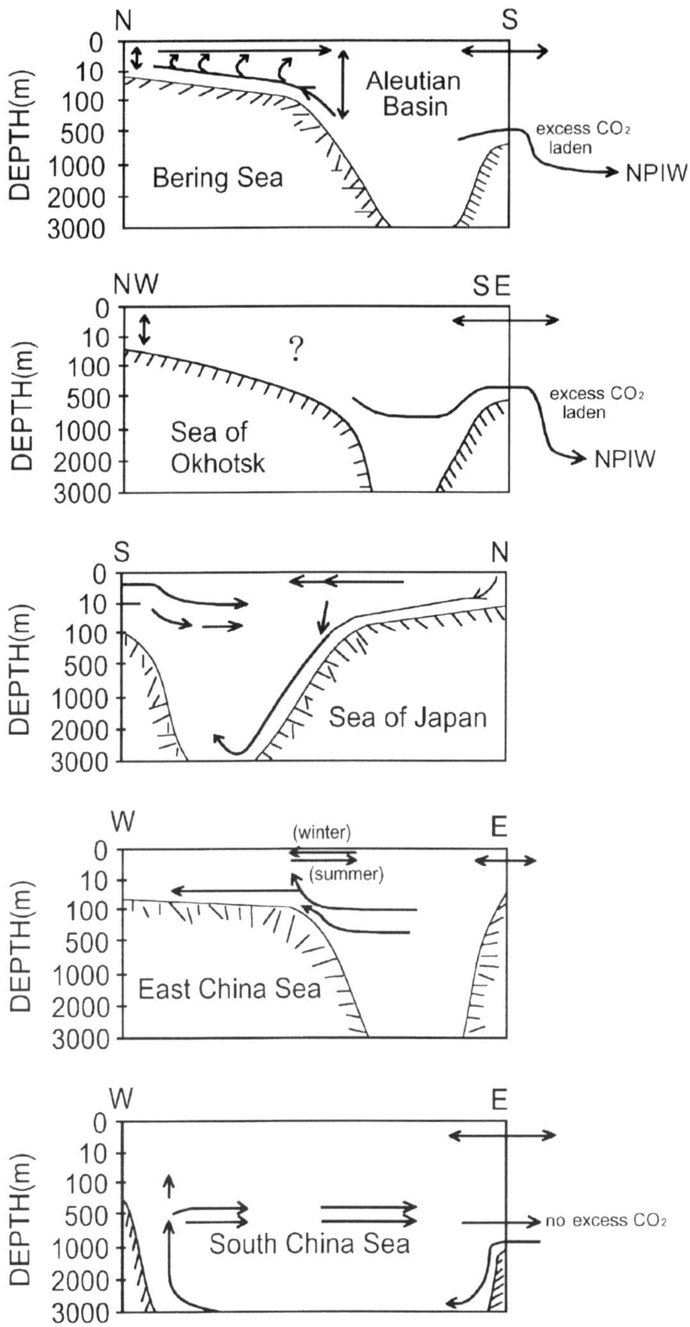

Figure 11.8 *Schematic diagrams of flow patterns in the Bering Sea, Sea of Okhotsk, Sea of Japan, East China Sea, and South China Sea.*

are unlikely to neutralize excess CO_2 in the next century, but carbonate is regenerated from organic matter at the bottom. This regenerated carbonate may be exported to, and stored in, the pelagic ocean (Tsunogai *et al.*, 1997).

The excess CO_2 budget in the global oceans

As the first attempt on a global scale, we have integrated the amount of excess CO_2 contained in each ocean (Chen, 1993*b*). The vertical integration, however, was from the sea surface to the depth where the excess CO_2 concentration equals 5 μmol kg^{-1}. We re-evaluated the carbon budget by integrating the excess CO_2 to where it is approximately zero, resulting in an increase of about 10%. We also added more data in the marginal seas. The results indicate that the Norwegian and Greenland Seas (between 65–80° N) contained 0.95 (± 0.2) Gt excess carbon in 1982 (Chen *et al.*, 1990). The rest of the North Atlantic Ocean and the South Atlantic Ocean contained 24 (± 4.5) and 13.4 (± 2.5) Gt excess carbon, respectively, around 1973. Overall, the Atlantic Ocean contained 38.4 (± 7) Gt excess carbon around 1973 (Table 11.3). This is slightly higher than the estimate of 33 (± 5) Gt excess carbon in 1983 estimated by Bolin *et al.* (1989).

The North Pacific Ocean (excluding the marginal seas) and the South Pacific Ocean contained 16 (± 4) and 18.6 (± 4) Gt excess carbon, respectively, around 1980. Overall, the Pacific Ocean contained 35.8 (± 6) Gt excess carbon, less than that found in the Atlantic Ocean despite of the much larger capacity (Table 11.3). This is mainly because the Pacific Ocean lacks deep water formation in the north and because of the higher upwelling rate.

The Red Sea contained 0.07 (± 0.02) Gt excess carbon in 1985 (Chen & Wu, 1991). The entire Indian Ocean contained 18.6 (± 4) Gt excess carbon around 1980. Overall the major oceans contained 76–110 Gt excess carbon around 1980 (Table 11.3). Since data from most marginal seas and the Arctic Ocean are lacking, and the excess CO_2 may penetrate deeper in these seas, the global oceans may actually contain 5% more excess CO_2 perhaps. The above values compare with 90–120 Gt net release of carbon from terrestrial ecosystems between 1850 and 1980 (Houghton & Skole, 1990) and 150–190 Gt C from total fossil fuel release between 1860 and 1982. The atmosphere contained 118–160 Gt excess carbon in 1980 (Table 11.4). These data indicate that the high estimates for the atmospheric and oceanic sinks (270 Gt C) could account for the low estimates for the fossil fuel and land change sources (240 Gt C). The sum of the high estimates for the sources (310 Gt C), however, is much higher than the combined atmospheric and oceanic sinks (194–270 Gt C). The carbon budget could not yet be balanced with any certainty.

Table 11.3 *Estimated amount of excess carbon in the oceans*

Area	Excess carbon (Gt C)
Greenland and Norwegian Seas (65 – 80° N)	0.95 (± 0.2)
North Atlantic (EQ – 65° N)	24 (± 4.5)
South Atlantic	13.4 (± 2.5)
Subtotal	38.4 (± 7)
Bering Sea	0.21 (± 0.05)
Sea of Okhotsk	0.18 (± 0.08)
Sea of Japan	0.31 (± 0.05)
East China and Yellow Seas	0.07 (± 0.02)
South China Sea	0.43 (± 0.1)
North Pacific (excluding the above marginal seas)	16 (± 4)
South Pacific	18.6 (± 4)
Subtotal	35.8 (± 6)
Red Sea	0.073 (± 0.02)
Indian (excluding the Red Sea)	18.6 (± 4)
Total (1980)	92.8 (± 17)

Table 11.4 *Estimated values of key sources and sinks in the global carbon cycle*

Parameter	Source (Gt C)	Sink (Gt C)
Total fossil fuel release 1860 – 1982 (Rotty & Masters, 1985)	150 – 190	
Total land change 1850 – 1980 (Houghton & Skole, 1990)	90 – 120	
Atmospheric excess carbon 1982 (Solomon *et al.*, 1985)		118 – 160
Oceanic excess carbon (*c.* 1980)		76 – 110
Total	240 – 310	194 – 270

Acknowledgments

We wish to thank T. Takahashi, K. Iseki and the Nagasaki Marine Observatory for assistance. An anonymous reviewer provided detailed, constructive comments which strengthened the manuscript. The ROC National Science Council supported the preparation of this manuscript (NSC 86-2611-M-110-007-W).

References

Anderson, L. G., Holby, O., Lindegren, R. & Ohlson, M. (1991). The transport of anthropogenic carbon dioxide into the Weddell Sea. *Journal of Geophysical Research*, **96**, 16 679–87.

Biscaye, P. E., Flagg, C. N. & Falkowski, P. G. (1994). The shelf edge exchange processes experiment, SEEP-II: an introduction to hypotheses, results and conclusions. *Deep-Sea Research*, **41**, 231–52.

Boden, T. A., Kaiser, D. P., Sepanski, R. J. & Stoss, F. W. (1994). Trends '93: a compendium of data on global change. Carbon Dioxide Information Analysis Center. 984 pp.

Bolin, B. (1977). Modeling the oceans and ocean sediments and their response to fossil fuel carbon dioxide emissions. In *The Fate of Fossil Fuel CO_2 in the Oceans*, ed. N. Andersen & A. Malahoff, pp. 81–95. New York: Plenum Press.

Bolin, B. (1983). Changing global biogeochemistry. In *Oceanography, The Present and Future*, ed. P.G. Brewer, pp. 305–26. New York: Springer-Verlag.

Bolin, B., Bjorkstrom, A. & Holmen, K. (1983). The simultaneous use of tracers for ocean circulation studies. *Tellus*, **35B**, 206–36.

Bolin, B., Bjorkstrom, A. & Moore, B. (1989). Uptake by the Atlantic Ocean of excess atmospheric carbon dioxide and radiocarbon, *Geophysical Monographs American Geophysical Union*, **52**, 57–78.

Brewer, P. G. (1978). Direct observation of the oceanic CO_2 increase. *Geophysical Research Letters*, **5**, 997–1000.

Brewer, P. G., Wong, G. T. F., Bacon, M. P. & Spenser, D. W. (1975). An oceanic calcium problem? *Earth Planetary Science Letters*, **26**, 81–7.

Brewer, P. G., Goyet, C. & Dyrssen, D. (1989). Carbon dioxide transport by ocean currents at 25° N latitude in the Atlantic Ocean. *Science*, **246**, 477–9.

Broecker, W. S. & Peng, T. H. (1992). Interhemispheric transport of carbon dioxide by ocean circulation. *Nature*, **356**, 587–9.

Broecker, W. S., Peng, T-H., Ostlund, G. & Stuiver, M. (1985). The distribution of bomb radiocarbon in the ocean. *Journal of Geophysical Research*, **90**, 6953–70.

Chen, C. T. (1982*a*). Oceanic penetration of excess CO_2 in a cross-section between Alaska and Hawaii. *Geophysical Research Letters*, **9**, 117–9.

Chen, C. T. (1982*b*). On the distribution of anthropogenic CO_2 in the Atlantic and Southern Oceans. *Deep-Sea Research*, **29**, 563–80.

Chen, C. T. (1984). Carbonate chemistry of the Weddell Sea, Department of Energy Technical Report, DOE/EU/10611–4, 118 pp.

Chen, C. T. A. (1987). On the depth of anthropogenic CO_2 penetration in the Atlantic and Pacific Oceans. *Oceanologica Acta.*, No. sp, 97–102.

Chen, C. T. A. (1990). Rates of calcium carbonate dissolution and organic carbon decomposition in the North Pacific Ocean. *Journal of Oceanographic Society of Japan*, **46**, 201–10.

Chen, C. T. A. (1992). Comparison of summer–winter carbonate and nutrient data in the southwest Indian Ocean. *Terrestrial, Atmospheric and Oceanic Sciences*, **3**, 59–80.

Chen, C. T. A. (1993*a*). Anthropogenic CO_2 distributions in the North Pacific Ocean. *Journal of Oceanography*, **49**, 257–70.

Chen, C. T. A. (1993*b*). The oceanic anthropogenic CO_2 sink. *Chemosphere*, **27**, 1041–64.

Chen, C. T. A. (1993*c*). Carbonate chemistry of the wintertime Bering Sea marginal ice zone. *Continental Shelf Research*, **13**, 67–87.

Chen, C. T. A. (1996). The Kuroshio Intermediate Water is the major source of nutrients on the East China Sea continental shelf. *Oceanologica Acta*, 1, 523–7.

Chen, C. T. & Drake, E. T. (1986). Carbon dioxide increase in the atmosphere and oceans and possible effects on climate. *Annual Review of Earth and Planetary Sciences*, **14**, 201–35.

Chen, D. W. & Chen, C. T. A. (1989). The anthropogenic CO_2 signals in the Indian Ocean. *Journal of Enviromental Protection Society of ROC*, **12**, 45–65 [in Chinese with English abstract].

Chen, C. T. A. & Huang, M. H. (1995). Carbonate chemistry and the anthropogenic CO_2 in the South China Sea. *Acta Oceanologica Sinica*, **14**, 47–57.

Chen, C. T. A. & Huang, M. H. (1996). A mid-depth front separating the South China Sea water and the west Philippine Sea water. *Journal of Oceanography*, **52**, 17–25.

Chen, C. T. & Millero, F. J. (1978). The gradual increase of oceanic carbon dioxide. *EOS*, **59**, 1101.

Chen, C. T. & Millero, F. J. (1979). Gradual increase of oceanic carbon dioxide. *Nature*, **277**, 205–6.

Chen, C. T. & Pytkowicz, R. M. (1979). On the total CO_2–titration alkalinity–oxygen system in the Pacific Ocean. *Nature*, **281**, 362–5.

Chen, C. T. A. & Wu, C. J. (1991). The anthropogenic CO_2 signals in the Red Sea and northwest Indian Ocean. *Oceanography and Limnology*, **22**, 434–42 [in Chinese with English abstract].

Chen, C. T., Millero, F. J. & Pytkowicz, R. M. (1982*a*). Comment on "calculating the oceanic CO_2 increase: a need for caution", by A. M. Shiller. *Journal of Geophysical Research*, **87**, 2083–5.

Chen, C. T., Pytkowicz, R. M. & Olson, E. J. (1982*b*). Evaluation of the calcium problem in the South Pacific. *Geochemical Journal*, **16**, 1–10.

Chen, C. T. A., Jones, E. P. & Lin, K. J. (1990). Wintertime total carbon dioxide measurements in the Norwegian and Greenland Seas. *Deep-Sea Research*, **37**, 1455—73.

Chen, C. T. A., Holligan, P., Hong, H. S., Iseki, K., Krishnaswami, S., Wollast, R. & Yoder, J. (1994). Land–ocean interactions in the coastal zone, JGOFS Report No. 15, SCOR, 20 pp.

Chen, C. T. A., Wang, S. L. & Bychkov, A. S. (1995*a*). Carbonate chemistry of the Sea of Japan. *Journal of Geophysical Research*, **100**, 13 737–45.

Chen, C. T. A., Ruo, R. Pai, S. C. Liu, C. T. & Wong, G. T. F. (1995*b*). Exchange of water masses between the East China Sea and the Kuroshio off northeastern Taiwan. *Continental Shelf Research*, **15**, 19–39.

Chen, C. T. A., Gong, G. C., Wang, S. L. & Bychkov, A. S. (1996). Redfield ratios and regeneration rates of particulate matter in the Sea of Japan as a model of closed system. *Geophysical Research Letters*, **23**, 1785–8.

Chen, M. P., Huang, C. C., Lin, F. L., Liu, C. C., L, H. W., Gong, G. C., Huang, J. J. & Lee, Y. L. (1992). The KEEP-MASS data report, 728 pp., NSC R.V. Ocean Researcher 1, Reg. Instrum. Cent., Taipei, Taiwan.

Christensen, J. P. (1994). Carbon export from continental shelves, denitrification and atmospheric carbon dioxide. *Continental Shelf Research*, **14**, 547–76.

Cline, J. D., Feely, R. A., Kelly-Hansen, K., Gendron, J. F., Wisegarver, D. P. & Chen, C. -T. (1985). *Current inventory of anthropogenic carbon dioxide in the North Pacific Gyre.* NOAA Technical Memorandum, ERL PMEL–60, 46 pp.

Codispoti, L. A. (1995). In the ocean losing nitrate? *Nature*, **376**, 724.

Degens, E. T., Kempe, S. & Richey, J. E. (1991). Summary: biogeochemistry of major world rivers. In *Biogeochemistry of Major World Rivers*, ed. E. T. Degens, S. Kempe & J. E. Richey, SCOPE Report 42, pp. 323–47. Chichester: John Wiley.

Delwiche, C. C. & Likens, G. E. (1977). Biological response to fossil fuel combustion products. In *Global Cycles and Their Alteration by Man*, ed. W. Stumm, pp. 73–88,

Dahlem Conferences, Berlin.

Denman, K., Hofmann, E. & Marchant, H. (1995). Marine biotic response to environmental change and feedbacks to climate, manuscript prepared for IPCC 95 report, 40 pp.

Falkowski, P. G. & Wilson, C. (1992). Phytoplankton productivity in the North Pacific Ocean since 1990 and implications for absorption of anthropogenic CO_2. *Nature*, **358**, 741–3.

Feely, R. A. & Chen, C. T. (1982). The effect of excess CO_2 on the calculated calcite and aragonite saturation horizons in the northeast Pacific. *Geophysical Research Letters*, **9**, 1294–7.

Galloway, J. N., Schlesinger, W. H., Levy, H. II., Michaels, A. & Schnoor, J. L. (1995). Nitrogen fixation: anthropogenic enhancement-environmental response. *Global Biogeochemical Cycles*, **9**, 235–52.

Gammon, R. H., Cline, J. D. & Wisegarver, D. P. (1982). Chlorofluoromethanes in the Northeast Pacific Ocean: measured vertical distribution and application as transient tracers of upper ocean mixing. *Journal of Geophysical Research*, **87**, 9441–54.

Garcon, V., Thomas, F., Wong, C. S. & Minster, J. F. (1992). Gaining insight into the seasonal variability of CO_2 at O.W.S.P. using an upper ocean model. *Deep-Sea Research*, **30**, 921–38.

Garside, C., Malone, T. C., Roels, O. A. & Sharfstein, B. A. (1976). An evaluation of sewage derived nutrients and their influence on the Hudson Estuary and New York Bight. *Estuarine Coastal Marine Science*, **4**, 281–9.

GESAMP (1987). Land/Sea Boundary Flux of Contaminants: Contributions from Rivers, IMCO/FAO/UNESCO/WMO/WHO/IAEA/UN/UNEP. Joint Group of Experts on the Scientific Aspects of Marine Pollution, Reports and Studies, **32**, 172 pp.

Gordon, D. C., Jr., Boudreau, P. R., Mann, K. H., Ong, J. E., Silvert, W. L., Smith, S. V., Wattayakorn, G., Wulff, F. & Yanagi, T. (1996). LOICZ Biogeochemical Modelling Guidelines. *LOICZ Report and Studies*, No. 5, 96 pp.

Goyet, C. & Brewer, P. G. (1993). Biogeochemical properties of the oceanic carbon cycle. NATO ASI Series, Vol. III. *Modeling Oceanic Climate Interactions*, ed. J. Willebrand & D. L. T. Anderson, pp. 271–97. Berlin: Springer–Verlag.

Gu, H. K. (ed.). (1991). *Marine Chemistry of the Bohai Sea, The Yellow Sea and the East China Sea*, 500 pp. Beijing: Science Press.

Guan, B. X. (1994). Patterns and structures of the currents in Bohai, Huanghai and East China Seas. In *Oceanology of China Seas*, Vol. 1, ed. D. Zhou, Y. B. Liang & C. K. Zeng, pp.17–26. The Netherlands: Kluwer.

Holligan, P. M. & Reiners, W. A. (1992). Predicting the responses of the coastal zone to global change. *Advances in Ecological Research*, **22**, 211–55.

Hong, G. H., Kim, S. H., Chung, C. S. & Pae, S. J. (1995). The role of the anthropogenic nutrient input in the carbon fixation of the coastal ocean Yellow Sea: a preliminary studey. In *Direct Disposal of Carbon Dioxide*, ed. N. Handa & T. Ohsumi, pp. 13–22. Tokyo: Terra Science Publishing Company.

Houghton, R. & Skole, D. L. (1990). The long-term flux of carbon between terrestrial ecosystems and the atmosphere as a result of changes in land use. 4 pp., Carbon Dioxide Information Analysis Center, ORNL.

IGBP (1994). Land-ocean interactions in the coastal zone, implementation plan. *Global Change Report*, **33**, 215 pp.

Inoue, H. Y., Matsueda, H., Iseki, M., Fushimi, K., Hirota, M., Asanuma, I. & Takasugi, Y. (1995). Long-term trend of the partical pressure of carbon dioxide (pCO_2) in surface waters of the western North Pacific, 1984–1993, *Tellus*, **47B**, 391 - 413.

Ito, T., Kaneko, A., Tsubota, H. & Gohda, N. (1994). The characteristic distribution of silica over the East China Sea Shelf slope. *Journal of Oceanography*, **50**, 465–77.

Joos, F., Sarmiento, J. L. & Siegenthaler, U. (1991*a*). Estimates of the effect of Southern Ocean iron fertilization on atmospheric CO_2 concentrations. *Nature*, **349**, 772–5.

Joos, F., Siegenthaler, U. & Sarmiento, J. L. (1991*b*). Possible effects of iron fertilization in the Southern Ocean on atmospheric CO_2 concentration. *Global Biogeochemical Cycles*,

5, 135–50.

Keeling, C. D. (1993). Surface ocean CO_2. In *The Global Carbon Cycle*, ed. M. Heimann, pp. 413–29. Heidelberg: Springer-Verlag.

Kempe, S. (1995). Coastal seas: a net source or sink of atmospheric carbon dioxide? LOICZ Report and Studies, No 1, 27 pp.

Kempe, S. & Pegler, K. (1991). Sinks and sources of CO_2 in coastal seas: the North Sea. *Tellus*, **43**, 224–35.

Kester, D. R. (1975). Dissolved gases other than CO_2. In *Chemical Oceanography*, ed. J. P. Riley & G. Skirrow, 2nd edn. Vol.1, pp 497–556. New York: Academic Press.

Krumgalz, B. S., Erez, J. & Chen, C. T. A. (1990). Anthropogenic CO_2 penetration in the northern Red Sea and in the Gulf of Elat. *Oceanologica Acta*, **13**, 283–90.

Krysell, M. & Wallace, D. W. R. (1988). Arctic Ocean ventilation studied using a suite of anthropogenic halocarbon tracers. *Science*, **242**, 746–9.

Laska, R., Schaefer, M., Myers, D., Parker, T., Wells, W., Dorset, P., McLean, J., Altscchuler, K., Wicker, M. & Scott-Walton, B. (1980). *Acid Rain*. Environmental Protection Agency, Washington, DC: US Government Printing Office.

Laws, E. A. (1983). Men's impact on the marine nitrogen cycle. In *Nitrogen in the Marine Environment*, ed. E. J. Carpenter & D. G. Capone, pp. 459 –485. New York: Academic Press.

Lerman, A., Mackenzie F. T. & Garrels, R. M. (1975). Modeling of geochemical cycles: phosphorus as an example. *Geological Society American Memoirs*, **142**, 205–17.

Liss, P. S. & Merlivat, L. (1986). Air–sea gas exchange rates: introduction and synthesis. In *The Role of Air-Sea Exchange in Geochemical Cycling*, ed. P. Buat-Menard, pp. 113–27. Dordrecht, Holland: D. Reidel Publ. Co.

Liu, K. K., Gong, G. C. Shyu, C. Z., Pai, S. C., Wei, C. L. & Chao, S. Y. (1992). Response of Kuroshio upwelling to the onset of the northeast monsoon in the sea north of Taiwan: observations and a numerical simulation. *Journal of Geophysical Research*, **97**, 12511–26.

McCarthy, J. J. & Carpenter, E. J. (1983). Nitrogen cycling in near-surface waters of the open ocean. In *Nitrogen in the Marine Environment*, ed. E. J. Carpenter & D. G. Capone, pp. 487–512. New York: Academic Press.

Martin, J. (1990). Glacial–interglacial CO_2 change: the iron hypothesis. *Paleoceanography*, **5**, 1–13.

Meybeck, M. (1982). Carbon, nitrogen and phosphorus transport by world rivers. *American Journal of Science*, **282**, 401–50.

Meybeck, M. (1993). Natural sources of C, N, P, and S. In *Interactions of C, N, P, and S Biogeochemical Cycles and Global Change*, ed. R. Wollast, pp. 163–93. Berlin: Springer-Verlag.

Moore, B. & Bolin, B. (1986). The oceans, carbon dioxide, and global change. *Oceanus*, **29**, 9–15.

Noriki, S. & Tsunogai, S. (1986). Particulate fluxes and major components of settling particles from sediment trap experiments in the Pacific Ocean. *Deep-Sea Research*, **33**, 903–12.

Okuda, K., Tsunogai, S., Watanabe, S., & Kusakabe, M. (1995). Anthropogenic CO_2 dissolved in the western North Pacific water estimated from $d^{13}C$ of dissolved inorganic carbon. Global flux of carbon and its related substances in the Coastal Sea–Ocean–Atmosphere System, ed. S. Tsunogai, K. Iseki, I. Koike & T. Oba, pp. 400 - 5. Yokohama: M and J International.

Orr, J. C. (1993). Accord between ocean models predicting uptake of anthropogenic CO_2. *Water, Air and Soil Pollution*, **70**, 465–81.

Peng, T. S. & Broecker, W. S. (1985). The utility of multiple tracer distributions in calibrating models for uptake of anthropogenic CO_2 by the ocean thermocline. *Journal of Geophysical Research*, **90**, 7023–35.

Poisson, A. & Chen, C. T. (1987). Why is there little anthropogenic CO_2 in the Antarctic Bottom Water? *Deep-Sea Research*, **34(2)**, 1255–75.

Quay, P. D., Tillbrook, B. & Wong, C. S. (1992). Oceanic uptake of fossil fuel CO_2: carbon-13 evidence. *Science*, **256**, 74–9.

Redfield, A. C., Ketchum, B. H. & Richards, F. A. (1963). The influence of organisms on the composition of seawater. In *The Sea*, ed. M. N. Hill, Vol. 2, pp. 26–77. New York: Wiley, Interscience.

Reid, J. L. (1965). Intermediate waters of the Pacific Ocean. *The Johns Hopkins Oceanographic Studies*, No. 2, 85 pp.

Reid, J. L. (1973). Northwest Pacific Ocean waters in winter. *The Johns Hopkins Oceanographic Studies*, No. 5, 96 pp.

Riebesell, U., Wolf-Gladrow D. A. & Smetacek, V. (1993). Carbon dioxide limitation of marine phytoplankton growth rates. *Nature*, **361**, 249–51.

Robertson, J. E. & Watson, A. J. (1992). Thermal skin effect of the surface ocean and its implication for CO_2 uptake. *Nature*, **35B**, 738–40.

Rotty, R. M. & Masters, C. D. (1985) Carbon dioxide from fossil fuel combination: trends, resources, and technical implications. In *Atmospheric Carbon Dioxide and the Globol Carbon Cycle*, ed. J. Trabalka, DOE/ER-0239, pp. 63–80. USDOE.

Sarmiento, J. L. & Sundquist, E. T. (1992). Revised budget for the oceanic uptake of anthropogenic carbon dioxide. *Nature*, **356**, 589–93.

Sarmiento, J. L., Orr, J. C. & Siegenthaler, U. (1992). A perturbation simulation of CO_2 uptake in an ocean general circulation model. *Journal of Geophysical Research*, **97**, 3621–45.

Schimel, D., Enting, I. G., Heiman M., Wigley, T. M. L., Raynaud, D., Alves, D. & Siegenthaler, U. (1995). CO_2 and the carbon cycle, climate change 1994: radiative forcing of climate change and an evaluation of the IPCC IS 92. *Emission Scenarios*, 35–71.

Schlesinger, W. H. & Melack, J. M. (1981). Transport of organic carbon in the world's rivers. *Tellus*, **33**, 172–87.

SCOR (1990). *Joint Global Ocean Flux Study Science Plan*, JGOFS Report No. 5, 61 pp.

SCOR (1992). *Joint Global Ocean Flux Study Implementation Plan*, JGOFS Report No. 9, 78 pp.

Sharp, J. H. (1983). The distributions of inorganic nitrogen and dissolved and particulate organic nitrogen in the sea. In *Nitrogen in the Marine Environment*, ed. E. J. Carpenter & D. G. Capone, pp. 1–35. New York: Academic Press.

Siegenthaler, U. & Sarmiento, J. L. (1993). Atmospheric carbon dioxide and the ocean. *Nature*, **365**, 119–25.

Simpson, H. J. (1977). Man and the global nitrogen cycle group report. In *Global Chemical Cycles and their Alternatives by Man*, ed. W. Stumm, pp. 253–74. Berlin: Dahlem Konferenzen.

Soderlund, R. & Svensson, B. (1976). Nitrogen, phosphorus and sulfur. *Global Cycles. Scope Report 7*, pp. 23–73. Stockholm: Swedish National Research Council.

Solomon, A. M., Trabalka, J. R., Reichle, D. E. & Voorhees, L. D. (1985). The global cycle of carbon. In *Atmospheric Carbon Dioxide and the Global Carbon Cycle*, pp. 13. US DOE Technical Report, DOE/ER-0329.

Spencer, C. P. (1975). The micronutrient elements. In *Chemical Oceanography*, 2nd edn., ed. J. P. Riley & G. Skirrow, vol. 2, pp. 245–300. New York: Academic Press.

Stocker, T. F., Broecker, W. S. & Wright, D. G. (1994). Carbon uptake experiments with a zonally-averaged ocean circulation model. *Tellus*, **46B**, 103–22.

Stuiver, M., Burk, R. L. & Quay, P. D. (1984). $^{13}C/^{12}C$ ratios and the transfer of biospheric carbon to the atmosphere. *Journal of Geophysical Research*, **89**, 1731–48.

Stumm, W. (1973). The acceleration of the hydrogeochemical cycling of phosphorus. *Water Resources*, **17**, 131–44.

Takahashi, T., Chipman, D. & Volk, T. (1983). Geographical, seasonal, and secular variations of the partial pressure of CO_2 in the surface waters of the North Atlantic Ocean: the results of the north Atlantic TTO program. *Proceedings Carbon Dioxide Research Conference*, pp. II. 123–II. 145, Berkeley Springs, W.Va., 1982.

Takahashi, T., Olafsson, J., Goddard, J. G., Chipman, D. W. & Sutherland, S. C. (1993). Seasonal variation of CO_2 and nutrients in the high-latitude surface oceans: a comparative study. *Global Biogeochemical Cycles*, **7**, 843–78.

Takahashi, T., Takahashi, T. T. & Sutherland, S. C. (1995). An assessment of the role of the North Atlantic as a CO_2 sink. *Philosophic Transactions Royal Society of London*, **348B**,

143–52.

Talley, L. D. (1991). An Okhotsk Sea water anomaly: implication for sub-thermocline ventilation of the North Pacific. *Deep-Sea Research*, **38** (Supplement) S171–90.

Talley, L. D., Nagata, Y., Fujimura, M., Iwao, T., Kono, T., Inagake, D., Hirai, M. & Okuda, K. (1995). North Pacific intermediate water in the Kuroshio/Oyashio mixed water region. *Journal of Geophysical Research*, **25**, 475–501.

Tans, P. P., Fung, I. Y. & Takahashi, T. (1990). Observational constraints on the global atmospheric CO_2 budget. *Science*, **247**, 1431–38.

Tans, P. P., Berry, J. A. & Keeling, R. F. (1993). Oceanic $^{13}C/^{12}C$ observations: a new window on ocean CO_2 uptake. *Global Biogeochemical Cycles*, **7**, 353–68.

Tokieda, T., Watanabe, S. & Tsunogai, S. (1996). Chlorofluorocarbons in the western North Pacific in 1993 and formation of North Pacific Intermediate Water. *Journal of Oceanography*, **52**, 475–90.

Tsunogai, S. (1979). Dissolved silica as the primary factor determining the composition of phytoplankton classes in the ocean. *Bulletin Faculty Fisheries Hokkaido University*, **30**, 314–22.

Tsunogai, S. (1991). Contribution of Kosa to oceanic biological production. In *Kosa: Science of the Atmosphere and Hydrosphere*, pp. 233–39. Nagoya University (Kokin Shoin, Tokyo): Water Research Institute [in Japanese].

Tsunogai, S. & Noriki, S. (1991). Particulate fluxes of carbonate and organic carbon in the ocean: is the marine biological activity working as a sink of the atmospheric carbon? *Tellus*, **43B**, 256–66.

Tsunogai, S. & Watanabe, Y. (1983). Role of dissolved silicate in the occurrence of phytoplankton bloom. *Journal of Oceanographical Society of Japan*, **39**, 231–39.

Tsunogai, S., Yamahata, H., Kudo, S. & Saito, O. (1973). Calcium in the Pacific Ocean. *Deep-Sea Research*, **20**, 717–26.

Tsunogai, S., Kusakabe, M., Iizumi, H., Koide, I. & Hattori, A. (1979). Hydrographic features of the deep water of the Bering Sea – the sea of silica. *Deep-Sea Research*, **26/6A**, 641–59.

Tsunogai, S., Suzuki, T., Kurata, T. & Uematsu, M. (1985). Seasonal and areal variation of continental aerosol in the surface air over the western North Pacific region. *Journal of Oceanographical Society of Japan*, **41**, 427–34.

Tsunogai, S., Noriki, S., Harada, K., Kurosaki, T., Watanabe, Y. & Maeda, M. (1986). Large but variable particulate flux in the Antarctic Ocean and its significance on the chemistry of the Antarctic water. *Journal of Oceanographical Society of Japan*, **42**, 83–90.

Tsunogai, S., Ono, T. & Watanabe, S. (1993*a*). Increase in the total carbonate in the western North Pacific water and a hypothesis on the missing sink of anthropogenic carbon. *Journal of Oceanography*, **49**, 305–15.

Tsunogai, S., Watanabe, Y. W., Harada, K., Watanabe, S., Saito, S. & Nakajima, M. (1993*b*). Dynamics of the Japan Sea deep water studied with chemical and radiochemical tracers. In *Deep Ocean Circulation, Physics and Chemical Aspects*, ed. T. Teramoto, pp. 105–19. Elsevier Oceanography Ser. Elsevier Sciences Publ.

Tsunogai, S., Watanabe, S., Honda, M. & Aramaki, T. (1995). North Pacific Intermediate Water studied chiefly with radiocarbon. *Journal of Oceanography*, **51**, 519–36.

Tsunogai, S., Watanabe, S., Nakamura, J., Ono, T., & Sato, T. (1997). A preliminary study of carbon system in the East China Sea. *Journal of Oceanography*, **53**, 9–17.

van Bennekom, A. J., & Salomons, W. (1981). Pathways of nutrients and organic matter from land to ocean through rivers. In *River Inputs to Ocean System*, ed. J. M. Martin, J. D. Burton & D. Eisma, pp. 33–51. United Nations Environment Programme.

Wallace, D. W. R., (1995). Monitoring global ocean carbon inventories, OOSDP Background Report No. 5. Ocean Observing System Development Panel, 54 pp.

Wallace, D. W. R., Beining, P. & Putzka, A. (1994). Carbon tetrachloride and chlorofluorocarbons in the South Atlantic Ocean, 19° S. *Journal of Geophysical Research*, **99**, 7803–19.

Wallace, D. W. R., Schlosser, P., Krysell, M. & Boenisch, G. (1992). Halocarbon ratio and tritium/^{3}He dating of water masses in the Nansen Basin, Arctic Ocean. *Deep-Sea*

Research, **39** (Supplement), S435–58.

Walsh, J. J., Rowe, G. T., Iverson, R. C. & McRoy C. P. M. (1981). Biological export of shelf carbon is a sink of the global CO_2 cycle. *Nature*, **291**, 196–201.

Walsh, J. J., McRoy, C. P., Coachman, L. K., Goering, L. K., Nihoul, J. J., Whitledge, J. J., Blackburn, T. E., Parker, T. H., Wirick, P. L., Shuert, C. D., Gremeir, P. G., Springer, J. M., Tripp, A. M., Hansell, R. D., Dienidi, D. A., Delersnijder, S., Henriksen, E., Lund, K., Andersen, B. A., Muller-Karger, P., F. E. & Dean, K. (1990). Carbon and nitrogen cycling within the Bering/Chukchi Seas: source regions for organic matter affecting AOU demands of the Arctic Ocean. *Progress in Oceanography*, **22**, 277–359.

Wanninkhof, R. (1992). Relationship between wind speed and gas exchange over the ocean. *Journal of Geophysical Research*, **97**, 7373–82.

Watanabe, Y. W., Harada, K. & Ishikawa, K. (1994). Chlorofluorocarbons in the central North Pacific and southward spreading time of North Pacific Intermediate Water. *Journal of Geophysical Research*, **99**, 25195–213.

Watson, A. J. (1993). Air–sea gas exchange and carbon dioxide. In *The Global Carbon Cycle*, ed. M. Heiman, pp. 397–411. Heidelberg: Springer-Verlag.

Weyl, P. (1981). Man's influence on river inputs to ocean system. In *River Inputs to Ocean Systems*, ed. J. M. Martin, J. D. Burton & D. Eisma, pp. 305–307. Proceedings of a SCOR/ACMRR/ECOR/IAHS/UNESCO/CMG/IABO/IAPSO Review and Workshop, Rome, 26–30 March, 1979.

Wollast, R., (1983). Interactions in estuaries and coastal waters in the major biogeochemical cycles and their interactions. In *SCOPE*, ed. B. Bolin & R. B. Cook, pp. 385—407. New York: John Wiley.

Wong, C. S. & Chan, Y-H. (1991). Temporal variations in the partial pressure and flux of CO_2 at ocean station P in the subarctic northeast Pacific Ocean. *Tellus*, **43B**, 206–23.

Wong, C. S., Chan, Y-H., Page, J. S., Bellegay, R. D. & Pettit, K. G. (1984). Trends of atmospheric CO_2 over Canadian WMO background stations at Ocean Weather Station P, Sable Island and Alert. *Journal of Geophysical Research*, **89**, 9527–39.

Wong, C. S., Chan, Y-H., Page, J. S., Smith, G. E. & Bellegay, R. D. (1993). Changes in equatorial CO_2 and nutrient levels in Pacific surface waters during the 1986/87 El Niño. *Tellus*, **45B**, 64–79.

Zhang, S. L. & Gu, H. K. (1994). Marine chemistry in northern seas of China. In *Oceanography of China Seas*, ed. D. Zhou, Y. B. Liang & C. K. Zeng, Vol. 1, pp. 157–70. Dordrecht: Kluwer Academic Publishers.

12 # Preliminary assessment of impacts of global change on Asia

C. B. Fu, J-W. Kim and Z. C. Zhao

Introduction

Asia has been one of the first priority regions in the global change studies for two major reasons: first, the unique Asian monsoon system is of great importance in global hydrological and biogeochemical processes of the Earth system; second, the ever-growing human activities, due to rapid economic growth and a large population, constitute the major driving force causing environmental changes, such as a rapid increase of anthropogenic emission of greenhouse gases, deforestation and desertification due to the industrialization and irrational use of land and biotic resources (Fu & Wang, 1992). A better assessment of impacts of global change on Asia, due to the vulnerability of economic infrastructure to global change in most developing countries in this continent, would be of benefit to the sustainable development of Asia if scientists could provide a scientific basis for decision makers in taking measures of adaptation and mitigation.

Despite much ongoing research in Asian countries on the assessment of global change impacts, this review focuses on the features of Asia as a whole, and sometimes references some individual countries as examples in the analysis of these general features.

The chapter first presents the main characteristics of Asian changes as a part of global change in this elapsing century, particularly in several recent decades. It then analyzes major forcing factors causing Asian changes, especially the changing Asia monsoon in response to global climate change and anthropogenic forcings. Based on available results of global climate model (GCM) simulations, the potential changes in Asia in the next 20–50 years under global warming scenarios are projected. Finally, the uncertainties in assessing the potential changes are discussed, with a focus on the defective performance of GCMs in simulation of regional climate. It is further proposed in this section that the studies of the atmosphere–biosphere interaction in the integrated Asia monsoon system and land use and land cover changes in Asia could be incorporated as Asian components within IGBP. Since these components consist of both biophysical and biogeochemical processes and social processes, such research will be related to several core projects of IGBP, WCRP and IHDP.

Asia changes as a part of global environmental change

Changes are taking place in all components of the environmental system in Asia, including air, water, soil and biota at various temporal and spatial scales. A probe into these changes, with a particular stress on the anthropogenic forcings, which appear to be the most striking forcings in recent decades, are presented below.

Regional climatic change

The climate in Asia from 1960–1990 is compared with that from 1930–1960 in order to illustrate the climatic changes in Asia in recent decades. The changes of annual mean temperature from 1930–1960 to 1960–1990 were mainly characterized by a warming of less than 0.5 °C in most parts of Asia, with the highest positive departure center of more than 1 °C in Siberia, Russia. The exception is that a large part of southern China was cooling by -0.5–1.0 °C (Fig. 12.1*a*). The significant warming area in northern Asia extended from 30° E to 130° E to the north of 40° N in winter (Fig. 12.1*b*), while the cooling area in summer extended to the north of 30° N with a center in China reaching more than -1.0 °C (Fig. 12.1*c*).

Fig. 12.2*a* presents changes of annual mean precipitation in Asia. The dry area less than -10% appeared to the south of 40° N and it was wet in 10–20% to its north and in southern India. The dry area extended in summer to cover most of Asia with centers in central China and northeastern China (-10–30%) except for southern India (Fig. 12.2*b*). In winter, although the dry area was reduced, the intensity of dryness was enhanced in western Asia (-10–30%) (Fig. 12.2*c*).

As will be discussed in a later part of this review, such regional climate change is closely linked with global climate change and to a certain extent with regional anthropogenic factors. It has significant impacts on water resources and agriculture in the region.

Rapid increase of anthropogenic emission of greenhouse gases and aerosols

CO_2

According to a recent report by the US Department Energy (Marland *et al.*, 1994), in the past 42 years, the global total CO_2 emission increased from 1638 MMT (million metric tonnes) of carbon to 6188 MMT with an annual rate of 6.5%. North America contributed one-fourth of the global total emission, much higher than all the Asian countries. It increased rapidly before the mid-1970s and then came into rather a stable state with inter-decadal oscillation for reasons that are not clear. The situation in the Asian countries is much different from the United States. All of them except Japan show a significant increase since the 1970s. Japan shows a rather different feature from most Asian countries. From 1950–

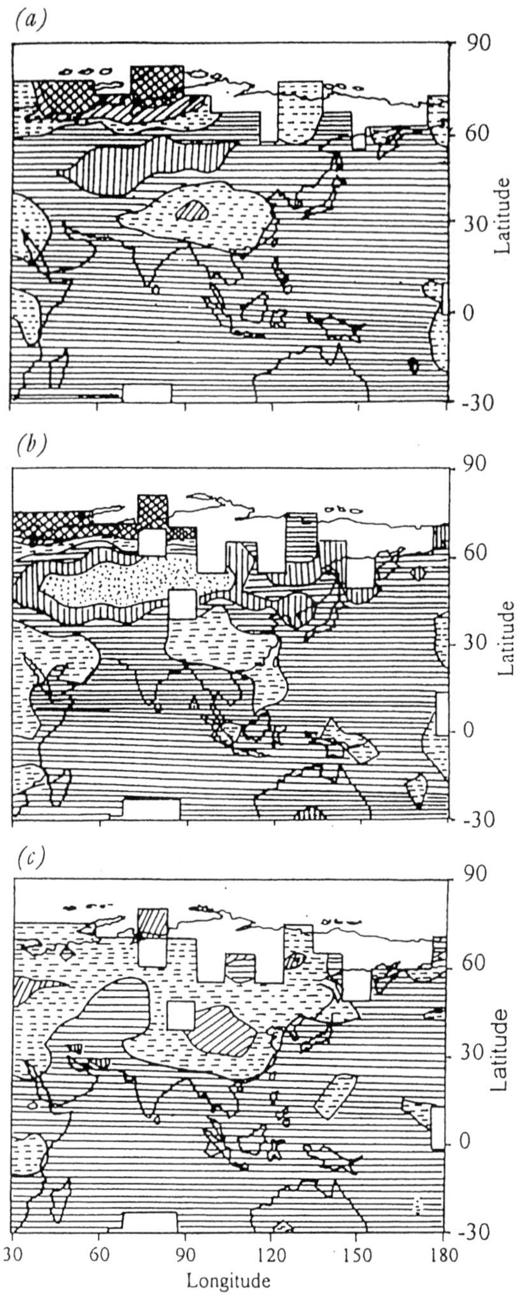

Figure 12.1 *Mean temperature changes 1931–1960 to 1961–1990: (a) annual; (b) winter, December–January–February; (c) summer, June–July–August. (Temperature change (°C): above 10, 0.5–1.0, 0.0–0.5, -0.5–0.0, -1.0– -0.5, below -1.0.)*

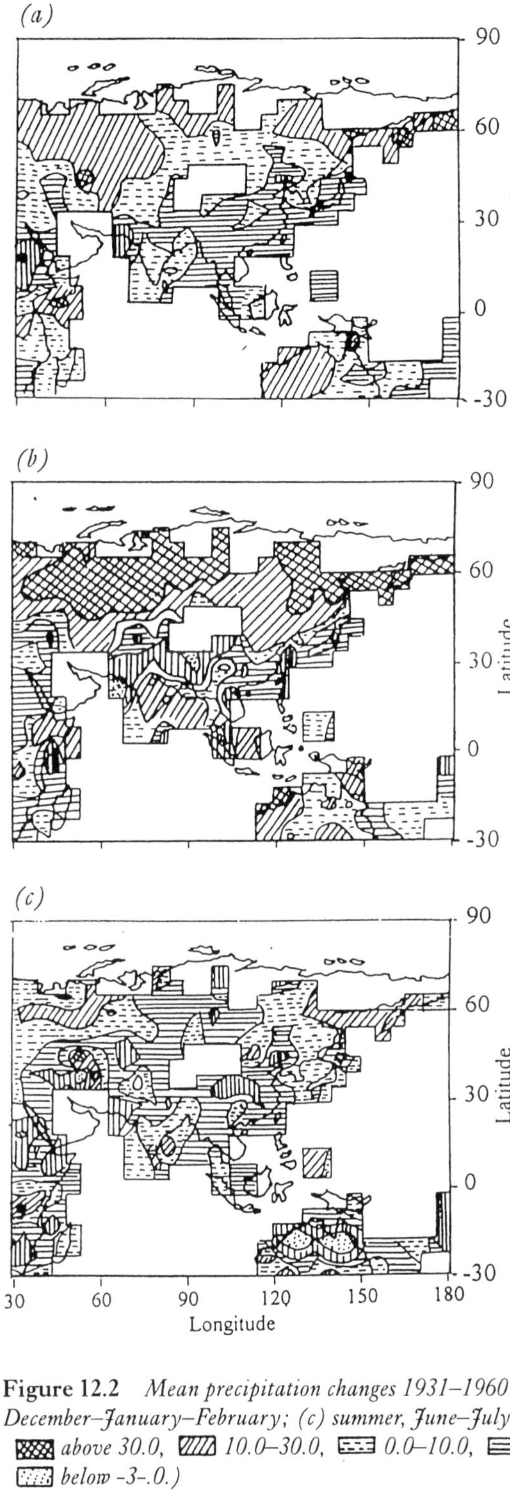

Figure 12.2 *Mean precipitation changes 1931–1960 to 1961–1990: (a) annual; (b) winter, December–January–February; (c) summer, June–July–August. (Precipitation change (%): above 30.0, 10.0–30.0, 0.0–10.0, -10.0–0.0, -30.0– -10.0, below -3-.0.)*

1973, post-war Japan showed a growth at 9.9% per year and since then the total emissions have been virtually constant, but there are recent signs of renewed growth. 1990 and 1991 were the two years of highest emissions. Up to 1990, Asia contributed 25.8% of global total industrial CO_2 emission, nearly the same as that of North America (Fig. 12.3).

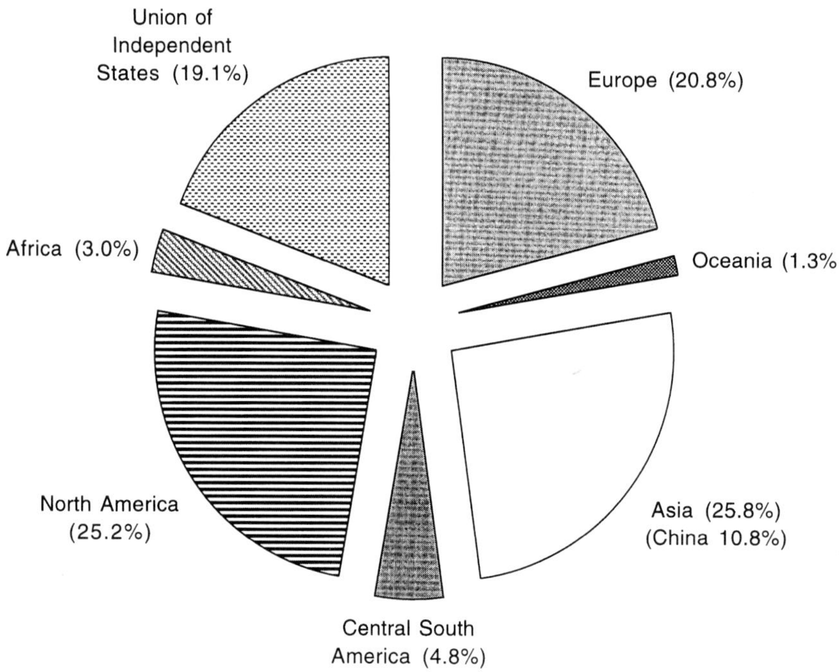

Figure 12.3 *Regional contributions of CO_2 emission. (Data from UNEP, 1991.)*

The features of per capita emission are quite different from the emission in total. Fig. 12.4 is the industrial CO_2 emission per capita in different regions of the world and in different countries of Asia (UNEP, 1991). Among all the regions, North America is the highest with 5.2 Gt C year^{-1}. Asia is the second lowest with 0.4 Gt C year^{-1}, 1/13 of North America, but its yearly increasing rate is the highest (4.9%), more than twice the global average (2.2%) and seven times that of North America (0.7%). Among Asian countries, Japan shows the highest emission per capita (2.2 Gt C year^{-1}), while Indonesia has the highest increasing rate (8.3%).

According to new projections released on April 24, 1995 by the International Energy Agency (IEA, 1995), by the year of 2010, global CO_2 emission will be 30–42% higher than they are today and most of the increase of CO_2 emission will come from the developing world, with India and China accounting for a increase larger than all OECD countries combined. With its potential fossil fuel consumption and growing population, it is possible that the total CO_2 emission by China will reach 1.6 billion tonnes by 2020 and 2.25 billion tonnes by 2050 (IEA, 1995).

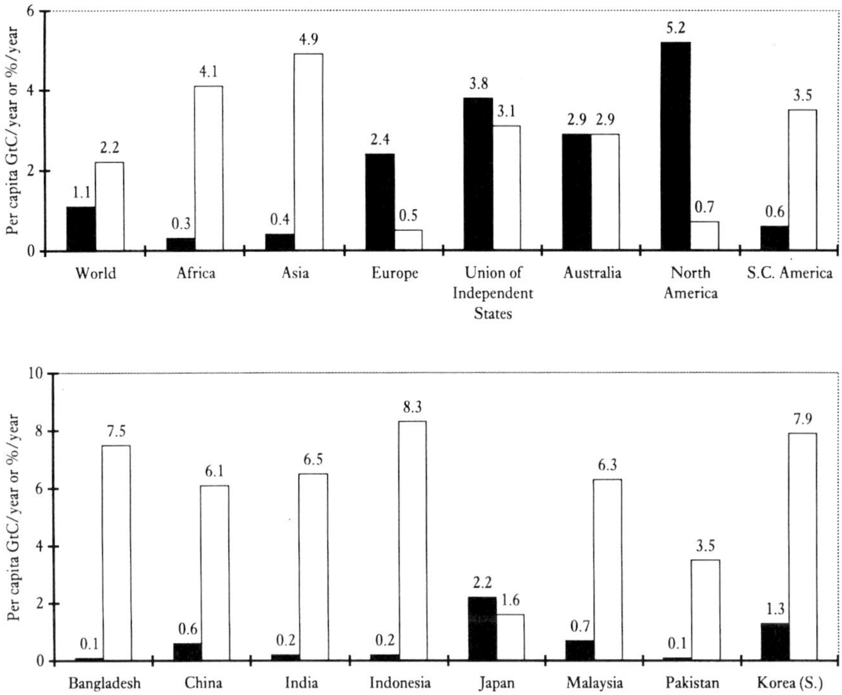

Figure 12.4 *The industrial CO_2 emission per capita (black columns) in different regions of world and different countries of Asia and their yearly increasing rates (white columns). (Data from UNEP, 1991, and revised from Hulme* et al., *1992.)*

SO_2 and NO_x

The emissions of SO_2 and NO_x in Asia, excluding Russia, were estimated, based on fuel consumption, sulfur content of each fuel type and emission factors for used fuels in each emission category (Akimoto & Narita, 1994). The total emissions of SO_2 in Asia increased from 9.1 Tg S in 1975 to 14.5 Tg S in 1987 at the annual rate of 5% and the total NO_x emissions in Asia increased from 2.9 Tg N in 1975 to 4.7 Tg N in 1987 at the rate of 5% too, taking up about 20% of the global total anthropogenic emissions. Fig. 12.5*a* is the distribution of NO_x emission in Asia. The intensive emissions are observed in the Pacific rim region of East Asia which are about 70% of the Asia total, especially in China, Japan, R. O. Korea and the Philippines (Akimoto & Narita, 1994). Fig. 12.5*b* is the distribution of SO_2 emission in East Asia, showing that the intensive emissions are observed in Northern China plain, Southwest China, South Korea, and so on (Huang *et al.*,1995).

Similarly to the CO_2 emission, in decades to come, the contributions of developing countries in Asia to the global total emissions of SO_2 and NO_x will increase will produce more tropospheric sulfate aerosols, which have negative radioactive

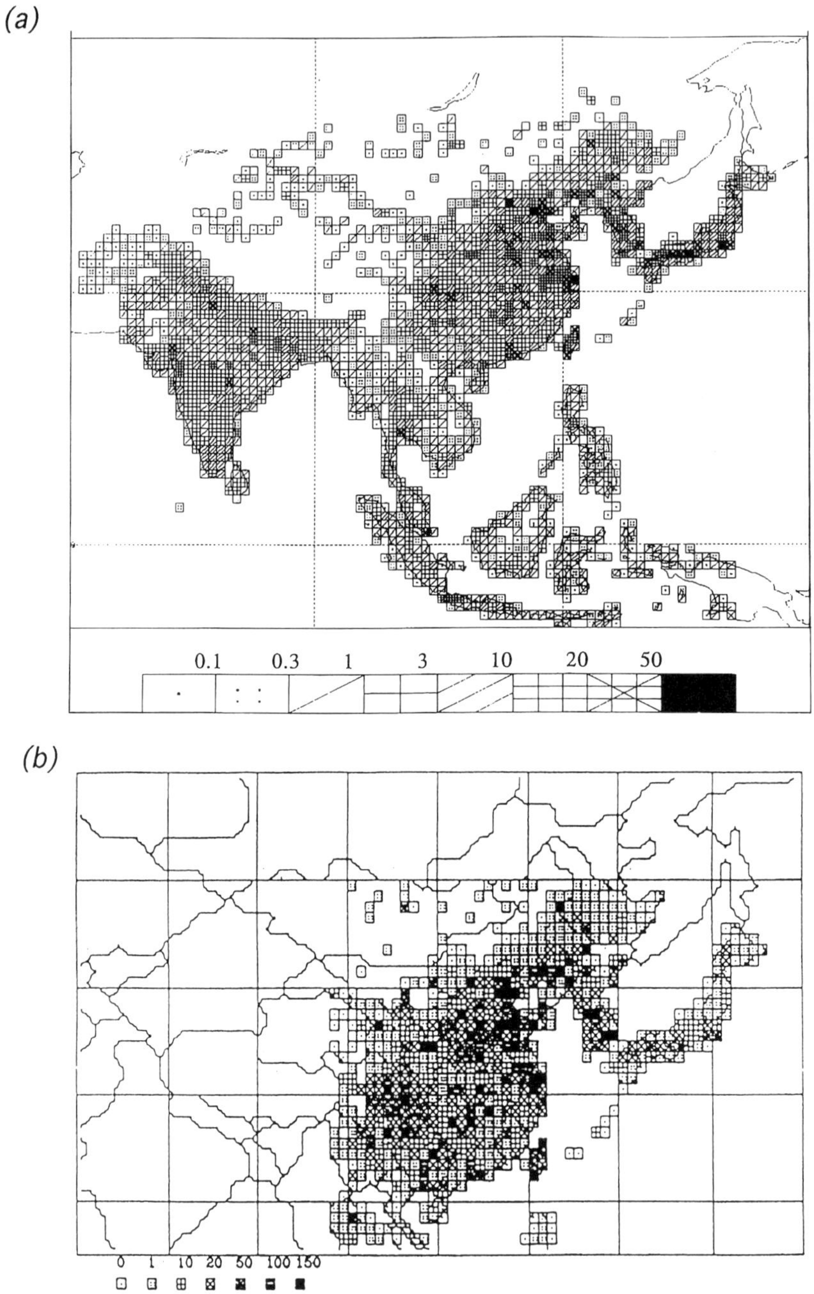

Figure 12.5 *(a) Distributions of NO_x emission (Gg N/yr) in Asia (Akimoto, 1994), and (b) distributions of SO_2 emission (Gg S/yr) in East Asia (Huang* et al., *1996).*

effects. Charlson *et al.* (1991, 1992) reported the distribution of calculated increase of reflected flux to space due to tropospheric sulfate aerosols derived from anthropogenic sources. There are three centers of such increase. The intensity of the one in East Asia is the second, but it is now increasing rapidly owing to the increasing emission of SO_2. The regional and global climatic impact of increasing tropospheric sulfate aerosols in Asia is a very interesting topic.

On the other hand, the atmosphere received a great portion of the anthropogenic reactive N which is converted by the non-reactive N, the N fixation, not only by energy consumption, but also by fertilizer production (see Mackenzie *et al.*, this volume). Fig. 12.6 presents the total fertilizer consumption for the period of 1961–1989 and projection to 2020 for the globe, Asia, Europe, North and Central America, the former Soviet Union, Africa and South America, respectively (Galloway *et al.*, 1995). It can be used as an indicator for the future trend of atmospheric concentration of NO_x. It is clear that the fertilizer consumption increased the most in Asia over the period of 1961–1991. The annual growth rate in Asia was about 10.8%, much higher than the global average rate of 6.6%. It is predicted that the anthropogenic N fixation rate will increase by about 60% by the year 2020, primarily due to increased fertilizer use and fossil fuel combustion. About two-thirds of the increase will occur in Asia, which by 2020 will account for over half of the global anthropogenic N fixation.

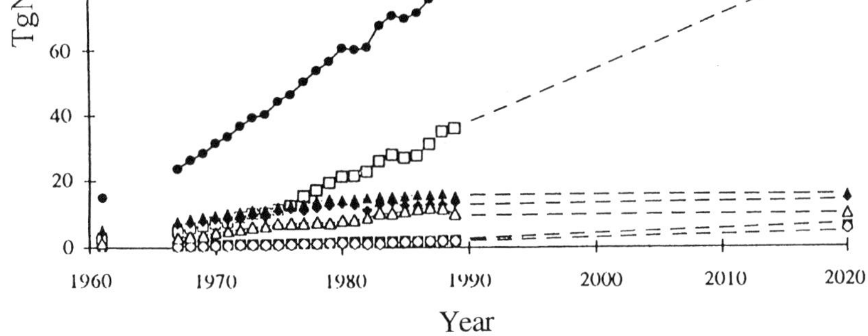

Figure 12.6 *The total fertilizer consumption (Tg N/yr) for the period of 1961–1989 and projection to 2020 for global (solid circle) and regions (Asia, open square; Europe, solid triangle; North and Central America, solid diamond; USSR, open triangle; Africa, solid square; South America, open diamond) (Galloway* et al., *1995).*

CH_4

The current emissions of methane by source show that natural sources may contribute less than 25% of the total emissions. The emissions from rice production, domestic animals and fossil fuel are three major anthropogenic sources. According to the EPA (1990), the top three producers of CH_4 for these sources are: rice production (India, China and Bangladesh); domestic animals (India, former USSR and Brazil) and fossil fuel production (United States, former USSR and China). Obviously, Asia is the major producer of anthropogenic CH_4, especially due to the largest area of rice paddies in the world (see Neue and Boonjawat, this volume). Fig. 12.7 presents the distribution of total harvest rice paddy areas, showing that eight countries of Asia (India, China, Bangladesh, Indonesia, Thailand, Vietnam, Burma and Japan) occupy nearly 87% of world's total.

The emission of CFCs and other trace gases in Asia are also increasing rapidly, but no systematic data can be used for this review. The deteriorating air quality due to many air pollutants created by the development of industry and transportation is now causing serious problems of human health in Asia, but the impacts of such regional emissions of greenhouse gases and aerosols on the regional as well as global climate require further study.

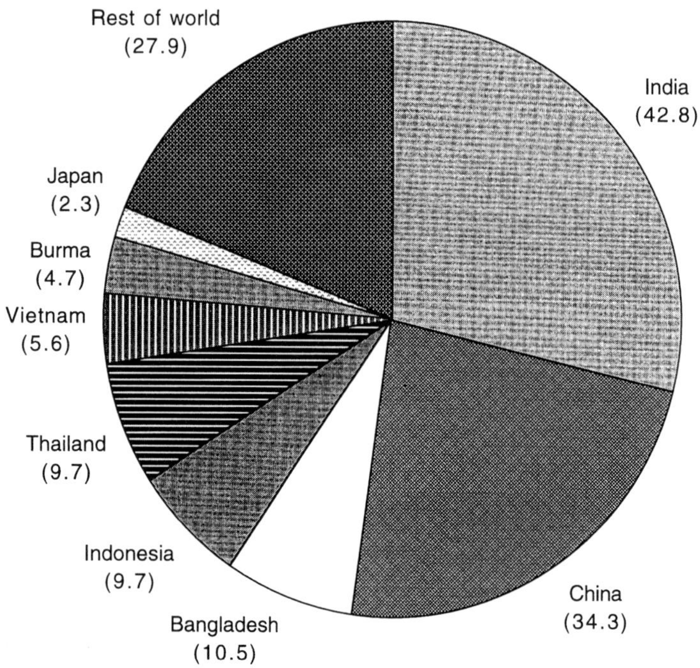

Figure 12.7 *Regional distribution of total rice harvest paddy area (units 10^6 ha) (EPA, 1990).*

Shortage of water resources

The renewable water resource per capita in Asia is 3.37×10^3 m^3, only 44% of the global average (7.69×10^3 m^3). Some countries of Asia are even below this level, for example, China (2.47×10^3 m^3), India (2.17×10^3 m^3) and R.O. Korea (1.45×10^3 m^3). Asia supplies 46% of the total world water use, while it shares only 26% of the resources (WRI, 1991). Table 12.1 presents the present and the projected water use in the year 2000 in three categories (irrigation, municipal and industrial use) in Asia and their comparison with the global. Asia supplies nearly 60% of the total world irrigation water at present and will slightly decrease in the year 2000. Water pollution in Asia has become more serious than before, due to the rapid increase of municipal and industrial waste water. The ratio of municipal and industrial water use in Asia to those of the world will be nearly twice the present level in the year 2000. The ratio for municipal and industrial waste water in Asia to the world total is expected to double in the year 2000. Water pollution is extraordinarily serious in some countries. For instance, in India, 70% of groundwater has been polluted. Among 3119 towns, only 209 have the facilities for treating waste water. Along the 48 km of the Yamuna river before it runs through New Delhi, the capital of India, it contains 700 colon bacillus per 100 mg water. In China, of 78 main rivers, 54 have been polluted by industrial and municipal water discharge. In the Huangpu river near Shanghai, nearly 20–25% of the total volume of water is pollutants. The underground water in 41 big cities (of a total of 44) in China has been polluted, mainly by heavy metal elements (SSTC, China, 1990). Therefore, the water supply shortage in Asia will be more serious in the coming decade. The shortage of water resources in Asia is partly related to the climate change. As seen earlier, a large part of East Asia has experienced a drying trend in the last 30 years.

Land use/land cover changes and land degradation

As the most populated region of the world, Asia has experienced significant changes in its traditional pattern of land use, particularly in the background of the rapid economic development in recent decades. Table 12.2 presents the global and regional changes of three main categories of land use/land cover: forest, grassland and cropland for 1700–1980. These are the major ecosystems playing significant roles in the hydrological and biogeochemical cycles of the earth system. As seen in Table 12.2 (Grubler, 1994), in Asia the decrease of forests and grasslands occurred significantly, with the most rapid change in the last 30 years. During the past 280 years, the total decrease of forests and grasslands in Asia was 313 million ha, the biggest decreasing region of the world in comparison with other regions. In the last 30 years, Asia, Africa and the Middle East ranked at the top in such a decrease, while there has been a trend of increase of croplands since 1700 with a peak in the last 30 years. It is also Asia that shows the most rapid increase of croplands in either 280 years or in the last 30 years.

Table 12.1 *Water use in Asia and the World in the 1980s with projections for 2000*

Categories	Irrigation		Municipal use		Industrial use		Municipal waste water		Industrial waste water	
	1980s	2000	1980s	2000	1980s	2000	1980s	2000	1980s	2000
Asia	1300	1500	88	200	118	320–340	35	100	88	255–270
World	2206	2585	263	456	760	1153–1310	153	282	662	963–993
A/W (%)	59	57	33	43	15	26–27	22	46	13	26–27

Units: km^3.
WRI (1992).

Table 12.2 *Global and regional land use and population change*

	1700–1800	1800–1850	1850–1920	1920–1950	1950–1980	Total 1700–1980
Europe						
Forest	-15	-10	-5	-1	+13	-18
Grassland	-15	-25	-11	-3	+2	-52
Forest+Grassland	-30	-35	-16	-4	+15	-70
Cropland	+30	+35	+15	+5	-15	+70
Population	+53	+63	+105	+79	+92	+392
North America						
Forest	-6	-39	-27	-5	+3	-74
Grassland	0	-1	-103	-22	+1	-125
Forest+Grassland	-6	-40	-130	-27	+4	-199
Cropland	+6	+41	+129	+27	-3	+200
population	+3	+20	+89	+52	+82	+246
USSR and Oceania						
Forest	-29	-42	-86	-38	-23	-218
Grassland	+2	+7	-12	-9	-22	-34
Forest+Grassland	-27	-35	-98	-47	-45	-252
Cropland	+27	+35	+97	+47	+47	+253
population	+19	+30	+62	+50	+95	+256
Africa and Middle East						
Forest	-11	-15	-68	-96	-118	-308
Grassland	0	+5	+23	+24	-9	+43
Forest+Grassland	-11	-10	-45	-72	-127	-265
Cropland	+11	+9	+47	+71	+127	+265
population	0/+1	+4	+39	+70	+250	+364
Latin America						
Forest	-6	-19	-51	-96	-122	-294
Grassland	+2	+11	+25	+54	+67	+159
Forest+Grassland	-4	-8	-26	-42	-55	-135
Cropland	+4	+7	+27	+42	+55	+135
population	+9	+15	+67	+63	+200	+354
Asia						
Forest	-38	-20	-50	-53	-89	-250
Grassland	-1	-8	-11	-12	-31	-63
Forest+Grassland	-39	-28	-61	-65	-120	-313
Cropland	+38	+29	+61	+65	+120	+313
population	+195	+171	+216	+372	+1190	+2144

Table 12.2 *(cont.)*

	1700–1800	1800–1850	1850–1920	1920–1950	1950–1980	Total 1700–1980
World						
Forest	-105	-145	-287	-289	-336	-1162
Grassland	-12	-11	-89	+32	+8	-72
Forest+Grassland	-117	-156	-376	-257	-328	-1234
Cropland	+116	+156	+378	+257	+331	+1236
population	+278	+603	+578	+686	+1909	+3755

In 10^6 ha and 10^6 people, positive or negative sign indicates direction of change.
Revised from Grubler (1994).

Deforestation, particularly in tropical areas, is an important issue causing global environmental effects, although the debate is still going on with respect to the rate of deforestation. Table 12.3 presents two estimates (FAO and World Resources Institute) over deforestation of closed tropical forests in nine selected, key countries (Williams, 1994, units: 10^3 ha) with six in Asia. Although the discrepancies are enormous, both show that Costa Rica is the highest. According to FAO, the averaged annual rate of loss in six Asian countries was 0.87%, higher than the Amazon (0.4%), but according to the WRI, it was 2.1%, much closer to the Amazon (2.2%) .

Table 12.3 *Deforestation estimates for closed tropical forests, selected countries*

Country	FAO estimate	Annual rate of loss (%)	Recent estimate	Annual rate of loss (%)	Period of recent estimate
Brasilian Amazon	1480	0.4	8000	2.2	1987
Cameroon	80	0.4	100	0.6	1976–86
Costa Rica	65	4.0	124	7.6	1977–83
India	147	0.3	1500	4.1	1975–82
Indonesia	600	0.5	900	0.8	1979–84
Myanmar	105	0.3	677	2.1	1975–81
Philippines	92	1.0	143	1.5	1981–88
Thailand	379	2.4	397	2.5	1978–85
Vietnam	65	0.7	173	2.0	1976–81
Total	3013		12 014		

Units: 10^3 ha.
Williams (1994).

The desertification of grassland is related mainly to land use, particularly to overgrazing and over-cropping. Table 12.4 presents the distribution of range lands among major geographic areas and the percentage of desertification in each region according to the UNEP (Graetz, 1994). The most serious desertification occurred in Sudano-Sahelian Africa. The desertified lands in western and southern Asia have reached 85% and those in China and Mongolia 70%. All these regions in Asia show higher rates than the global average (62%). Another study (Table 12.5, WRI, 1990) shows human-induced land degradation (UNEP and ISRIC) all over the world. It is very important to note that the land degradation is more serious in Asia (31%) than in any other region, while the degradation caused by overgrazing is the most serious in Oceania.

Table 12.4 *Distribution of rangelands among major geographic areas and percentage of desertification in each region*

Rangelands	Area ($\times 10^6$ km^2)	Desertified (%)
Sudano-Sahelian Africa	3.80	90
Southern Africa	2.50	80
Mediterranean Africa	0.80	85
Western Asia	1.16	85
Southern Asia	1.50	85
USSR in Asia	2.50	60
China and Mongolia	3.00	70
Australia	4.50	22
Mediterranean Europe	0.30	30
South America and Mexico	2.50	72
North America	3.00	42
Total	25.56	62

Source: Graetz (1994).

Urbanization

Actually the changing pattern of human settlement is a part of land use/land cover change issues, but urbanization is mainly related to industrialization rather than to the development of agriculture as other land use changes discussed above. Recent estimates suggest that, by 1900, only 14% of the world population were living in towns, then the urban numbers increased rapidly after 1950 and now the urban areas contain 43% of the world population. Although they occupy only 1% of the total land surface, their environmental effects are highly significant.

The growth of city population and population movement from rural areas to cities has been most rapid in Asia in recent decades. Among the 12 most populated

Table 12.5 *World land degradation*

	Used area (A)	Degradation area (B)	B/A (%)	Degradation by overgrazing (%)
World	8056	1943	24	35
Africa	1799	494	27	49
N.C. America	1044	158	15	24
S. America	1386	244	18	28
Asia	2440	745	31	26
Oceania	544	103	19	80

Units: 10^6 ha.
Graetz (1994).

cities of the world, 8 are located in Asia. (see Table 12.6; Douglas, 1994). According to UN projection (World urbanization prospects, the 1992 revision, New York, 1993), by the end of this century, cities with a population greater than 10 million will number 21, and 13 of them will be located in Asia as shown in Fig. 12.8. The growing of semiurbanized corridors between major cities, known as "Metro-Agro plexes", is now a world-wide phenomenon (Chameides, 1994). This is a land use unit rapidly growing in Asia, particularly in China, where the township enterprises bloom day by day, such as in the areas of the Yangtze River (Changjiang) delta, Pearl river delta, and so on. The growth of Metro-Agro plexes increases the demand for food, energy and water supply, and promotes science and technology flows and information flows. However, it creates a series of environmental influences too, such as air, water and soil pollution from the township enterprises, it also decreases croplands and agricultural population. This is a new issue challenging the sustainable development of Asia.

Major driving factors for Asia changes

Changing Asian monsoon in response to global climate

The observed global mean temperature records have shown a general warming trend with the amplitude of about 0.3–0.6 °C in the last century (IPCC, 1992), although it is still too early to conclude whether this warming trend is a part of natural variability or if it could be attributed to the effects of increasing atmospheric CO_2 and other trace gases.

Analysis of the monsoon climate has clearly shown the sensitivity of the monsoon system in response to global mean temperature changes. For example,

Table 12.6 *Major city population growth*

	1950	1960	1970	1980	2000
Mexico City, Mexico	2.9	5.1	8.9	15.0	31.0
São Paolo, Brazil	2.4	4.4	8.0	13.5	25.8
Shanghai, China	5.8	7.4	10.0	13.4	22.7
Beijing, China	2.2	4.4	7.0	10.7	19.9
Rio de Janeiro, Brazil	2.9	4.4	7.1	10.6	19.0
Bombay, India	2.9	4.0	5.8	8.3	17.1
Calcutta, India	4.4	5.5	6.9	8.8	16.7
Jakarta, Indonesia	1.7	2.7	4.5	7.3	16.5
Cairo, Egypt	2.5	3.7	5.5	7.4	13.1
Madeira, India	1.4	1.7	3.0	5.4	12.9
Manila, Philippines	1.6	2.3	3.6	5.7	12.3
Bangkok, Thailand	1.4	2.2	3.2	4.9	11.9

Units: 10^6.
Douglas (1994).

the decreasing rainfall in a large part of Asia over the last several decades is closely linked with global mean temperature changes. Take China for example: Fig. 12.9 presents a changes of aridity index of East China, superimposed on the annual mean temperature over the Northern Hemisphere land since 1880, showing a close linkage between two curves in the following aspects: the long-term trend, non-periodic oscillation and the abrupt features on decadal scale changes (Fu, 1994). There is a significant aridity trend in China (weakening of the summer monsoon) under the global warming trend. The monsoon climate in China exhibited an abrupt change around the early 1920s, when a rapid warming occurred over the Northern Hemisphere land area, as a Mann–Kendall rank Statistic test was applied for two time series. That means the China summer monsoon entered into a weak regime abruptly around the1920s when there was a global rapid warming, indicating clearly the sensitivity of the Asian monsoon system to the global warming, no matter what causes the warming. This relationship indicates that the shortage of water resources in a large part of Asia as mentioned in the previous section may, to a certain extent, be related to global warming.

There is also a good historical analog in China on a longer time scale, showing the changes of the monsoon system, including the broad-scale distribution of ecosystem under the global temperature changes (An, 1990). Fig. 12.10*c* shows the present distribution of ecosystems over China, where the heavy solid line marks the boundary of humid monsoon climate and dry inland climate. It could be looked upon as the northern edge of the summer monsoon in general. During a

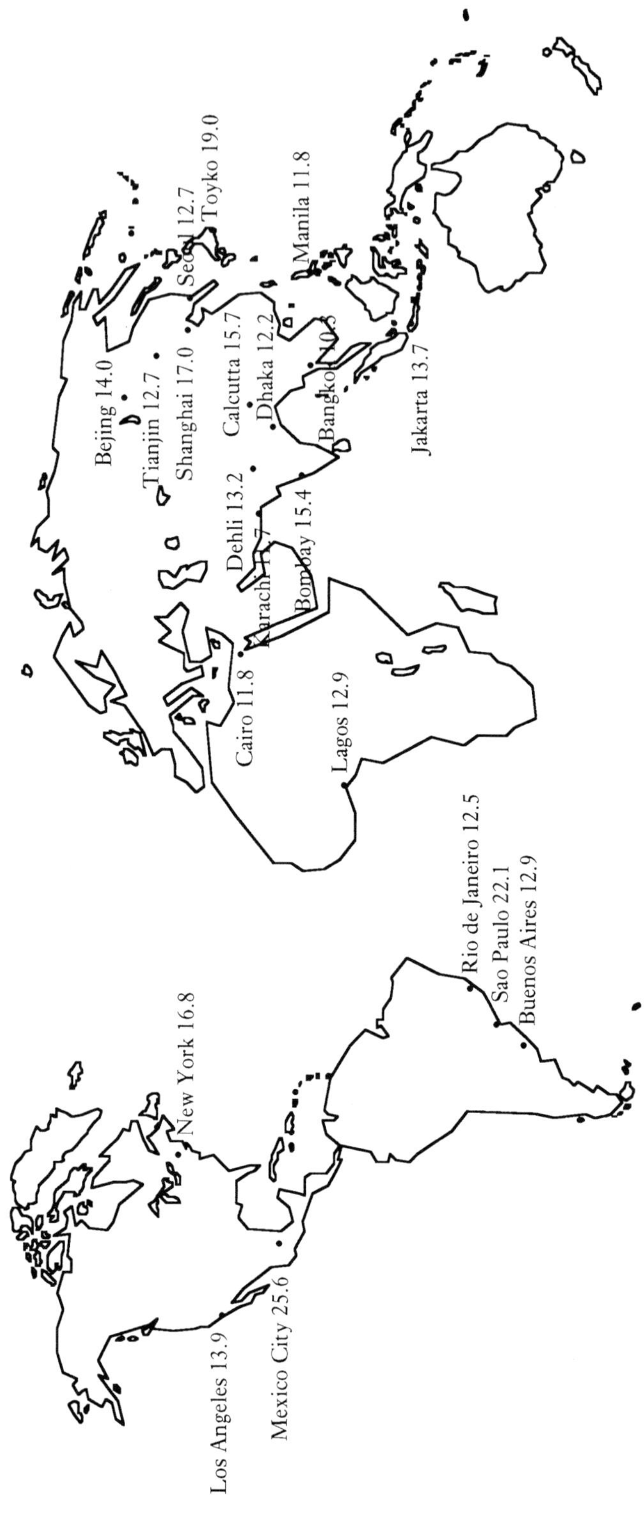

Figure 12.8 ***Cities with population greater than 10 million in 2000 (UN, 1993).***

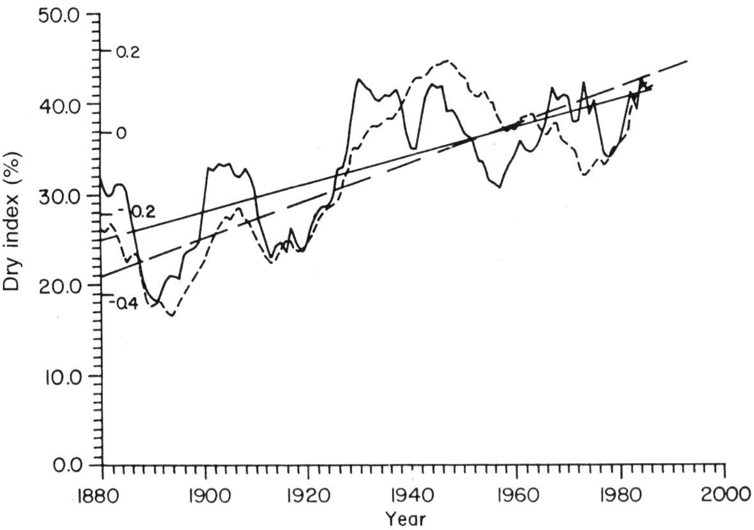

Figure 12.9 *10-years running means of drought index of East China (solid line), superimposed with that of Northern Hemisphere air temperature over land (dashed line) (°C) (Fu, 1994).*

cold period (15 000–18 000 BP) (Fig. 12.10*a*), the pattern of ecosystem distribution changed significantly, particularly the southward shift of the northern edge of summer monsoon, while during a warming period of global climate (6000–9000 BP), the strong summer monsoon advanced northward and the entire structure of ecosystem distribution over East China also experienced a significant change (Fig. 12.10*b*).

The India monsoon, however, responds to global warming in an opposite direction to that of the East Asia monsoon. It was found that the Indian summer monsoon had an abrupt change from an inactive period to an active one around 1920s when the globe warmed rapidly (Fu, 1992).

These two examples show the close relationship between the Asia monsoon climate and global mean temperatures and provide excellent evidence of the regional response in Asia to global warming, and can be used as a historical analog for projecting potential changes of the Asia monsoon climate under future global warming.

Rapid economic development and growing population

Asia accounts for the largest share of the world's population (3.32 billion, *c.* 57.2%), with China (1.17 billion) and India (860 million) the most populous countries. At present (1990–1995), the fastest rate of population increase is in Africa at an average rate of 3% per year, and the average growth rate in Asia also reached 1.8% per year (1990–1995). The growing population is a major driving

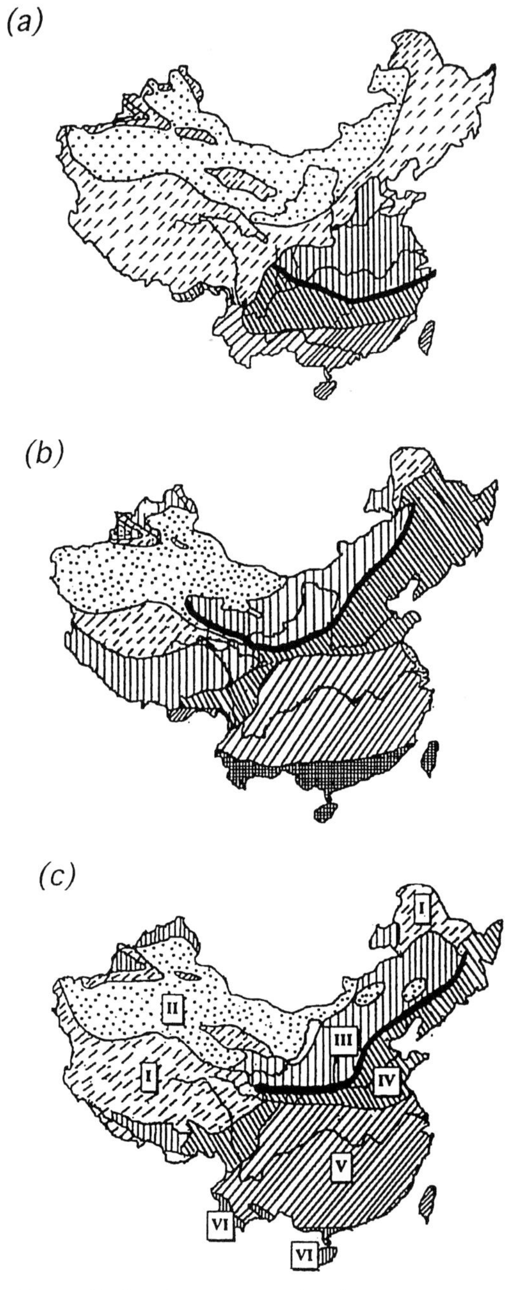

Figure 12.10 *Distributions of major ecosystems over China: (a) 15 000 – 18 000 BP (cold period), (b) 6000 – 9000 BP (warm period), (c) present. (Key: I, perma frost; II, desert; III, range-grassland; IV, temporate forest; V, subtropical forest; VI, tropical rainforest.) (An, 1990).*

force for the Asia changes. As shown in Fig. 12.11, there is a close positive correlation between the population and areas of croplands, and a negative correlation between the population and the areas of forests and grasslands in Asia over the last 280 years.

Rapid industrial and agricultural development is most significant in Asia in recent decades, particularly since the middle 1970s. Fig. 12.12*a* shows the change of the agricultural index of Asia and its comparison with the globe and North America. Agricultural development in Asia is much higher than the global average, while North America is much below the global average. In Asia (Fig. 12.12*b*), the development of agriculture is most rapid in Malaysia, China, Indonesia, India and Pakistan, with an average rate of more than 10% per year (*The Yearbook of World Economy*, 1994).

Fig. 12.13 shows the changes in the industrial index of five countries of Asia (China, India, Japan, R. O. Korea, Singapore) and their comparison with the United States and the global average. China is at the top with an average rate of

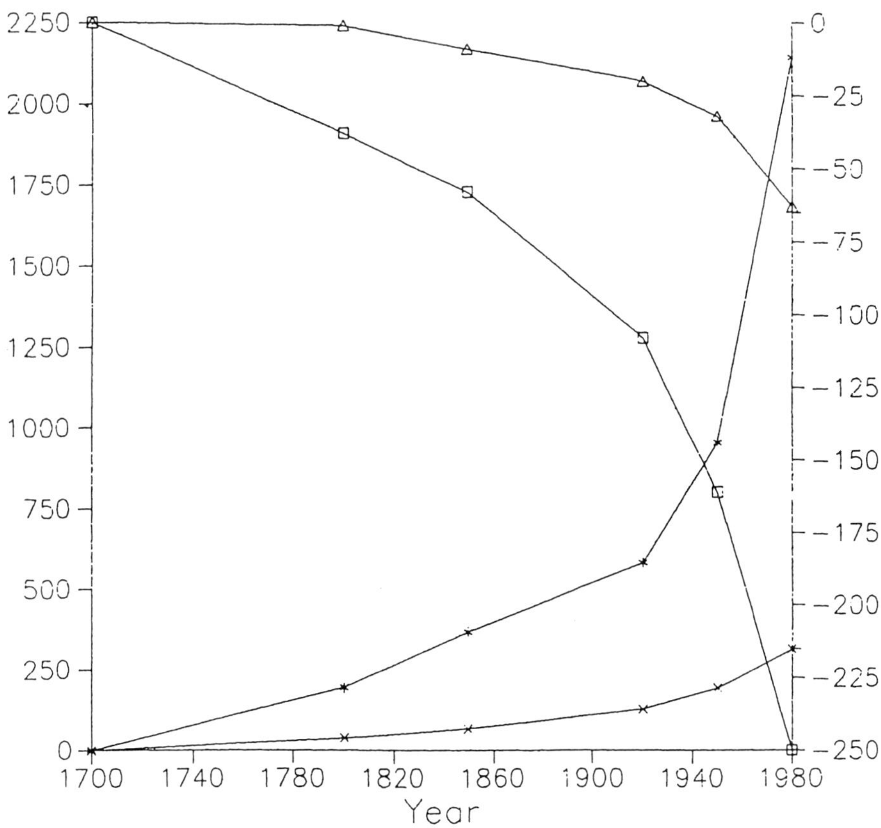

Figure 12.11 *Changes in land use relating to population growth: forest, 10^6 ha (□□□□); grassland, 10^6 ha (ΔΔΔΔ); cropland, 10^6 ha (××××); and population, 10^6 (****).*

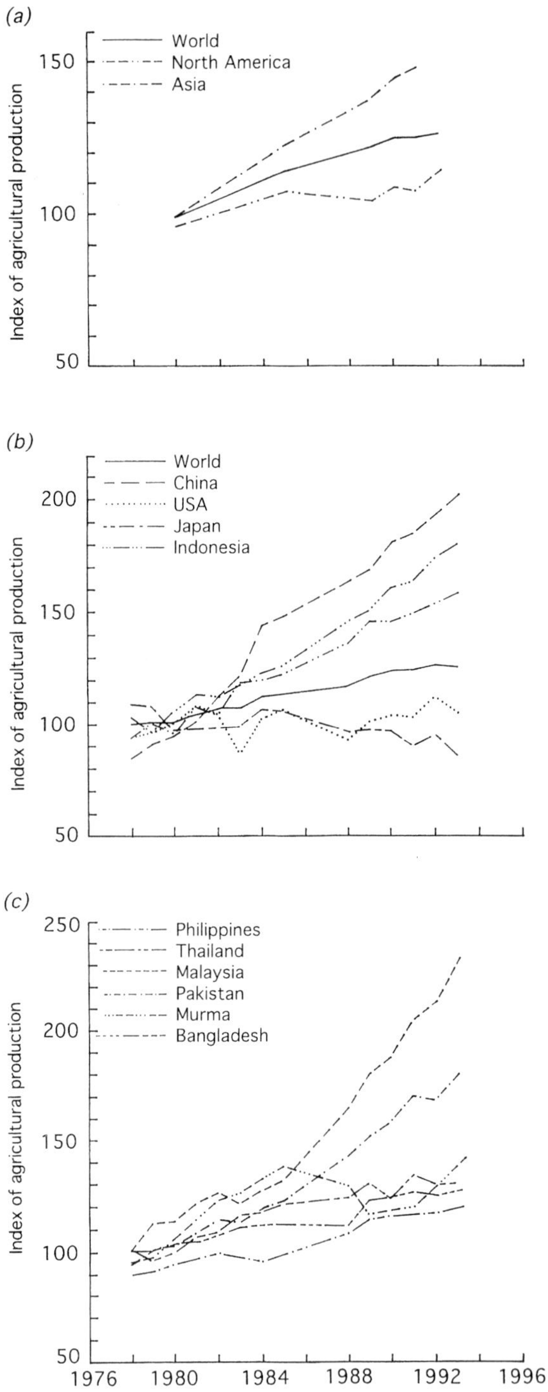

Figure 12.12 *Changes of agriculture index of Asia and its comparison with global and North America (a) and some major countries of Asia (b) and (c). (Note: the index of 1979–1981 is equal to 100.)*

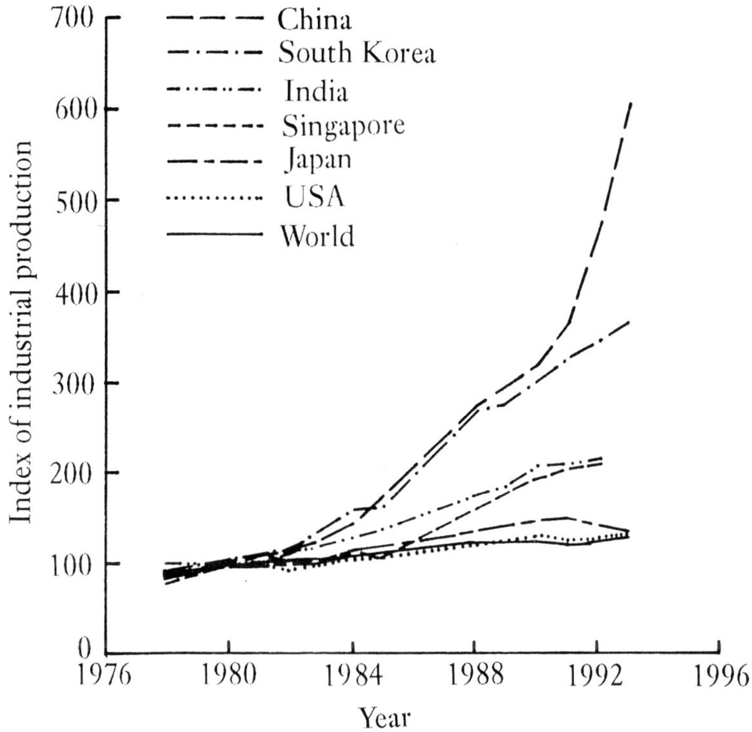

Figure 12.13 *Changes of industrial index of five countries of Asia (China, South Korea, India, Singapore, and Japan) and their comparison with the USA and global. (Note: the index of 1980 is equal to 100.)*

33% per year. Singapore, Korea, R. O. and India are all above the global average, while the United States and Japan are close to the global average (1% per year) (*The Yearbook of World Economy*, 1994).

It is clear that all the major Asia changes, such as the rapid increase of anthropogenic emission of greenhouse gases and sulfate aerosols, shortage of water resources discussed earlier can be attributed mainly to the rapid development of the economy and the ever-growing population, which are the major driving forces of Asia environmental changes in recent decades.

Potential changes in Asia in the next 20–50 years

There is increasing demand to make a better projection of global change and the impacts on different regions of the world over the next 20–50 years. The IPCC has made great efforts to make such a projection to meet the immediate requirement of human society. However, our present knowledge is too little to make a confident projection for such a complicated earth system as one consisting of atmosphere, hydrosphere, lithosphere and biosphere, regulated by the non-linear interaction of physical, chemical and biological processes and strongly interfered with by human society.

So far, global climate models (GCMs), including the atmospheric global circulation model (AGCM) and the oceanic global circulation model (OGCM) and the coupled atmospheric–ocean model (A–O GCM), have been the major tools for making projections of global climate change in the future, although great uncertainties remain in the results of such projections. The outputs of GCMs are used to drive various kinds of subsystem models, such as hydrological models, ecosystem models, crop models, and so on, to make further assessments of impacts when needed practically. Under such a restriction, caution must be taken in using conclusions from this area.

Potential climate changes in Asia with increasing atmospheric radioactive trace gases

Fig. 12.14*(a)–(d)* presents the climate changes in 2050 in East Asia, simulated from the averaged seven GCMs outputs (UKMO-L, UKMO-H, GFDL, GISS, OSU, LLNL and MPI) (Hulme *et al.*, 1992). Among these models, only the MPI is a complete ocean–atmospheric coupled model, and simulation is undertaken under a stable transitional increase of greenhouse forcing. The simulation is made according to IS92 emission scheme with the increase of CO_2, feedback of O_3 and sulfate aerosols negative impact. The results show that the warming in East Asia will be about 1.5–2.0 °C in winter and 1.0–1.5 °C in summer, respectively. There will be an increasing trend of precipitation of about 5–10% to the north of 20° N, but a decreasing trend in southern China and northwestern India in winter, and in summer, the precipitation will be increasing about 5–10% in southern part of the Asia and 5% in northern East Asia, but it will be dry in the most of middle latitude regions.

Another set of experiments were made under two different conditions: one for greenhouse gas forcing only and the other for both greenhouse gas forcing and sulfate aerosol effects (IPCC, 1995). As shown in Table 12.7, when both effects are included, the mean area climate change over Asia in the years 2030–2050 will be characterized by a warming of about 1.7 and 2.8 K in winter and summer, respectively, much less than those for greenhouse gas forcing only (3.7 K and 3.5 K for winter and summer, respectively), and decreasing precipitation mainly in summer (-0.6 mm/day) in contrast with increasing precipitation for greenhouse only (0.2 mm/day).

Impact of climate changes on water resources

As discussed earlier, there is an increasing demand for water resource in Asia, due to population growth and economic development. Water resources in the future

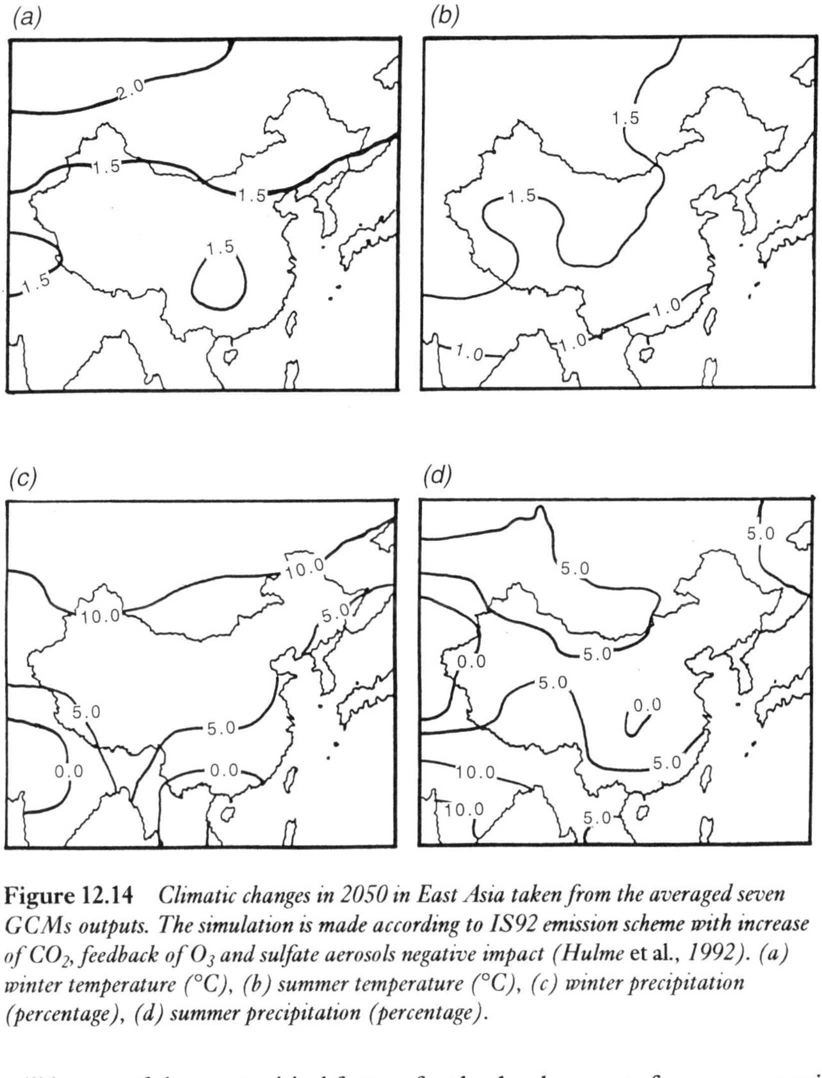

Figure 12.14 *Climatic changes in 2050 in East Asia taken from the averaged seven GCMs outputs. The simulation is made according to IS92 emission scheme with increase of* CO_2*, feedback of* O_3 *and sulfate aerosols negative impact (Hulme* et al.*, 1992). (a) winter temperature (°C), (b) summer temperature (°C), (c) winter precipitation (percentage), (d) summer precipitation (percentage).*

will be one of the most critical factors for the development of economy, particularly for agriculture.

The projections of water resources in Asia in the next 20–50 years are made by climate–hydrological models and climate water resources models using the outputs of GCM simulated climate under 2 × CO_2 (e.g. Morita *et al.*, 1993; Lal, 1994; Neilson & Danny, 1994). The changes of soil moisture in Asia under the greenhouse effect simulated by 5 GCMs (GFDL, GISS, NCAR, OSU, UKMO) indicate 0.5–1.0 cm drier in the southern part of Asia and nearly no changes in other parts in winter, while in summer most parts of the continent will be 0.5–1.5 cm drier except slightly wetter in the Indian peninsula as shown in Fig. 12.15 (Zhao &

Table 12.7 *Projections of climate changes in Asia for 2030 – 2050*

	Temperature (K)		Precipitation ($mm\ day^{-1}$)		Soil moisture (cm)	
	DJF	JJA	DJF	JJA	DJF	JJA
Greenhouse gases only	3.7	3.5	0.2	0.3	0.4	-0.6
Greenhouse gases and aerosols	1.7	2.8	0.0	-0.6	-0.2	-1.4

DJF: winter, December–January–February
JJA: summer, June–July–August

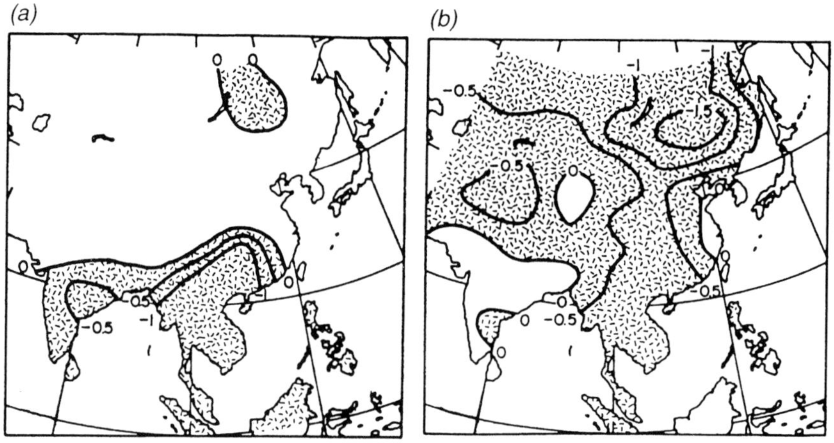

Figure 12.15 *The averages of changes in soil moisture shown by five models (Zhao* et al., *1988) (a) winter (b) summer. (Units: cm)*

Kellogg, 1988). The potential changes of area mean soil moisture over Asia were projected under greenhouse gas forcing only, and under both effects of greenhouse gases and sulfate aerosol, as seen in Table 12.7. It is obvious that, when the effect of sulfate aerosol is added, the soil becomes drier.

The averaged changes of global surface runoff simulated from 5 GCMs scenarios (GFDL, GFDL-Q, GISS, OSU and UKMO) of the double CO_2 with and without increasing water use efficiency (Neilson & Danny, 1994) show that the runoff might decrease over the most parts of Asia, except for India and Indochina Peninsula. Fig. 12.16 gives the simulated changes of runoff in South Asia using MPI model's output. A moderate decrease may occur over a large area of southeastern China and northern India but a slight increase will occur in southern India and in the Indochina Peninsula in winter. In summer, most of Asia will have less surface runoff except for a narrow zone around 25° N and near the Equator (Lal, 1994).

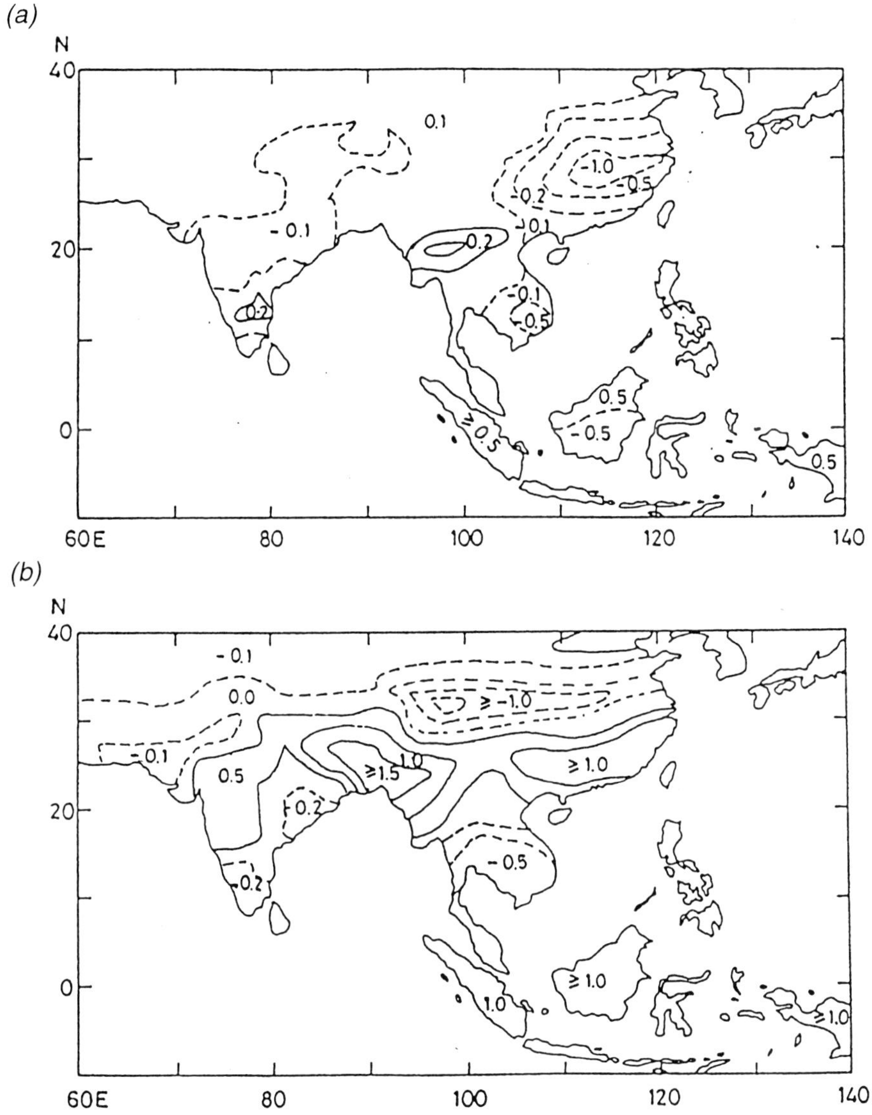

Figure 12.16 *Simulated changes of surface runoff (mm day^{-1}) over South Asia expected due to greenhouse warming (Lal, 1994). (a) Northern Hemisphere, winter (b) Northern Hemisphere, summer.*

Impact of climate changes on vegetation and ecosystem distributions

The changes of temperature, precipitation, radiation, soil moisture and other climate variables under greenhouse effect simulated by GCMs could be used as the inputs to drive the models of ecosystem, vegetation, and so on, to estimate their future changes with the assumption that the relationship between the vegetation and climate variables will remain the same with the warming earth.

Several such simulations (e.g. Leemans, 1989; Morita *et al.*, 1993; Neilson & Danny, 1994; Handerson-Sellers & Mcguffie, 1994) focus on the Asia region. In general, the impacts of global warming on the vegetation in Asia are significant, with more changes in middle and higher latitudes than in the tropics and subtropics. According to Neilson & Danny's (1994) results simulated by the biogeographical model, the areas of boreal forest over Siberia might increase, but the areas of temperate forests in China and the areas of tropical dry forest and savannas might decrease under increasing CO_2. The result of Handerson-Sellers and McGuffie (1994) are simulated under an instantaneous double CO_2 and indicate that the areas of deciduous and conifer trees might reduce, while the approximate agricultural areas might extend northward due to the greenhouse effect, as shown in Fig. 12.17.

Impact of climate changes on agriculture in Asia

Being the largest agricultural continent, Asia has 60–90% of the agricultural population. It produces 80–90% of the world total productions of rice, tea and some industrial crops such as natural rubber, quinine, Malaysia gummy, shaddock,

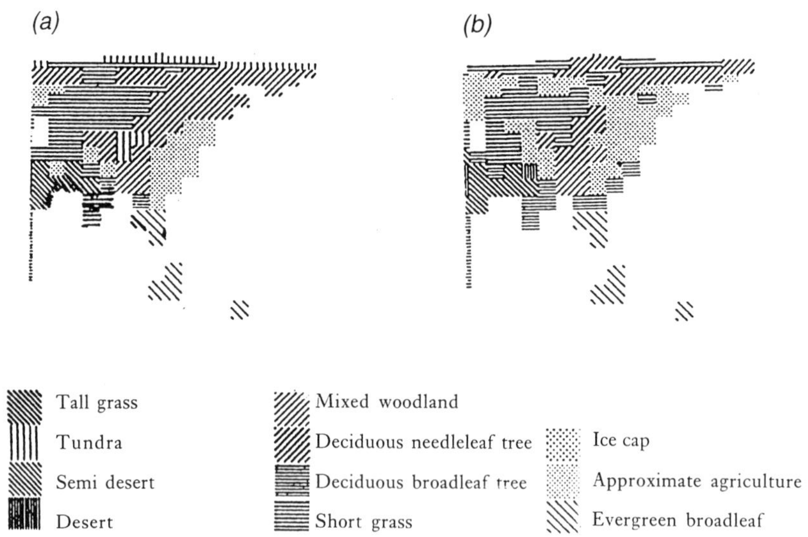

Figure 12.17 *Simulated vegetation distributions of Asia under (a) 1 × CO_2 and (b) 2 × CO_2 (Handerson-Sellers* et al., *1994).*

paper, jute and copra, and so on. The productions of peanut, sesame, tobacco and rapeseed in Asia are 45% of world totals and that of soybean and cotton are 30–40% of the world totals. Great concern is given to the impact assessment of global warming on the agriculture in Asia. As seen from the models simulation (e.g. Rosenzweig *et al.*, 1992; Hulme *et al.*, 1992; Zhao & Wang, 1994), such potential impacts are remarkable. In general, the growing seasons might be extended owing to warm temperature. Fig. 12.18 gives an example of cropping system change in China under projected climate change of composite GCM scenario for 2050, showing the distinct expansion of triple cropping system area (Zhao & Wang, 1994). However, the warm temperature will also increase the evaporation and dry the soil which has negative effects on the crops.

Rosenzweig *et al.* (1992) estimated that the grain yield might decrease by 10–30% over India and the Arabian Peninsula and increase by 10–30% in other areas of Asia under the double CO_2 climate change scenarios of three GCMs (GFDL, GISS and UKMO) with the direct effects of CO_2 . He also reported that the total crop yield might decrease by 10–30% over most of the Asia region except for China. Table 12.8 gives the changes of wheat yield in four countries of Asia under the above scenarios with and without the direct effects of CO_2. The wheat yield will decrease in all four countries, and under all three scenarios, if no direct effects are taken into account, while it will slightly decrease in most countries or slightly increase in China and occasionally in India and Pakistan in some models.

Further development of Asia component of IGBP

As seen from the above analysis, Asia is changing rapidly, mainly due to the strong anthropogenic forcing, which is superimposed on the natural evolution of the environment. The assessment of global change impact on Asia in the future largely depends on the projection of climatic changes at both the global and regional scale. The uncertainties of climatic change projection at present come from several sources: (1) incomplete understanding of the complicated climate system, particularly the non-linear interactions of major processes, such as cloud–radiation interaction, atmosphere–biosphere interaction, etc. in which physical, chemical and biological processes are interacting with each other; (2) uncertainties in the projection of human driving forces, such as population growth, and development of the economy which depend on social and political factors and to some extent is unpredictable.

The difficulty in the projection of regional impacts also comes from the imperfect performance of present GCMs in simulating the regional climate and climatic changes mainly due to their coarse resolution and lack of better parameterization of the meso-scale forcing, such as vegetation and land cover classification, etc.

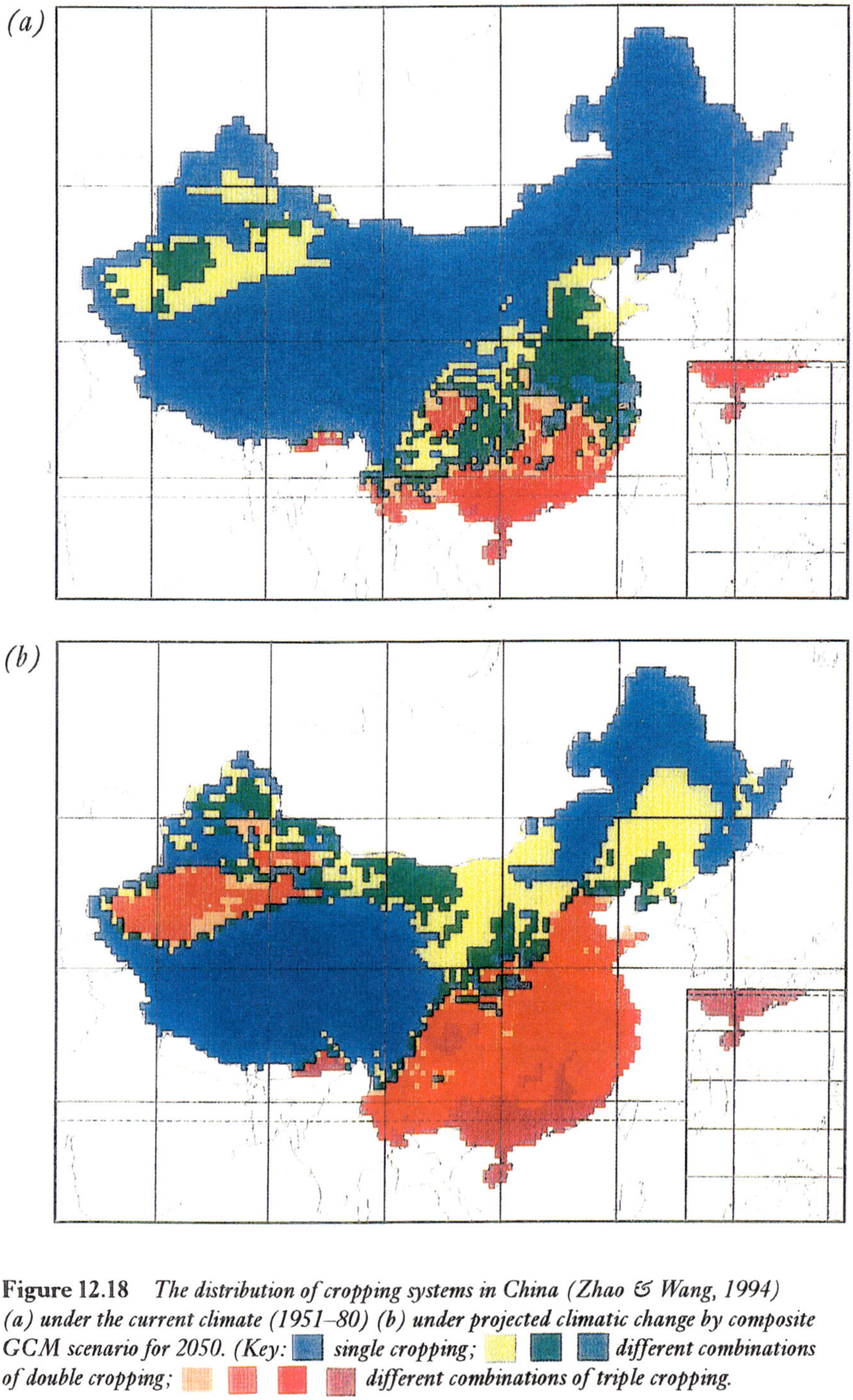

Figure 12.18 *The distribution of cropping systems in China (Zhao & Wang, 1994) (a) under the current climate (1951–80) (b) under projected climatic change by composite GCM scenario for 2050. (Key:* ▪ *single cropping;* ▪ ▪ ▪ *different combinations of double cropping;* ▪ ▪ ▪ ▪ *different combinations of triple cropping.*

Table 12.8 *Changes in wheat yields under three climate change scenarios, with and without the direct effects of* CO_2

Country	Current production			Changes in simulated yield (without)			Changes in simulated yield (with)		
	Yield (t ha^{-1})	Area (ha × 10^3)	Product (t × 10^3)	GISS	GFDL (%)	UKMO	GISS	GFDL (%)	UKMO
China	2.53	29 092	73 527	-5	-12	-17	16	8	0
India	1.74	22 876	39 703	-32	-38	-56	3	-9	-33
Japan	3.25	237	772	-18	-21	-40	-1	-5	-27
Pakistan	1.73	7478	12 918	-57	-29	-73	-19	31	-55

Rosenzweig *et al.* (1992).

Since there are a number of researches devoted to studying the uncertainties of climatic projection, some ongoing researches in core projects of IGBP, WCRP and HDP are addressing these issues in order to reduce uncertainties on a global scale. This chapter only gives some preliminary thoughts related to Asian development of IGBP components.

As evident in a recent scientific conference of AMIP (Atmospheric Model Intercomparison Project) of WCRP, so far, most GCMs fail to simulate the significant regional structure of the Asian monsoon. The reasons are not clear yet. The effect of the Tibetan plateau has been thought to be one of the possible factors (e.g. see An & Thompson, this volume). However, its effect could be on large-scale circulation patterns rather than on the regional structure as seen in the rainfall distribution in detail.

From some preliminary studies, it is suggested that further research on the atmosphere–biosphere interaction in the Asia monsoon region could be one of the critical issues for improving monsoon simulation. Asia is covered from south to north by tropical rainforests, tropical monsoon forests, subtropical monsoon evergreen forests, deciduous broad-leaved forests, mixed broad-leaved and coniferous forest and cool, temperate coniferous forest, large areas of grassland and various cropland. These major ecosystems play significant roles in the hydrological and biogeochemical cycles of the monsoon system, which should have strong impacts on the formation of the monsoon system and its variation.

On the other hand, the Asian monsoon climate is characterized by a high rate of change at various time scales, such as seasonal abrupt change, high interannual variability, and distinct climate regimes and the abruptness of their transition (Fu & Zheng, 1997). All these are strong forces imposing on the ecosystem. Therefore, the variability of ecosystems in the monsoon region shows close linkage with the monsoon climate variability. This is a concept we refer to as "monsoon-driven ecosystems" (Fu, 1995). The monsoon-driven ecosystems should also have significant feedback effects on the monsoon climate. The integrated monsoon system should be considered to consist not only of the physical component of the monsoon as it used to be, but also the monsoon-driven ecosystem, the biological component of the monsoon system. A better simulation of the integrated monsoon system and its changes under the large-scale forcing such as greenhouse gases forcing and sulfate aerosol could provide better inputs for impact assessment in Asia. This requires enhancing study of the atmosphere–biospheric interaction in the Asia monsoon system (Fig. 12.19).

Another issue which is also critical for Asia is land use/land cover changes. As the most populated area of the world, the major format of human activities in Asia is the changing land use patterns, which is mainly related to rapid development of industry, agriculture and urbanization and growing population. It is expected that the changing land use/land cover patterns will have significant impacts on the

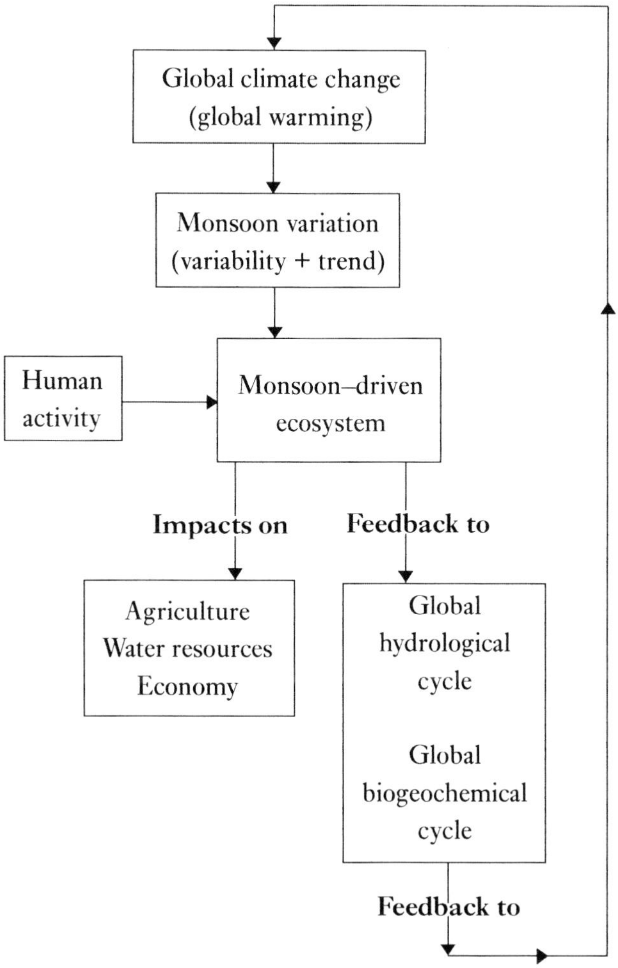

Figure 12.19 *A schematic diagram for proposed study on integrated monsoon system.*

environment at regional as well as global scales. The study on the impact of changing land use/ land cover on the physical and biogeochemical processes on land surface in Asia and their feedback to regional and global climate system, including the modification of the monsoon system, should be strengthened as Asia component of IGBP and HDP (Fig. 12.20). This diagram shows that the issue of land use/land cover change links not only the LUCC of IGBP/HDP (IGBP report No.35 and HDP report No.7, 1995) but also links to many core projects of IGBP and WCRP. A integrated study on land use/land cover changes in Asia could be developed as a cross-IGBP/HDP/WCRP project at a regional level.

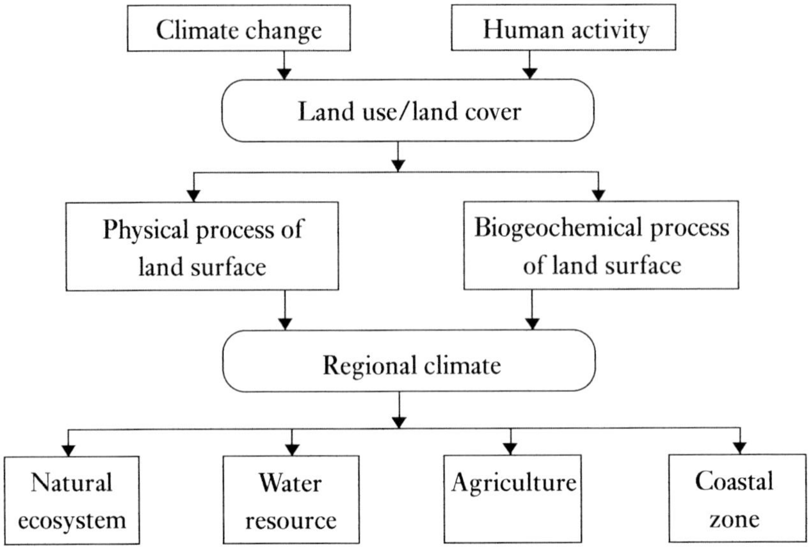

Figure 12.20 *A schematic diagram for proposed study on land-use/land-cover in Asia global change.*

Conclusions

Significant environmental changes have taken place in Asia in the last century, particularly in last several decades. The major portion of such changes which superimpose on the nature changes could be attributed to the anthropogenic forcing factors, mainly the ever-growing population and recent rapid development of the economy. Although the potential changes of Asia in the next several decades are presented in this chapter, caution must be taken in using the conclusions from that section, because great uncertainties remain in such projections which are based on the climate changes simulated by GCMs under the scenarios of global social development. Further development of global change research in Asia is strongly recommended to reduce the uncertainties. The atmosphere–biosphere interactions in the monsoon system and the regional anthropogenic forcing, such as the land use change, could be the major areas of further research.

References

Akimoto, H. & Narita, H. (1994). Distributions of SO_2, NO_x and CO_2 emission from fuel combustion and industrial activities in Asia with 1° × 1° resolution. *Atmospheric Environment*, 28, 213–25.

An, Z. S. (1990). Paleo-environment change in China in last 20 000 years. In *Loess-Quaternary Geology Global Change*, ed. D. S. Liu, pp. 8–18. Beijing: Science Press.

Chameides, W. L., Kasibhatla, P. S., Yienger, J.,

Levy II, H., (1994). Growth of continental scale metro-agro-plexes, regional ozone pollution and world food production. *Science*, **264**, 74–5.

Charlson, R. J., Langner, J., Rodhe, H., Leovy, C. B. & Warren, S. G. (1991). Perturbation of Northern Hemisphere radiative balance by backscattering from anthropogenic sulfate aerosols. *Tellus*, **43A-B(4)**, 152–63.

Charlson, R. J., Schwartz, S. E., Hales, J. M., Cess, R. D., Coakley Jr., J. A., Hansen, J. E. & Hofmann, D. J. (1992). Climate forcing by anthropogenic aerosols. *Science*, **255**, 423–30.

Douglas, I. (1994). Changes in land use and land cover, human settlements. In *Change in Land Use and Land Cover, A Global Perspective*, eds. W. Meyer, and B. Turner II, pp. 149–69. Cambridge: Cambridge University Press.

EPA (1990). Policy options for stabilizing global climate, report to Congress, ed. Daniel A. Lashof & Dennis A. Tirpak, Climate Change Division (PM-221), Office of Policy Planning and Evaluation, US Environmental Projection Agency.

Fu, C. B. (1992). A progress report on global change studies in China, *Proceedings of Asian Symposium on Global Environmental Change*, Tokyo, **1–2**, 27–9.

Fu, C. B. (1994). An aridity trend in China in association with global warming. In *Climate Biosphere Interaction: Biogenic Emissions and Environmental Effects of Climate Change*, ed. R. G. Ztpp, pp. 1–17. Chichester: John Wiley.

Fu, C. B. (1995). Dynamics of monsoon-driven ecosystem: concept, preliminary evidence and proposal for further research. In *Proceedings of the Second International Study Conference on GEWEX in Asia and GAME*, pp. 68–70.

Fu, C. B. & Wang, Q. (1992). On the abrupt change in South Asian summer monsoon climate and its synchronism with the global rapid warming. *Science in China*, Series B, 35, 859–69.

Fu, C. B. & Zheng, Z. M. (1997). Monsoon region, the highest rate in global climate change – observed from global precipitation analysis. Chinese Science Bulletin, 42(16), 4–6 [in English].

Galloway, J. N., Schlesinger, W. H., Levy II, H., Michales, A. & Schnoor, J. L. (1995). Nitrogen fixation: anthropogenic enhancement – environmental response. *Global Biogeochemical Cycles*, **9**, 235–52.

Graetz, D. (1994). Changes in land use and land cover, grasslands. In *Change in Land Use and Land Cover, A Global Perspective*, ed. W. Meyer & B. Turner II, pp. 125–47. Cambridge: Cambridge University Press.

Grubler, A. (1994). Human driving force: technology. In *Change in Land Use and Land Cover, A Global Perspective*, ed. W. Meyer & B. Turner II, pp. 287–328. Cambridge: Cambridge University Press.

Handerson-Sellers, A. & Mcguffie, K. (1994). Land-surface characterization in greenhouse climate simulations. *International Journal of Climatology*, **14**, 1065–94.

Huang, M. Y., Wang, Z., He, D., Xu, H & Zhou, L. (1995). Modeling studies on sulfur deposition and transport in East Asia. *Water, Air and Soil Pollution*, **85(1)**, 921-26.

Hulme, M., Wigley, T., Jiang, T., Zhao, Z. C., Wang, F. T., Ding, Y. H., Leemans, R. & Markham, A. (1992). Climatic change due to greenhouse effect and implication for China, Banson Production, 57 pp.

IEA (1995). World Energy Outlook, 1995 Edition, quoted from Global Environmental Change Report, vol. 7, No. 8, p3, Published by Cutter Information Corp., USA.

IGBP and HDP (1995). Land-use and land cover change, science/ research plan, IGBP report No. 35, and HDP report No. 7, 132 pp.

IPCC (1992). Climate Change, 1992, The supplementary report to the IPCC Scientific Assessment, ed. J. T. Houghton, B. A. Callemder & S. K. Varney, 200 pp. Press Syndicate of the University of Cambridge.

IPCC (1995). Climate Change, 1995, The science of climate change: contribution of working group I to the second assessment report of the intergovernmental panel on climate change, ed. J. T. Houghton, L. G. Meira Filho, B. A. Callander, N. Harris, A. Kattenberg & K. Maskell. Cambridge: Cambridge University Press.

Lal, M. (1994). Water resources of the South-East Asian region in a warmer atmosphere. *Advances in Atmosphere Science,* **11**, 2.

Leemans, R. (1989). Possible changes in natural vegetation patterns due to a global warming, *Der Treibhauseffect: Das Problem-Mögliche Folgen-Erforderliche Massnahmen*, ed. Hackl, pp. 105–21. Laxenberg, Austria: Akademie für Umvelt und Energie.

Marland, G. R., Andres, J. & Boden, T. A. (1994). Global, regional and national CO_2 emissions. In *Trends=9293*: A compendium of data on global change. ORNL/CDIAC-65, ed. T. A. Boden, D. P. Kaiser, R. J. Sepanski & F. W. Stoss, pp. 505–84. Oak Ridge, TN: Carbon Dioxide Information Analysis Center, Oak Ridge National Laboratory.

Morita, T., Matsuoka, T. Y., Kainuma, M., Harasawa, H. & Kai, K. (1993). AIM-Asia-Pacific integrated model for evaluating policy options to reduce GHG emissions and global warming impacts, proceedings of Workshop on Global Warming Issues in Asia, ed. S. C. Bhattacharya, AIT Press.

Neilson, R. P. & Danny, M. (1994). A global perspective of regional vegetation and hydrologic sensitivities from climatic change. *Journal of Vegetation Science*, 5, 715–30.

Rosenzweig, C., Parry, M., Fischer, G. & Frohberg, K. (1992). Climate change and world food supply, 40 pp. Oxford, UK: Environmental Change Unit, University of Oxford.

SSTC China (1990). Climate (abridged), *Blue Book of Science and Technology of China*, No. 5, 53 pp. The State Science and Technology Commission, China: China Meteorological Press.

The Yearbook of World Economy (1994). World Economy Press.

UN (1993). *UN World Urbanization Prospects*, the 1992 revision, New York.

UNEP (1991). *Environmental Data Report*, Vol. 3, 1991/92, 408 pp. Alden Publish.

Williams, M. (1994). Changes in land use and land cover, forest and tree cover. In *Change in Land Use and Land Cover, A Global Perspective*, ed. W. Meyer & B. Turner II, pp. 97–124. Cambridge: Cambridge University Press.

WRI. (1992). *World Resources 1990-1991.* A Report by the World Resources Institute in collaboration with the United Nations Environment Programme and the United Nations Development Programme, 679 pp. Peking: Peking University Press [in Chinese].

Zhao, Z. C. & Wang, F. (1994). Climate change and cropping systems in China. *Bulletin of CNC-IGBP*, 3, 52–62.

Zhao, Z. C. & Kellogg, W. W. (1988). Sensitivity of soil moisture to doubling of carbon dioxide in climate model experiments, part II: the Asian monsoon region. *Journal of Climate*, **1**, 367–78.

Symposium summary

D. Ye and M. Yoshino

Introduction

The centerpiece of the Scientific Advisory Council IV meeting held in Beijing, China, 3–27 October 1995 was the scientific symposium co-chaired by Drs James Galloway and Jerry Melillo. The symposium was distinguished by numerous presentations which highlighted current IGBP and other global change research topics. The focus of the symposium dealt with natural and anthropogenic changes in the global biogeochemical cycles. Changes to global biogeochemical cycles have occurred in more developed parts of the world, owing to increased industrialization, fossil fuel combustion, and intensification of agricultural practices. However, because of recent changes taking place in Asia relative to the expanding industrialization and intensification of agriculture in the region, there have been significant impacts to biogeochemical cycles. Hence the papers of the symposium have reviewed not only different aspects of the IGBP research activities and Core Projects but also have provided relevant regional aspects of changes taking place in Asia. These papers (11 contained in this volume) presented at the symposium were augmented by a series of poster sessions (total of 64 posters presented) that focused primarily on global research studies conducted in Asia (IGBP *Book of Abstracts*, 1995).

These papers and posters dealt with a variety of global change topics including biogeochemical cycles of elements (CONPS), trace-gas emissions, land use changes, interaction of soil–vegetation–atmosphere, global change and sea-level rise, paleoclimatic change, climate and land use impact studies, freshwater and river systems, and marine and coastal aspects of global biogeochemical cycles. The material presented covered topics relevant to IGBP, IHDP and the WCRP.

Main findings

Emission of CH_4 in the ricefield and its reduction

One of main sources of CH_4 is the ricefield of which Asia is the most important region of its plantation. It is thus important to know the total amount of emission of CH_4 from the ricefield in Asia. According to Neue and Boojawat (this volume), its annual amount to the atmosphere ranges from less than 20 to 100 million tonnes. Within this amount the contribution from China is about 13.3 (11.4–15.2) Tg CH_4 year^{-1} according to Wang and Xingjian (1995), 15.3 (7.7–24.6) Tg CH_4 year^{-1} according to Yao *et al.* (1995), or 24.1 (19–29) Tg CH_4 year^{-1} by the same later group of scientists by a different method of estimation. The above estimation gives an idea about the order of magnitude of how much of the annual amount of CH_4 the ricefields can emit to the atmosphere.

How to reduce the CH_4 emission from the ricefield is a concern of the scientists. Neue and Boojawat (this volume), Wassman *et al.* (1995), and Wang and Xingjian (1995) all discussed this problem. According to them, the mid-season drainage can reduce it, because CH_4 emission is high in good water conditions. A shift from organic manure to mineral fertilizer is an option, and some high-yielding rice varieties may have a lower methane emission potential than the traditional ones. However, there is a relation between CH_4 and N_2O emissions in ricefields, as observed by Chen *et al.* (1995). N_2O is another important greenhouse gas. Therefore, to discuss the methods of reducing CH_4 emission, attention should be paid in parallel to N_2O emission. Further, the yield of rice should also be considered. No farmers will accept the method of reducing the CH_4 emission at the expense of yield.

Neue and Boojawat (this volume) also pointed out that, due to the increase of population, the required increase in rice production up to 1 million tonnes in the next century will increase concomitant emission further if current technologies are continued. This will be a big problem demanding various national and international studies. The most comprehensive one is an inter-regional research program with the participation of China, India, Indonesia, Philippines and Thailand, aiming to determine CH_4 sources, strength of major rice ecologies, its controlling factors and processes and mitigation options.

Gaps in CO_2 sources study and the problem of "missing sink" of CO_2

At present, the most predominant greenhouse gas is CO_2. The study of its cycle is well received by scientists. But Malingreau and Zhuang (this volume) point out that people are only beginning to realize the importance of biomass burning in Asia and Africa in the CO_2 cycle. Biomass burning includes the agricultural residue and fuelwood combustion in household stoves, crop stalk incineration in the open fields, forest fires, land use changes, and so on. The study by Dadhwal *et al.* (1995) of the changes in the carbon cycle for the Indian agroecosystem may be cited as another example. Their study indicates an increase in crop NPP from 1.19 Mg C ha^{-1} in 1950–1951 to 2.88 Mg C ha^{-1} in 1989–1990. The major fluxes for 1989–1990 were human consumption, livestock consumption and biomass fuel.

In the study of the CO_2 cycle, the "missing sink" of CO_2 is an important problem and in this symposium the presentations of several authors are related to it. The review paper by Chen and Tsunogai (this volume) pointed out that the seawater in the marginal seas in North Pacific alone may have taken up over 1 Gt of excess carbon. The poster by Yuan Daoxian shows that one of the biggest carbon reservoirs, i.e. the carbonate rocks, is still active in the modern cycle and not negligible because dissolution of every tonne of limestone will remove 120 kg of carbon from the atmosphere. He estimated that the annual removal of carbon by such dissolution on the Kweichow (Guizhou) Plateau alone could reach 0.415 mt C.

Contribution to PAGES

One of the important contributions to PAGES was the study of paleomonsoon changes in China using various kinds of paleoclimate data and the comparison of these data with other regional data. An and Thompson (this volume) found similarities of precipitation variability and decade-to-century scale fluctuations in ice core records between China and Peru during last 1500 years. Yao (1995) additionally found similarity between the dust records of the ice cores in the Tibetan Plateau and those in Greenland. An and Thompson (this volume) also discussed the role of the Tibetan Plateau in global change. Over the last 20 000 years, the cold and warm events seemed to have occurred earlier in the Tibetan Plateau than in other regions. Similar findings were obtained by other Chinese scientists. These findings indicate that the Tibetan Plateau might be a place for early signals of climate change.

Another unique natural environmental feature in Asia is the loess which contains rich paleoclimate data. Indeed, Chinese scientists have made use of loess-derived paleodata to study past climate changes. Shen *et al.* (1995), using ^{10}Be dating techniques, produced continuous time scales of loess strata at Luochuan, NW China, with a precision of 5–10 Ka for the last 0.75 Ma, and 3–5 Ka for the last 0.15 Ma. Because this dating is independent of the local sediment rate, it may form a base for properly comparing the regional paleo-environmental changes and the global glacial cycle.

The study of the loess deposits and subsequent erosion can be used to assess the impact of human activities on river sedimentation. Saito *et al.* (1995) estimated that about 1000–2000 years ago, the suspended sediment discharge of the Yellow

River (Huanghe) was about one-tenth of the present level. The cultivation of the Loess Plateau has been the major cause of this increase of sediment discharge.

Land use/land cover changes and land surface processes

Quite a few plenary reviews and posters related to this topic. The importance of land cover change in the release of carbon to the atmosphere has been discussed. Here, we would like to mention the result of a model simulation by Chen (1995). He showed that the local climate could be altered considerably by land cover change and suggested that the greenland area in the plan of new cities should not be less than 50% in order to retard the unfavorable change of local ecological climate.

Related to this topic is a new model given by Ji and Hu. Based on the feedback mechanism between the plant physiological processes and abietic environment, a new approach to the linkage between physical and biogeochemical processes at land surface, the atmosphere–vegetation interaction model (AVIM) is developed. The AVIM consists of two compartments: the plant physiological growth model and the land surface physical transfer model. All physiological processes are affected by physical climate conditions. Meanwhile, the physical transfers of heat and water depend on vegetation morphological parameters. This AVIM has been applied to mixed forest and agricultural ecosystems in temperate China.

Impact of global changes

Asia is a continent of very dense population and with a very high rate of economic development. So, a big concern is the impact of global change on Asia. Therefore quite a few plenary reviews and posters discussed this problem. Here we will mention two works. The first one is a new concept, the monsoon-driven ecosystem, developed by Fu *et al.* (this volume). They emphasized that, in Asia, the ecosystems are formed and changed under the strong forcing of the monsoon climate. The seasonal jumps, high interannual variability, and rapid change on decadal or longer time scales of the monsoon strongly affect the ecosystems. On the other hand, the ecosystems in the Asian monsoon region play an important role in regulating the regional and even the global hydrological processes and biogeochemical cycles. The dynamics of monsoon-driven ecosystems has already been accepted as one of the foci of the research programs of TEACOM.

The second work to be mentioned is the Northeast China Transect (NECT) proposed by Zhang (1995) and accepted by IGBP (IGBP Report No. 36, 1995) as one of the four major transects in the world. It extends from the Plateau of Inner Mongolia, through the Songliao Plain to the North Changbai Mountains along 43°30'N. From east to west it passes through temperate desert, temperate meadow, temperate deciduous broad-leaved brush, farmland, and temperate deciduous broad-leaved forest. It possesses a strong underlying gradient between different ecosystems. The impact of global change in this region can be studied thoroughly. This study can contribute to IGAC, BAHC, LUCC and DIS a great deal.

Before ending, we would like to add a few words to the impact of human activities on the global climate and environment. Most of the reviews and posters in one way or another have discussed this important problem. Here, we would like to single out what Hu *et al.* (this volume) have pointed out. They said that about 13% of global fluvial discharge is presently dammed in dams or reservoirs and that, by the year after 2000, 66% of the world total stream flow to the ocean margins could be controlled by dams. What a large number! Suppose that half of this continental water were used in one way or another and not allowed to flow into the ocean for many years, its impact could be imagined. Its influence will not only be limited to ocean water density, but it will also disturb the marine ecosystems in the marginal sea (nutrients from the continents to oceans would be greatly reduced). Therefore, the hydrological cycle as well as the biogeochemical cycles and other processes will greatly be affected at least in coastal zones and marginal seas.

Future global change research in Asia

Why does Asian change play an important role in the impacts on the global change? This is because Asia has two kinds of important unique features, namely the natural features and the social features. There are three (positional, topographical and climatical) components in the natural features: (1) Asia occupies a peculiar position, which is the boundary between the Eurasian continent, the largest landmass, and the Pacific Ocean, the largest sea surface, in the Northern Hemisphere. (2) Asia's topography is peculiar: the Tibetan Plateau, the high mountain mass, whose thermal and mechanical effects are definite on general circulation in the atmosphere. On the other hand, Southeast and East Asia is composed of so many islands and peninsulas, having very long coastlines in total. This means that the problems of sea level rise in the 21st century and of the sediment/nutrient transport to the coastal zone are of importance. (3) Climatically, the world's strongest westerlies, monsoons in summer and winter, severe tropical depressions such as typhoons and cyclones, and early summer rainy season named Meiyu (plum rains) in China, Mae-ue in South Korea or Bai-u in Japan, are striking in Asia. On the other hand, arid and semiarid regions are broad in Inner Asia to Northeast Asia. River discharge in Asia is significant for material transport, hydrological power and water use management.

The most important social features in Asia are its high population density and the high rate of economic development. Related to them is the anthropogenic forcing of global change which can be summarized as follows: (1) land use/land cover change, (2) intensive cultivation of paddy rice, which is connected to the

NH_4 problem, (3) urbanization and CO_2 increase, (4) industrialization and air pollution, (5) emission increasing rate, and (6) deforestation or cutting forests with CO_2 problems and with wild fire and burning. However, uncertainty in the projection of human driving forces is making the studies difficult.

Problems to be studied in the coming decades are summarized in the context of global change as follows. From the viewpoint of impacts on global biogeochemical cycles, NH_4, C, N, S, P and K cycles are the first priority. Missing carbon, CO_2 and C in soil should be measured accurately and taken into consideration. Among the important problems in human dimensions, rice production not only from the standpoint of agricultural biotechnology but also from the socio-economic development with the labor force, village system, customs of minorities, national and regional policy, etc. should be studied intensively in relation to fluctuation of the monsoon activities in Asia. Because of the high land use rate, a slight change of the environment tends to result in occurrence of an unexpected forcing. Changes in the growing season of crops causes changes of the crop calendar and cropping system from region to region. For example, in a shifting cultivation systems, areas and cycle times should be changed strikingly.

The estimation of Chinese and Indian population in the future is one of the severe tasks for the global change studies. GNP in the various countries in Asia is now increasing remarkably and this relates to the urbanization and industrialization in the respective regions. These will affect the emission of pollutants and effects of pollution through long-distance transportation by the tropospheric westerlies. Mitigation options of NH_4 should be studied more in Asia: water management, fertilizer sources, new cultivators, minimizing addition of fresh organic matter, etc.

For reconstructing paleo-environment, China provides rich sources of proxy data such as old documents, because of its long history. These should be a strong background for our research not only on paleomonsoon reconstruction, but also on the relationships between the human activities and physical environment during the historical period.

Adding to these programs we propose the following.

Human impact on the global climate and environment

Many research problems may be suggested for this topic, but here we would only emphasize the land use/land cover change (including desertification which unfortunately was not discussed in the symposium). Since Asia is the most densely populated and long civilized region, it is an ideal place for studying the historical change of land use and land cover, present status, the future projection and the effect of land cover change on various natural processes. Because land use and its change happen mostly on a local or regional scale but have global significance through cumulative effects, it is a problem how to incorporate the diversity of small-scale land use into global assessment.

Researches on the early signals of the global climate and environmental change

One of the important practical purposes of the global change research is its prediction. So far, the method of prediction is far from being found. One practical way is to watch for early signs of changes. The sign of change should appear earlier in places which are most sensitive to changes – places usually occurring at the transitional zones between two climate regions, two ecosystems (ecotons) or any regions where the changes usually occurred earlier in the past. Asia is one of the ideal regions for this kind of study. First of all, the study of variation of the intensity of the monsoon and the limit of its northern reach will serve the purpose because in global warming the continent will warm more rapidly than the ocean, thus enhancing the temperature contrast in summer between the land and ocean. The Tibetan Plateau may also be a region to show earlier changes, because in the past the warming or cooling usually occurred earlier here than in other regions. Further, in Asia, there are quite a few large-scale zones between different climate or ecosystem regimes.

Sustainable development

Asia is not only a place of very dense population but also a place with a high rate of economic development. How to get a sustainable development is a big problem, but up to now, to the knowledge of the authors, a proper definition of sustainable development has not existed. Sustainable development relates the development to the environment. Evidently, the developed countries will dwell on good environment while the developing countries will dwell more on development and therefore sustainable development will have different meanings in different countries. Even in a country such as China or India, the stage of economic development is different in different regions, and so one definition cannot be applied everywhere even in one country. The natural environment is also different from region to region. Although the problem of sustainable development is an extremely difficult one, it should be attacked in Asia.

Public education

The practical purpose of the global change research is to provide the decision-makers with the scientific strategy for maintaining a good environment, along with the economic development. No matter how good the scientific strategy is, however, it should be implemented through the hands of the general population. Unless they understand it thoroughly, they will not implement it successfully. Public education is therefore of urgent importance. This is especially so in Asia where the culture level of the population is not yet high and living standards are low in general (except Japan and others). The people urgently need to improve their standard of living and, at present, try to do this at the expense of the environment. It is therefore an urgent matter to educate the population to understand the scientific meaning of global change and its practical importance and to educate them not to be short-sighted.

In conclusion, it should be stressed that Asian change may cause a significant effect on the global change in the coming 25 years. Co-operation of scientists of the various countries in Asia and also of the various continents of the world is needed for work on these topics to continue effectively. International scientific programs/networks already established, such as IGBP and HDP, under which TEACOM, SASCOM and ASEAN-Group function, are for this purpose. The Asia–Pacific Network (APN) will also have a strong function in development in this direction.

References

Chen, W. (1995). A study on the local ecological climatic effects of land use. In *Book of Abstracts*, IGBP SAC IV Meeting, Beijing 23–25 October 1995, p. 85.

Chen, G., Huang, G, Huang, B. W. J., Yu, K., Xu, H. & Wang, Z. (1995). Greenhouse gases emissions by plant and rice field. In *Book of Abstracts*, IGBP SAC IV Meeting, Beijing 23–25 October 1995, p. 54.

Dadhwal, V. K., Shah, A. K. & Vora, A. B. (1995). Changes in carbon cycle for agro-ecosystems and its implications: a case study of India for 1950–51 to 1989–90. In *Book of Abstracts*, IGBP SAC IV Meeting, Beijing 23–25 October 1995, p. 91.

IGBP Report No. 36. (1995). *The IGPB Terrestrial Transects: Science Plan*, ed. G. W. Koch, R. J. Scholes, W. L. Steffen, P. M. Vitousek & B. H. Walker, 61 pp., Stockholm.

Saito, Y., Katayama, H., Ikehara, K., Matsumoto, E. & Kato, Y. (1995). Cultivation history of the loess plateau in China recorded in sediments of the East China Sea. In *Book of Abstracts*, IGBP SAC IV Meeting, Beijing 23–25 October 1995, p. 83.

Shen, C., Beer, J., Liu, T., Bonani, G. & Suter, M. (1995). ^{10}Be and the dating of loess strata. In *Book of Abstracts*, IGBP SAC IV Meeting, Beijing 23–25 October 1995, p. 46.

Wang, M. & Xingjian, S. (1995). Factors regulating methane emission from rice fields. In *Book of Abstracts*, IGBP SAC IV Meeting, Beijing 23–25 October 1995, p. 51.

Wassmann, R., Neue, H. U., Lantin, R. S., Papen, H., Renneberg, H., & Seiler. (1995). Advanced agricultural technologies affecting methane emission from rice fields. In *Book of Abstracts*, IGBP SAC IV Meeting, Beijing 23–25 October 1995, p. 52.

Yao, T. D. (1995). Ice core records on the Qinghai-Xizang (Tibetan) plateau and its relation to global changes. In *Book of Abstracts*, IGBP SAC IV Meeting, Beijing 23–25 October 1995, p. 45.

Yao, H., Zhuang, Y., Chen, Z. (1995). Estimation of methane emission from rice paddies in mainland China. In *Book of Abstracts*, IGBP SAC IV Meeting, Beijing 23–25 October 1995, p. 53.

Zhang, X. (1995). Chinese terrestrial ecosystems in response to global change. In *Book of Abstracts*, IGBP SAC IV Meeting, Beijing 23–25 October 1995, p. 102.

Index